Direction Dependence in Statistical Modeling

Direction Dependence in Statistical Modeling

Methods of Analysis

Edited by

Wolfgang Wiedermann
Daeyoung Kim
Engin A. Sungur
Alexander von Eye

Registered Office
John Wiley & Sons, Inc., 111 River Street, Hoboken, NJ 07030, USA

Editorial Office
111 River Street, Hoboken, NJ 07030, USA

For details of our global editorial offices, customer services, and more information about Wiley products visit us at www.wiley.com.

Wiley also publishes its books in a variety of electronic formats and by print-on-demand. Some content that appears in standard print versions of this book may not be available in other formats.

Library of Congress Cataloging-in-Publication Data

Names: Wiedermann, Wolfgang, 1981- editor. | Kim, Daeyoung, editor. | Sungur, Engin, editor. | Eye, Alexander von, editor.
Title: Direction dependence in statistical modeling : methods of analysis / edited by Wolfgang Wiedermann, Daeyoung Kim, Engin Sungur, Alexander von Eye.
Description: Hoboken, NJ : Wiley, 2021. | Includes bibliographical references and index.
Identifiers: LCCN 2020015364 (print) | LCCN 2020015365 (ebook) | ISBN 9781119523079 (cloth) | ISBN 9781119523130 (adobe pdf) | ISBN 9781119523147 (epub)
Subjects: LCSH: Dependence (Statistics)
Classification: LCC QA273.18 .D57 2020 (print) | LCC QA273.18 (ebook) | DDC 519.5–dc23
LC record available at https://lccn.loc.gov/2020015364
LC ebook record available at https://lccn.loc.gov/2020015365

Cover design by Wiley
Cover image: © zhengshun tang/Getty Images

Set in 9.5/12.5pt STIXTwoText by SPi Global, Chennai, India

Printed in the United States of America

SKY10022145_110220

Contents

About the Editors

Wolfgang Wiedermann

Wolfgang Wiedermann is Associate Professor at the University of Missouri, Columbia. He received his PhD in Quantitative Psychology from the University of Klagenfurt, Austria in 2012. His primary research interests include the development of methods for causal inference, methods to determine the causal direction of dependence in observational data, and methods for person-oriented research settings. He has edited books on advances in statistical methods for causal inference (with von Eye, Wiley) and new developments in statistical methods for dependent data analysis in the social and behavioral sciences (with Stemmler and von Eye). His work appears in leading quantitative methods journals, including *Psychological Methods*, *Multivariate Behavioral Research*, *Behavior Research Methods*, and the *British Journal of Mathematical and Statistical Psychology*. He currently serves as an associate editor for *Behaviormetrika* and the *Journal for Person-Oriented Research*.

Daeyoung Kim

Daeyoung Kim is Associate Professor of Mathematics and Statistics at the University of Massachusetts, Amherst. He received his PhD from the Pennsylvania State University in Statistics in 2008. His original research interests were in likelihood inference in finite mixture modeling including empirical identifiability and multimodality, development of geometric and computational methods to delineate multidimensional inference functions, and likelihood inference in incompletely observed categorical data, followed by a focus on the analysis of asymmetric association in multivariate data using (sub)copula regression. He also has active collaborations with colleagues in food sciences at the University of Massachusetts, Amherst, focusing on the use of statistical models to analyze data for colon cancer, obesity and diabetes.

Engin A. Sungur

Engin A. Sungur has a BA in City and Regional Planning (Middle East Techni-cal University, METU, Turkey), MS in Applied Statistics, METU, M.S. in Statistics (Carnegie-Mellon University, CMU) and PhD in Statistics (CMU). He taught at Carnegie-Mellon University, University of Pittsburg, Middle East Technical Uni-versity, and University of Iowa. Currently, he is a Morse-Alumni distinguished professor of statistics at University of Minnesota, Morris. He has been teaching statistics for more than 38 years, 29 years of which at the University of Minnesota, Morris. His research areas are dependence modeling with emphasis on directional dependence, modern multivariate statistics, extreme value theory, and statistical education.

Alexander von Eye

Alexander von Eye, PhD, is Professor Emeritus of Psychology at Michigan State University. He received his PhD in Psychology, with minors in Education and Psy-chiatry, from the University of Trier, Germany, in 1976. He is known for his work on statistical modeling, categorical data analysis, methods of analysis of direction dependence hypotheses, person-oriented research, and human development. He authored, among others, texts on Configural Frequency Analysis, the analysis of rater agreement (with Mun), and on log-linear modeling (with Mun; Wiley), and he edited, among others, two books on latent variables analysis (the first with Clogg, the second with Pugesek and Tomer), and one on Statistics and Causality (with Wiedermann; Wiley). His over 400 articles appeared in the premier journals of the field, including, for instance, *Psychological Methods*, *Multivariate Behavioral Research*, *Child Development*, the *Journal of Person-Oriented Research*, the *American Statistician*, and the *Journal of Applied Statistics*.

Notes on Contributors

Patrick Blöbaum

Patrick Blöbaum studied Cognitive Computer Science and Intelligent Systems at Bielefeld University (Germany) from 2009 to 2014 and received his PhD in Engineering (Machine Learning) from Osaka University (Japan) in 2019 with a research focus on causality. In addition to his PhD studies, he worked as an assistant researcher and machine learning engineer in Japan. In 2019, Patrick joined the newly founded causality team at the Amazon research and development center in Tübingen, Germany, which focuses on the development and application of novel causality algorithms.

G. Anne Bogat

G. Anne Bogat, PhD is a professor of clinical psychology at Michigan State University. Her research centers on intimate partner violence (IPV), including a focus on daily experiences of IPV among college students; how IPV during pregnancy affects women, children, and the mother–child relationship; and how bonding between mothers and infants is affected by pregnancy and postpartum IPV. In addition, she has written about and employed person-oriented methods in her research.

Yadolah Dodge

Yadolah Dodge was born in Abadan, Iran, and is a Swiss citizen. Along with a full-time position as Professor and Chair of Statistics at the University of Neuchâtel, Switzerland, his dedication to photography, painting, and film-making continued, resulting in three long documentaries: Turicum: This is Zurich (2014), Dear Son (2018), and Moving Heart (2019). He is author, co-author, and editor of over 20 books by Oxford University Press, Springer and John-Wiley, Dunod and North-Holland and several papers.

Regina García-Velázquez

Dr. Regina García-Velázquez has a background in Psychology and received her master's degree in Methodology of Behavioural and Health Sciences from the Autonomous University of Madrid. She received her PhD from the University of Helsinki. She is a post-doctoral researcher at the University of Helsinki and teaches courses on Psychometrics. She is interested in measurement issues applied to psychopathology, particularly on classification, validity, and statistical modelling. In her current research, she focuses on internalizing disorders.

Jade E. Kobayashi

Jade E. Kobayashi, MA is a clinical psychology graduate student in the doctoral program at Michigan State University. Her research interests include adult romantic attachment, interpersonal conflict and intimate partner violence (IPV), and intensive longitudinal and dyadic analytic methods.

Alytia A. Levendosky

Alytia A. Levendosky, PhD is a Professor in the Department of Psychology at Michigan State University. Her research over the past 25 years has focused on the effects of intimate partner violence (IPV) on mothers and children. Currently, her primary research interests are in the role of IPV as a stressor during the perinatal period. Her work has helped elucidate how IPV during pregnancy affects the very beginnings of motherhood through women's developing representations/schemas about their unborn child which later affect their parenting behaviors during early childhood.

Xintong Li

Xintong Li is Senior Research Analyst at the Assessment Resource Center at the University of Missouri, and he is an experienced researcher specialized in quantitative methods and educational research. He received his PhD in statistics, measurement and evaluation in education at the University of Missouri. His major research interests include causal inference with non-experimental data, educator effectiveness, and motivation in education. He is skilled and experienced in advance statistical modeling, programming, large-scale simulations, and large database management. He has multiple publications on methodological foundations and applications of direction dependence principals published in, e.g. *Multivariate Behavioral Research*, *Behavior Research Methods*, and *Prevention Science*.

Joel T. Nigg

Joel T. Nigg, PhD is a clinical psychologist and professor of Psychiatry and Behavioral Neuroscience and Director of the Center for ADHD Research at Oregon Health & Science University. His research on ADHD and related conditions has been funded by NIH continuously for over 20 years. His work focuses on refining the phenotype related to cognition and emotion and examining environmental and genetic etiology.

Tom Rosenström

Dr. Tom Rosenström has obtained his education in psychology (MA, PhD) and applied mathematics (MSc) at the University of Helsinki, Finland. He conducts mental health research, applying and developing mathematical and statistical models within the field. In addition, he has worked on theoretical biology at the University of Bristol and on behavior genetics at the Norwegian Institute of Public Health, and is currently employed by the Helsinki University Hospital, where he also conducts clinical patient work.

Valentin Rousson

Valentin Rousson was born in 1967 in Neuchâtel, Switzerland, where he got a PhD in Statistics in 1998. He then spent some time at the Australian National University in Canberra as a postdoc, and at the University of Zurich, sharing his time between statistical consulting and research. In 2007, he was named Associate Professor in Biostatistics at the University of Lausanne, where he is currently working and teaching.

Shohei Shimizu

Shohei Shimizu is a Professor at the Faculty of Data Science, Shiga University, Japan and leads the Causal Inference Team, RIKEN Center for Advanced Intelligence Project. He received a PhD in Engineering from Osaka University in 2006. His research interests include statistical methodologies for learning data generating processes such as structural equation modeling and independent component analysis and their application to causal inference. He received the Hayashi Chikio Award (Excellence Award) from the Behaviormetric Society in 2016. He is a coordinating editor of Behaviormetrika since 2016 and is an associate editor of Neurocomputing since 2019.

Diane D. Stadler

Diane Stadler has a PhD in Human Nutrition and is a registered dietitian with expertise in maternal and infant nutrition and providing care for children with metabolic disorders and developmental disabilities. She directs the Graduate Programs in Human Nutrition at Oregon Health & Science University in Portland, Oregon and is a leader in OHSU's nutrition education initiatives and research mentoring programs. She also oversees OHSU's clinical nutrition specialist training program and research initiatives in Lao People's Democratic Republic to support health care providers in addressing the country's high rates of childhood malnutrition.

Santi Tasena

Since 2011, Santi Tasena has been working at Chiang Mai University, Thailand. Being in love with mathematics, he enjoys discussing any topic related to mathematics. His research interests include mathematical analysis and related fields. His work includes heat kernel analysis on metric spaces, (sub)copulas and measures of dependence, and construction of aggregation and related functions. He received grants from the Commission on Higher Education, Thailand, the Centre of Excellence in Mathematics (CHE), Thailand, the Data Science Research Center, and the Center of Excellence in Mathematics and Applied Mathematics, Chiang Mai University, Thailand.

Tonghui Wang

Tonghui Wang is currently a full professor of statistics in the Department of Mathematical Sciences, New Mexico State University. He received his PhD degree from the University of Windsor, Canada in May, 1993. His research interests are multivariate linear modes under skew normal settings; copulas and their associated measures with applications; and big data analysis and statistical learning with applications.

Zheng Wei

Zheng Wei is currently an assistant professor of statistics in the Department of Mathematics and Statistics at the University of Maine. He served as a visiting assistant professor at Department of Mathematics and Statistics, University of Massachusettes Amherst from May 2015 to August 2017. He developed research in Bayesian statistical methods for data science, big data and analytics, the copula theory and its applications. He completed the PhD at the New Mexico State University in May 2015.

Phillip K. Wood

Phil K. Wood is a professor of Quantitative Psychology at the University of Missouri. He specializes in structural equation modeling, growth curve modeling and factor analysis, with particular emphasis on techniques for the analysis of longitudinally intensive data such as dynamic factor models. His substantive areas of interest include the cognitive outcomes of higher education and longitudinal inter-individual differences in behaviors during young adulthood such as problematic alcohol use, tobacco and other drug usage and risky sexual behaviors.

Xiaonan Zhu

Xiaonan Zhu is an assistant professor in the Department of Mathematics at University of North Alabama. Before joining UNA in Fall 2019, he obtained his PhD and MS in Mathematical Statistics from the Department of Mathematical Sciences at New Mexico State University in 2019 and 2014, respectively. His research interests include sampling distributions of skew normal distributions, distribution of quadratic forms under closed skew normal settings, construction of copulas, (local) dependence of random vectors, measures of dependence through (sub-)copulas.

Acknowledgments

There are numerous people to thank in regard of the preparation of this volume. First and foremost, we offer our deepest thanks to our contributing authors with whom we share a dedication to the development and application of statistical methods in the context of direction of dependence and causality. This volume would not have been possible without their excellent work.

We are also grateful to Wiley publishers for their interest in the topic and their support. This applies in particular to Sari Friedman, Kathleen Santoloci, Mindy Okura-Marszycki, Elisha Benjamin, Sechin Nithya, Amudhapriya Sivamurthy, and Ezhilan Vikraman who have supported and guided us from the very first contact to the completion of this book. Thank you all!

Most important, we are grateful for the love and support of our respective families. The first editor wants to emphasize that he is grateful to be allowed to experience a causal mechanism that does not require any empirical evaluation – the dependence between W's happiness and the existence of Anna and Linus. No statistical modeling is needed to show that $\{Anna, Linus\} \rightarrow (W = happy)$ holds in an unconfounded manner. The second editor would like to express gratitude and sincere thanks to his wife, Shu-Min, and his son, Minjun, for their tremendous support and love.

Preface

Questions concerning causation are omnipresent in the empirical sciences. In non-experimental research, however, it is often hard to determine the status of variables as cause and effect. Temporal order alone is of limited use, unless one observes antecedents and the beginning of a chain of events. That is, even when a putative explanatory variable (x) is measured earlier in time than the (putative) outcome (y), one cannot rule out that an outcome, measured at an earlier point in time, may have caused x. Similarly, temporality alone does not prevent causal effect estimates from being biased unless one is able to adjust for all relevant (potentially time-varying) confounders (Bellemare, Masaki, & Pepinsky, 2017). Cross-sectional research has often been looked-down upon because it is deemed of little use for the analysis of hypotheses that are compatible with (possibly competing) theories of causality. Based on cross-sectional data alone, for example, one is not able to distinguish whether a relation between x and y is observed because of an underlying causal model of the form $x \rightarrow y$ (i.e. x causes y), the reverse-causal model $y \rightarrow x$ (y causes x), or whether the observed relation is spurious due to (total or partial) confounding, $x \leftarrow u \rightarrow y$.

Limitations of longitudinal and cross-sectional observational research are (partly) rooted in the limitations of the statistical methods that are routinely applied to analyze dependence structures. In both research designs, covariance-based methods (such as correlational, linear regression, and structural equation modeling techniques) are de rigueur. Although, these methods can be useful in the estimation of the magnitude of causal effects (provided that certain unconfoundedness conditions are fulfilled, see, e.g. Pearl, 2009), they do not help to empirically distinguish between cause and effect. For example, in the standardized case, linear regression parameters for the model $x \rightarrow y$ are identical to the ones that are estimated for the reverse regression, $y \rightarrow x$ (von Eye & DeShon, 2012). These symmetry properties of the linear regression model have been known since its early origins (Galton, 1886). In fact, the observation that regression is inherently symmetric was one of the reasons why Francis Galton (the "founding father" of linear regression) changed his characterization of the phenomenon that previously suppressed hereditary traits can re-appear from a phenomenon of *reversion* to a phenomenon of *regression* (Gorroochurn, 2016). In other words, symmetry properties influenced how linear regression was conceptualized as a statistical tool. Similarly, symmetry properties of conventional representations of the Pearson product-moment correlation (for an overview of various facets of the Pearson correlation see, for example, Rodgers and Nicewander (1988), Rovine and von Eye (1997), Falk and Well (1997), and Nelsen (1998)) certainly contributed to the widespread and well-known mantra that *correlation does not imply causation* and to the belief that the means of statistic cannot be used to establish the causal direction of dependence.

Fortunately, this state of affairs has changed recently. It did take statisticians until the beginning of the new millennium to get a handle on the issue of direction dependence. But in 2000, Dodge and Rousson derived, within the framework of the linear regression model, the relation between cause and effect variables, for the (not so) particular case in which the cause variable is asymmetrically distributed. Specifically, these authors showed that variable information beyond means, variances, and covariances (e.g. skewness and co-skewness) can be used to empirically determine which of two variables, is more likely to be the cause and which is more likely to be the effect. Focusing on asymmetry properties of the linear regression and the Pearson correlation, the work by Dodge and Rousson (2000) initiated a new topic and line of statistical research, that of the development and application of methods for the analysis of direction dependence and causal hypotheses. Dodge and Rousson (2000) focused on asymmetry that emerges from marginal variable distributions. Asymmetry properties based on error distributions have later been proposed by Wiedermann, Hagmann, and von Eye (2015), Wiedermann and von Eye (2015b), and Wiedermann and Hagmann (2016). Extensions to measurement error models were recently discussed in Wiedermann, Merkle, and von Eye (2018). The second seminal paper in this new line of research was published in 2005 by Engin A. Sungur (see Sungur (2005a); a discussion of copulas in the regression context is given by Sungur (2005b)). While Dodge and Rousson's (2000) initial work focused on determining the direction of dependence through studying the marginal behavior of distributions, Sungur (2005a) proposed to study the behavior of joint variable distributions by making use of copulas. This copula-based direction dependence approach constitutes a second line of research that allows researchers to analyze cause-effect properties of variables while accounting for potential differences in marginal distributions. Copula-based directional dependence analysis has experienced rapid development. Various extension have been proposed by, e.g. Kim and Kim (2014, 2016), Wei and Kim (2017, 2018), and Kim and Hwang (2019) – more recent applications of the approach are given by Lee and Kim (2019) and Kim, Lee and Xiao (2019).

The third seminal paper in the development of methods to distinguish between cause and effect variables was published by Shimizu and colleagues in 2006 proposing the linear non-Gaussian acyclic model (LiNGAM) – a causal machine learning algorithm for non-normal variables that is closely related to independent component analysis (Hyvärinen, Karhunen, & Oja, 2001). LiNGAM rapidly developed in the area of machine learning research and has been extended to nonlinear variable relations (Zhang & Hyvärinen, 2016), models with hidden common causes (Hoyer, Shimizu, Kerminen, & Palviainen, 2008; Shimizu & Bollen, 2014), and mixed (continuous and categorical) data (Yamayoshi, Tsuchida, & Yadohisa, 2020), to name a few. For an overview of recent advances in causal machine learning, see Guyon, Statnikov, and Batu (2019).

The present book is concerned with novel statistical approaches to the analysis of the causal direction of dependence of variables in both, exploratory (i.e. learning the causal structures from observational data without background knowledge) and confirmatory (i.e. testing a priori existing competing causal theories) research scenarios, and presents original work in four modules. In the first module, *Fundamental Concepts of Direction Dependence*, Dodge and Rousson (Chapter 1) introduce the well-known Pearson correlation coefficient as an asymmetric concept of two variables which (as discussed above) served as a starting point for several lines of direction dependence research. Further, the authors provide a reminder that working with non-normality of variables (as a key requirement to derive asymmetry properties in the linear case) bears challenges in practice (e.g. distinguishing between non-normality as a characteristic of the construct under study versus non-normality due to outliers and suboptimal measurement). In Chapter 2, Wiedermann, Li, and von Eye then continue the discussion of asymmetry properties of the linear regression model and introduce three asymmetry concepts (summarized in a framework termed Direction Dependence Analysis (DDA), cf. Wiedermann and von Eye, 2015a; Wiedermann & Li, 2018) that can be used to detect potential confounding and distinguish between the two causally competing models $x \to y$ and $y \to x$. Applications of DDA in the context of mediation and moderation models are discussed. Chapter 3, by Engin A. Sungur, is devoted to the use of copulas in direction dependence modeling. This chapter introduces definitions and fundamental principles to model directional dependence of variables using asymmetric copulas and regression, and describes various copula-based directional dependence measures to perform model selection in both, continuous and categorical data settings.

The second module is devoted to *Direction Dependence in Continuous Variables*. Chapter 4, by Wolfgang Wiedermann, discusses asymmetry properties of the partial correlation coefficient in the research tradition of Dodge and Rousson (2000). Asymmetric facets of the partial correlation coefficient are presented which enable one to test causally competing models while adjusting for relevant background variables. Parameter recovery and accuracy of model selection is evaluated using Monte-Carlo simulation experiments. Chapter 5, by Shimizu and Blöbaum, gives an overview of recent advances in the development of algorithms for unsupervised causal learning. The authors start by introducing the standard LiNGAM and present extensions to structural vector autoregressive models for the analysis of time series data, models with hidden common causes, and methods for causal learning under nonlinearity of variable relations. In Chapter 6, Phillip K. Wood takes a regression diagnostic perspective and discusses the importance of evaluating the assumptions of the statistical models that are used to learn the causal structure of observational data. The author uses data from a longitudinal study on motives for alcohol consumption (cf. Sher & Rutledge, 2007) and compares the use of manifest variable composites, factor scores within a state-trait model, and latent difference factor scores in the evaluation of directional dependence hypotheses. The last chapter of this module (Chapter 7) by Santi Tasena, reviews definitions and basic properties of measures of complete dependence. The author gives examples of calculating complete dependence measures in the case of the multivariate Gaussian distribution and presents open problems and potential future directions.

In the third module, methods of direction dependence are extended to the categorical variable domain. Chapter 8, by von Eye and Wiedermann, introduces an event-based perspective in the analysis of hypotheses compatible with direction dependence. The authors introduce two-valued statement calculus to derive composite causality statements and use a design matrix approach to evaluate event-based direction dependence hypotheses. Three methods are compared with respect to their capability to test direction of dependence in categorical data, log-linear modeling, configural frequency analysis, and prediction analysis. Chapter 9 contributed by Zhu, Wei, and Wang, is devoted to a copula-based approach to measure associations in contingency tables. The authors start with reviewing some recently developed measures for the analysis of asymmetric associations in two-way or three-way contingency tables. Then, they propose two new measures of complete dependence on three-way contingency tables and present corresponding nonparametric estimators. Chapter 10, by Kim and Wei, investigates a subcopula-based asymmetric association measure for the analysis of dependence structures in three-way ordinal contingency tables. Their asymmetric measure utilizes sub-copula regressions obtained under the hypothesized dependence relations.

The fourth module is then devoted to *Applications and Software*. In Chapter 11, Rosenström and Regina García-Velázquez make use of LiNGAM in the context of psychiatric epidemiology. Specifically, the authors use distribution-based indicators to test the causal direction of the association between sleeping problems and depressive symptoms using data from the Swedish Adoption/Twin Study on

Aging (Pedersen, 2005). In addition, the authors provide application guidelines for epidemiologists, present a novel Monte-Carlo-based sensitivity analysis approach to evaluate the robustness of LiNGAM results, and integrate distribution-based causality approaches in the process of causal triangulation in etiologic epidemiology. Chapter 12, by Nigg, Stadler, von Eye, and Wiedermann, provides an application of direction dependence analysis in the context of determining risk factors of attention-deficit/hyperactivity disorder (ADHD). Specifically, direction dependence methods for linear models are used to evaluate the causal structure of the association between breastfeeding duration and ADHD. The authors use one of the largest well-characterized samples currently available and demonstrate DDA results can be affected by rater effects when measuring ADHD. Further an attempt is presented to account for potential ceiling/floor effects that can artificially increase the magnitude of non-normality of variables. In Chapter 13, Bogat, Levendosky, Kobayashi, and von Eye then take a longitudinal data perspective in the discussion of causal effect directionality. The authors use daily diary data to assess longitudinal dynamics of the causal structure of intimate partner violence and mood lability in young adult couples. Granger causality models (a causal prediction approach in which one tests whether the inclusion of past information of one variable (e.g. x_{t-1}) is useful in predicting another variable y_t above and beyond the information that is contained in y_{t-1}; Granger, 1969) are applied to test whether intimate partner violence is more likely to cause mood lability or vice versa. In the final chapter, by Li and Wiedermann (Chapter 14), a software implementation of direction dependence methods is presented. The authors introduce SPSS Custom Dialogs to perform DDA and use data from the High School Longitudinal Study 2009 (Ingels et al., 2011) for illustrative purposes. Specifically, the authors present a step-by-step tutorial to evaluate the causal direction of effect of academic achievement and intrinsic motivation in 9th grade Asian students.

Within the last two decades, tremendous progress has been made in the area of direction dependence modeling. We believe that this volume makes a timely and important contribution to the ongoing development of methods of direction dependence and we hope that this contribution will advance the statistical tools empirical sciences can use to better explain causal phenomena.

Wolfgang Wiedermann, University of Missouri, Columbia
Daeyoung Kim, University of Massachusetts, Amherst
Engin A. Sungur, University of Minnesota, Morris
Alexander von Eye, Michigan State University, East Lansing

References

Bellemare, M. F., Masaki, T., & Pepinsky, T. B. (2017). Lagged explanatory variables and the estimation of causal effect. *Journal of Politics, 79*, 949–963. doi:10.2139/ssrn.2568724

Dodge, Y., & Rousson, V. (2000). Direction dependence in a regression line. *Communications in Statistics-Theory and Methods, 29*(9–10), 1957–1972. doi:10.1080/03610920008832589

Falk, R., & Well, A. D. (1997), "Many faces of the correlation coefficient," *Journal of Statistics Education, 5*. Retrieved from http://www.amstat.org/publications/jse/v5n3/falk.html.

Galton, F. (1886). Family likeness in stature. *Proceedings of the Royal Society of London, 40*(242–245), 42–73. doi:10.1098/rspl.1886.0009

Gorroochurn, P. (2016). On Galton's change from "reversion" to "regression". *The American Statistician, 70*(3), 227–231. doi:10.1080/00031305.2015.1087876

Granger, C. W. J. (1969). Investigating causal relations by econometric models and cross-spectral methods. *Econometrica, 37*(3), 424–438. doi:10.2307/1912791

Guyon, I., Statnikov, A., & Batu, B. B. (Eds.) (2019). *Cause effect pairs in machine learning.* doi:10.1007/978-3-030-21810-2

Hoyer, P. O., Shimizu, S., Kerminen, A. J., & Palviainen, M. (2008). Estimation of causal effects using linear non-Gaussian causal models with hidden variables. *International Journal of Approximate Reasoning, 49*(2), 362–378. doi:10.1016/j.ijar.2008.02.006

Hyvärinen, A., Karhunen, J., & Oja, E. (2001). *Independent component analysis.* New York, NY: Wiley & Sons.

Ingels, S. J., Pratt, D. J., Herget, D. R., Burns, L. J., Dever, J. A., Ottem, R., … LoGerfo, L. (2011). *High School Longitudinal Study of 2009 (HSLS: 09): Base-year data file documentation.* Washington, DC: U.S. Dept. of Education, Institute of Education Sciences, National Center for Education Statistics.

Kim, D., & Kim, J.-M. (2014). Analysis of directional dependence using asymmetric copula-based regression models. *Journal of Statistical Computation and Simulation, 84*(9), 1990–2010. doi:10.1080/00949655.2013.779696

Kim, S., & Kim, D. (2016). *Directional dependence analysis using skew-normal copula-based regression.* In W. Wiedermann & A. von Eye (Eds.), *Statistics and causality: Methods for applied empirical research* (pp. 131–152). Hoboken, NJ: Wiley and Sons.

Kim, J.-M., & Hwang, S. Y. (2019). The copula directional dependence by stochastic volatility models. *Communications in Statistics - Simulation and Computation, 48*(4), 1153–1175. doi:10.1080/03610918.2017.1406512

Lee, N., & Kim, J.-M. (2019). Copula directional dependence for inference and statistical analysis of whole-brain connectivity from fMRI data. *Brain and Behavior*, *9*(1), e01191. doi:10.1002/brb3.1191

Nelsen, R. B. (1998). Correlation, regression lines, and moments of intertia. *American Statistician*, *52*, 343–345.

Pearl, J. (2009). *Causality: Models, reasoning, and inference* (2nd ed.). New York, NY: Cambridge University Press.

Pedersen, N. L. (2005). Swedish Adoption/Twin Study on Aging (SATSA), 1984, 1987, 1990, 1993, 2004, 2007, and 2010 [Data set]. doi:10.3886/ICPSR03843.v2

Rodgers, J. L., & Nicewander, W. A. (1988). Thirteen ways to look at the correlation coefficient. *American Statistician*, *42*, 59–66.

Rovine, M. J., & von Eye, A. (1997). A 14th way to look at a correlation coefficient: Correlation as the proportion of matches. *American Statistician*, *51*, 42–46.

Sher, K. J., & Rutledge, P. C. (2007). Heavy drinking across the transition to college: Predicting first-semester heavy drinking from precollege variables. *Addictive Behaviors*, *32*, 819–835.

Shimizu, S., & Bollen, K. A. (2014). Bayesian estimation of causal direction in acyclic structural equation models with individual-specific confounder variables and non-Gaussian distributions. *Journal of Machine Learning Research*, *15*, 2629–2652.

Shimizu, S., Hoyer, P. O., Hyvärinen, A., & Kerminen, A. (2006). A linear non-Gaussian acyclic model for causal discovery. *The Journal of Machine Learning Research*, *7*, 2003–2030.

Sungur, E. A. (2005a). A note on directional dependence in regression setting. *Communications in Statistics - Theory and Methods*, *34*(9–10), 1957–1965. doi:10.1080/03610920500201228

Sungur, E. A. (2005b). Some observations on copula regression functions. *Communications in Statistics - Theory and Methods*, *34*(9–10), 1967–1978. doi:10.1080/03610920500201244

von Eye, A., & DeShon, R. P. (2012). Directional dependence in developmental research. *International Journal of Behavioral Development*, *36*(4), 303–312. doi:10.1177/0165025412439968

Wei, Z., & Kim, D. (2017). Subcopula-based measure of asymmetric association for contingency tables. *Statistics in Medicine*, *36*, 3875–3894. doi:10.1002/sim.7399

Wei, Z., & Kim, D. (2018). On multivariate asymmetric dependence using multivariate skew-normal copula-based regression. *International Journal of Approximate Reasoning*, *92*, 376–391. doi:10.1016/j.ijar.2017.10.016

Wiedermann, W., & Hagmann, M. (2016). Asymmetric properties of the Pearson correlation coefficient: Correlation as the negative association between linear regression residuals. *Communications in Statistics: Theory and Methods*, *45*(21), 6263–6283. doi:10.1080/03610926.2014.960582

Wiedermann, W., & Li, X. (2018). Direction dependence analysis: A framework to test the direction of effects in linear models with an implementation in SPSS. *Behavior Research Methods, 50*(4), 1581–1601. doi:10.3758/s13428-018-1031-x

Wiedermann, W., & von Eye, A. (2015a). Direction-dependence analysis: A confirmatory approach for testing directional theories. *International Journal of Behavioral Development, 39*(6), 570–580. doi:10.1177/0165025415582056

Wiedermann, W., & von Eye, A. (2015b). Direction of effects in multiple linear regression models. *Multivariate Behavioral Research, 50*, 23–40.

Wiedermann, W., Hagmann, M., & von Eye, A. (2015). Significance tests to determine the direction of effects in linear regression models. *British Journal of Mathematical and Statistical Psychology, 68*, 116–141.

Wiedermann, W., Merkle, E. C., & von Eye, A. (2018). Direction of dependence in measurement error models. *British Journal of Mathematical and Statistical Psychology, 71*, 117–145.

Yamayoshi, M., Tsuchida, J., & Yadohisa, H. (2020). An estimation of causal structure based on latent LiNGAM for mixed data. *Behaviormetrika, 47*(1), 105–121. doi:10.1007/s41237-019-00095-3

Zhang, K., & Hyvärinen, A. (2016). Nonlinear functional causal models for distinguishing cause from effect. In W. Wiedermann & A. von Eye (Eds.), *Wiley series in probability and statistics* (pp. 185–201). doi:10.1002/9781118947074.ch8

Part I

Fundamental Concepts of Direction Dependence

1

From Correlation to Direction Dependence Analysis 1888–2018

Yadolah Dodge[1] and Valentin Rousson[2]

[1]*Institute of Statistics, University of Neuchâtel, Neuchâtel, Switzerland*
[2]*Division of Biostatistics, Center for Primary Care and Public Health (Unisanté), University of Lausanne, Lausanne, Switzerland*

1.1 Introduction

The Pearson product-moment correlation coefficient is one of the most popular statistical measure to summarize an association between two (continuous) variables X and Y. As suggested by Rodgers and Nicewander (1988), it should actually be renamed the "Galton–Pearson" correlation coefficient since both men played a significant role in the development and promotion of this coefficient in statistics. The concept of correlation was introduced by Francis Galton in 1888 (Blyth, 1994; Galton, 1888), although it was already presented in 1885 in relation to regression (Galton, 1885; Rodgers & Nicewander, 1988), while Karl Pearson (1895) provided the mathematical formula. See e.g. Stigler (1989) for some detailed historical account. Although Pearson (1930) quoted in Aldrich (1995) wrote that "up to 1889 men of sciences had thought only in terms of causation," it was clear from the very beginning that "correlation does not imply causation." For example, Aldrich (1995) mentioned that Francis Galton (1888) was well aware that "the correlation between two variables measures the extent to which they are governed by common causes." Thus, establishing a correlation between X and Y does not imply that one variable is the cause and the other is the (direct or indirect) consequence, just that the two variables are associated, due perhaps to the existence of a third variable Z which would be a common cause of both X and Y. In fact, even if one could rule out completely the possibility of the existence of such a variable Z, there would be no way to conclude from a correlation which of X and Y is the cause and which is the consequence, since the formula provided by Karl Pearson is perfectly (and beautifully) symmetric in X and Y.

Direction Dependence in Statistical Modeling: Methods of Analysis, First Edition.
Edited by Wolfgang Wiedermann, Daeyoung Kim, Engin A. Sungur, and Alexander von Eye.
© 2021 John Wiley & Sons, Inc. Published 2021 by John Wiley & Sons, Inc.

1.2   Correlation as a Symmetrical Concept of *X* and *Y*

Given a sample of n observations (X_i, Y_i) $(i = 1, \ldots, n)$ from a bivariate variable (X, Y), the (Pearson product-moment) correlation (coefficient) can be calculated as:

$$r_{XY} = \frac{\sum_{i=1}^{n}(X_i - \overline{X})(Y_i - \overline{Y})}{\sqrt{\sum_{i=1}^{n}(X_i - \overline{X})^2 \sum_{i=1}^{n}(Y_i - \overline{Y})^2}} \tag{1.1}$$

where $\overline{X} = \sum_{i=1}^{n} X_i/n$ and $\overline{Y} = \sum_{i=1}^{n} Y_i/n$ denote the sample means of X and Y. This is also the covariance between X and Y divided by the product of their standard deviations. Obviously, one has $r_{XY} = r_{YX}$. As mentioned in Section 1.1, correlation is intimately related to regression. Let us consider the regression equation with Y as the response variable and X as the predictor:

$$Y_i = a_{YX} + b_{YX}X_i + \varepsilon_i \tag{1.2}$$

as well as the regression equation with X as the response variable and Y as the predictor:

$$X_i = a_{XY} + b_{XY}Y_i + \varepsilon_i'. \tag{1.3}$$

If the goal is to get residuals ε_i and ε_i' with zero mean and with the smallest possible variances, the regression coefficients are obtained via the least squares criterion, which for the slopes are given by:

$$b_{YX} = \frac{\sum_{i=1}^{n}(X_i - \overline{X})(Y_i - \overline{Y})}{\sum_{i=1}^{n}(X_i - \overline{X})^2} \tag{1.4}$$

and by:

$$b_{XY} = \frac{\sum_{i=1}^{n}(X_i - \overline{X})(Y_i - \overline{Y})}{\sum_{i=1}^{n}(Y_i - \overline{Y})^2}. \tag{1.5}$$

Thus, the correlation is also the geometrical mean of the slopes in Eqs. (1.2) and (1.3):

$$r_{XY} = \sqrt{b_{YX} \cdot b_{XY}}. \tag{1.6}$$

Again, this is a symmetrical formula in X and Y. Many other ways to calculate or to interpret a correlation have been provided in the statistical literature. In particular, Rodgers and Nicewander (1988) identified 13 ways to look at the correlation, whereas a 14th way has been added to the list by Rovine and von Eye (1997), and even more by Falk and Well (1997). However, all these formulas, when involving two continuous variables, are symmetrical in X and Y.

1.3   Correlation as an Asymmetrical Concept of *X* and *Y*

When one considers a linear regression model (1.2) or (1.3), one usually assumes residuals which are normally distributed and with the same variance (homoscedasticity), yielding independence between the predictor and the residual variable (the residual distribution is the same whatever the value of predictor), which is also what is assumed in what follows. In that case, models (1.2) and (1.3) cannot hold simultaneously unless the distribution of (X, Y) is bivariate normal. In particular, if one considers that both X and Y are non-normal, at most one of (1.2) and (1.3) may hold. It is in such a context of non-normal X and Y that Dodge and Rousson (2000, 2001) introduced further formulas to interpret a correlation. Under model (1.2), and using basic properties of cumulants (see e.g. Kendall & Stuart, 1963), which differ from zero in case of non-normal variables, they noted that:

$$\text{cumulant}_m(Y) = r_{YX}^m \cdot \text{cumulant}_m(X) \tag{1.7}$$

where $\text{cumulant}_m(V)$ denotes the mth (standardized) cumulant of a random variable V, where $m \geq 3$, yielding the skewness coefficient for $m = 3$. One has thus for example:

$$r_{XY}^3 = \frac{\text{skewness}\ (Y)}{\text{skewness}\ (X)}. \tag{1.8}$$

Thus, although it is well known that the square of a correlation between two variables is the percentage of variance of one variable that one can linearly predict from the other, this holding also in the case where (1.2) and/or (1.3) do not hold, the cube of the correlation coefficient can be interpreted as the ratio of skewnesses of the response and of the predictor. On the other hand, if (1.3) holds, one will similarly get:

$$r_{XY}^3 = \frac{\text{skewness}\ (X)}{\text{skewness}\ (Y)}. \tag{1.9}$$

Thus, contrary to the numerous interpretations of a correlation mentioned above, the interpretation of the cube of a correlation is not symmetric in X and Y, (1.8) holding if (1.2) holds, and (1.9) holding if (1.3) holds. This may provide a criterion to choose between models (1.2) and (1.3), where one could select that model for which the response variable has the smallest skewness (i.e. select model (1.2) if Y has a smaller skewness than X, and select model (1.3) if X has a smaller skewness than Y).

1.4 Outlook and Conclusions

Formulas (1.8) and (1.9), although easily derived, might be surprising at first sight to many statisticians, including to some professional reviewers, since we are so much used to see the correlation presented as a symmetric concept of the variables involved. Discovering that it was actually possible (under some conditions) to statistically distinguish between models (1.2) and (1.3) based on the correlation led to a new domain of research, now called "direction dependence analysis." In particular, Alexander von Eye and Wolfgang Wiedermann initiated an impressive work in this new area, looking for further asymmetries in linear models and extending these concepts to other models, including multiple regression and categorical variables, among others (e.g. von Eye & Wiedermann, 2016; Wiedermann, Artner, & von Eye, 2017; Wiedermann & von Eye, 2015a, b, c; Wiedermann, Hagmann, & von Eye, 2014). While the very term of "direction dependence analysis" sounds as a promising step toward the golden concept of "causality," which is perhaps the ultimate goal of science, one should of course remain cautious. One issue are the outliers, whose detection might be even more problematic with skewed data, although new approaches have been recently proposed (Kimber, 1990; Walker, Dovoedo, Chakraborti, & Hilton, 2018). Dodge and Rousson (2016), discussing the influence of outliers on the testing procedures developed for direction dependence analysis, noted that "removing these observations may bring us back towards the normality which we wish to avoid," before concluding that "at the end, proving statistically the direction of a regression line, let alone causality, certainly remains a major challenge." Thus, despite the discovery of asymmetrical formulas for the correlation, one will still hear for a while that "correlation does not imply causation."

References

Aldrich, J. (1995). Correlations genuine and spurious in Pearson and Yule. *Statistical Science, 10,* 364–376.

Blyth, S. (1994). Karl Pearson and the correlation curve. *International Statistical Review, 62,* 393–403.

Dodge, Y., & Rousson, V. (2000). Direction dependence in a regression line. *Communications in Statistics: Theory and Methods, 29,* 1957–1972.

Dodge, Y., & Rousson, V. (2001). On asymmetric properties of the correlation coefficient in the regression setting. *The American Statistician, 55,* 51–54.

Dodge, Y., & Rousson, V. (2016). Statistical inference for direction of dependence in linear models. In W. Wiedermann & A. von Eye (Eds.), *Statistics and causality: Methods for applied empirical research* (pp. 45–62). Hoboken, NY: Wiley.

Falk, R., & Well, A. (1997). Many faces of the correlation coefficient. *Journal of Statistics Education, 5*(3), 1–18.

Galton, F. (1885). Regression towards mediocrity in hereditary stature. *Journal of the Anthropological Institute, 15*, 246–263.

Galton, F. (1888). Co-relations and their measurement, chiefly from anthropological data. *Proceedings of the Royal Society of London, 45*, 135–145.

Kendall, M., & Stuart, A. (1963). *The advanced theory of statistics* (2nd ed.). New York, NY: Hafner.

Kimber, A. (1990). Exploratory data analysis for possibly censored data from skewed distributions. *Applied Statistics, 39*, 21–30.

Pearson, K. (1895). Notes on regression and inheritance in the case of two parents. *Proceedings of the Royal Society of London, 58*, 240–242.

Pearson, K. (1930). *The life, letters and labours of Francis Galton* (Vol. *3A*). Cambridge, UK: Cambridge University Press.

Rodgers, J., & Nicewander, W. (1988). Thirteen ways to look at the correlation coefficient. *The American Statistician, 42*, 59–66.

Rovine, M., & von Eye, A. (1997). A 14th way to look at a correlation coefficient: Correlation as the proportion of matches. *The American Statistician, 51*, 42–46.

Stigler, S. (1989). Francis Galton's account of the invention of correlation. *Statistical Science, 4*, 73–79.

von Eye, A., & Wiedermann, W. (2016). Direction of effects in categorical variables: A structural perspective. In W. Wiedermann & A. von Eye (Eds.), *Statistics and causality: Methods for applied empirical research* (pp. 107–130). New York, NY: Wiley.

Walker, M., Dovoedo, Y., Chakraborti, S., & Hilton, C. (2018). An improved boxplot for univariate data. *The American Statistician, 72*, 348–355.

Wiedermann, W., Artner, R., & von Eye, A. (2017). Heteroscedasticity as a basis of direction dependence in reversible linear regression models. *Journal of Multivariate Behavioral Research, 52*, 222–241.

Wiedermann, W., Hagmann, M., & von Eye, A. (2014). Significance tests to determine the direction of effects in linear regression models. *British Journal of Mathematical and Statistical Psychology, 68*, 116–141.

Wiedermann, W., & von Eye, A. (2015a). Direction of effects in mediation analysis. *Psychological Methods, 20*, 221–244.

Wiedermann, W., & von Eye, A. (2015b). Direction of effects in multiple linear regression models. *Journal of Multivariate Behavioral Research, 50*, 23–40.

Wiedermann, W., & von Eye, A. (2015c). Direction-dependence analysis: A confirmatory approach for testing directional theories. *International Journal of Behavioral Development, 39*, 570–580.

2

Direction Dependence Analysis

Statistical Foundations and Applications

Wolfgang Wiedermann[1], Xintong Li[2], and Alexander von Eye[3]

[1]*Department of Educational, School, and Counseling Psychology, College of Education, & Missouri Prevention Science Institute, University of Missouri, Columbia, MO, USA*
[2]*Assessment Resource Center, University of Missouri, Columbia, MO, USA*
[3]*Department of Psychology, Michigan State University, East Lansing, MI, USA*

In this chapter, we introduce principles of direction dependence and present a unified statistical framework – direction dependence analysis (DDA) (Wiedermann & Li, 2018; Wiedermann & Sebastian, in press-a; Wiedermann & von Eye, 2015c) – to discern the causal direction of effects in linear models from observational data. Within the last decades, tremendous progress has been made in causal inference research. The potential outcome framework (Holland, 1986; Neyman, 1923; Rubin, 1974) provides a conceptual and formal basis for causal inference. Further, the development of causal graph theory (Greenland, Pearl, & Robins, 1999; Greenland & Robins, 1986; Pearl, 2009) enables researchers to evaluate whether a causal effect can be uniquely identified based on the hypothesized causal structure of variables. Covariance-based methods such as linear regressions and structural equation modeling (SEM) allow researchers to quantify the magnitude of hypothesized causal effects. However, these approaches are of limited use when establishing the causal direction of effects between variables, that is, whether $x \rightarrow y$ or $y \rightarrow x$ correctly describes the causal flow between two variables x and y. This can be explained by the inherent symmetry of intersection probabilities of events (cf. Pearl, 2009), i.e. the intersection probability $P(A \text{ and } B)$ for two events A and B is identical to the intersection probability $P(B \text{ and } A)$. Thus, measures of dependence such as the covariance are entirely symmetric (i.e. $\text{cov}(x, y) = \text{cov}(y, x)$) and cannot be used to discern statements about the causal direction of effects. In contrast, the presented DDA framework makes use of asymmetry properties of the linear model under variable non-normality (i.e. when variables deviate from the Gaussian distribution) which allow researchers to critically

Direction Dependence in Statistical Modeling: Methods of Analysis, First Edition.
Edited by Wolfgang Wiedermann, Daeyoung Kim, Engin A. Sungur, and Alexander von Eye.
© 2021 John Wiley & Sons, Inc. Published 2021 by John Wiley & Sons, Inc.

evaluate the causal precedence of variable pairs and the presence of influential confounders.

Causally competing theories are omnipresent in the educational, behavioral, and developmental sciences. In educational research, for example, one may theorize that math achievement causally affects math anxiety (*achievement → anxiety*; i.e. previous experiences of low math performance may lead to the development of math anxiety; Ma & Xu, 2004). This causal mechanism implies that one can expect that math anxiety can (to a certain degree) be reduced by providing additional math training. In contrast, as an alternative causal model, one can hypothesize that it is math anxiety that lowers math achievement (i.e. *anxiety → achievement*; Ramirez, Chang, Maloney, Levine, & Beilock, 2016). In this case, one would expect changes in math achievement after providing proper stress management training. From a developmental perspective, one could hypothesize a reciprocal causal relation in which achievement and anxiety affect each other in feedback loops that develop over time (Gunderson, Park, Maloney, Beilock, & Levine, 2018). In this case, the causal direction of effect for one of the two models may be more pronounced at a specific point in time than the causal effect for the competing model. In other words, the causally dominant mechanism varies with the snapshot of measurement.

Longitudinal data are usually used to evaluate unidirectional and reciprocal relations of variables. Here, however, it is important to realize that temporality alone is not sufficient to "exogenize" a tentative cause variable. That is, even when one variable (x_t; where the subscript t indicates the measurement occasion) is measured prior to the other variable (y_{t+1}), one has not gathered sufficient information to guarantee that the assumed causal mechanism $x \to y$ reflects the "true" underlying data-generating mechanism. The reason for this is that one cannot rule out that y active at an earlier point in time may have affected x_t. In addition, when estimating the lagged effect of the causal model $x_t \to y_{t+1}$ as an unbiased "conservative" estimate of the causal effect $x_t \to y_t$ (i.e. lag identification can, at best, estimate the contemporaneous effect weighted by the autocorrelation in x) one still has to assume that unconsidered confounders are absent. Using this lagged regression approach in the presence of a confounder u opens a "back door" path through $u_t \to x_t$ and $u_t \to u_{t+1} \to y_{t+1}$ (cf. Bellemare, Masaki, & Pepinsky, 2017). Similarly, when one directly compares the two cross-lagged paths $x_t \to y_{t+1}$ and $y_t \to x_{t+1}$ confounding can bias one path more toward zero than the other, which may lead to erroneous conclusions concerning the causal directionality of the *x–y* relation. In this chapter, we focus on cross-sectional data scenarios with the understanding that principles of direction dependence can also be applied in lagged regression contexts.

The chapter is structured as follows: In Section 2.1, we provide a brief summary of several research areas that show connections with the presented DDA framework. Such areas include methodological developments in theoretical statistics, econometrics, psychometrics, and machine learning research. Section 2.2 introduces fundamental asymmetry properties of cause–effect relations. In Section 2.3, we introduce the statistical foundations of DDA through describing several asymmetry properties of the linear model. These asymmetry properties constitute the basis to test the superiority of one causal model over a set of alternative explanatory models and can be observed for (i) distributional characteristics of observed variables, (ii) distributional characteristics of errors, and (iii) independence properties of predictors and errors in causally competing models. We will show that, taken together, these three components are sufficient to uniquely identify the "true" causal model (if it exists). Model selection guidelines and methods of statistical inference are discussed. Sections 2.4 and 2.5 of the chapter are devoted to testing the direction of dependence in mediation (i.e. in the presence of a third intervening variable that transmits the effect from x to y) and moderation (i.e. in the presence of causal effect heterogeneity). Section 2.6 summarizes previous applications and software implementations. The chapter closes with a discussion of potential pitfalls and directions for future research (Section 2.7).

2.1 Some Origins of Direction Dependence Research

Non-normality of observed variables and errors constitutes the key ingredient to derive statements about the direction of dependence of variable relations. Under non-normality, the linear regression model has various asymmetry properties that allow one to evaluate the causal direction of effects and to identify potential influential confounders. However, putting non-normality of variables to use in statistical modeling is certainly not new. Non-normality of observed variable distributions has been known to be highly prevalent for decades (Blanca, Arnau, López-Montiel, Bono, & Bendayan, 2013; Cain, Zhang, & Yuan, 2017; Cook, 1959; Ho & Yu, 2015; Lord, 1955; Micceri, 1989) and has motivated methodological research in areas such as theoretical statistics, econometrics, psychometrics, and machine learning. Early research largely focused on the negative consequences of non-normality in the context of testing hypotheses of location parameters (e.g. testing the difference in means of two groups) using, for example, Student's t-test or the Wilcoxon–Mann–Whitney U-test (Bartlett, 1935; Boneau, 1960; Posten, 1978, 1982; Rasch & Guiard, 2004; Wiedermann & Alexandowicz, 2007, 2011). This line of robustness research contributed to the dominant view among practitioners that non-normality is a threat to statistical inference (Stigler, 2010). The main message of this line of research is that parametric tests (such as t- and F-tests) are remarkably robust against non-normality (Box, 1953; Box & Watson,

1962). Distorted Type I error and power rates can only be expected under extreme distributions (e.g. mixture and extremely skewed distributions).

In a parallel development, researchers started to explore whether variable non-normality carries valuable information that can be useful in data modeling. In the context of error-in-variable (EIV) models, Geary (1949), Scott (1950), Reiersøl (1950), Drion (1951), Durbin (1954), Spiegelman (1979), Pal (1980), and Cragg (1997) proposed higher moment estimators when the predictor is subject to measurement error. An overview of higher moment estimators is given in Gillard (2014). Direction dependence properties of EIV models have been discussed by Wiedermann, Merkle, and von Eye (2018). In econometrics, higher-moment based internal instrumental variables (IVs) have been proposed to identify the parameters of EIV models (Dagenais & Dagenais, 1997; Erickson, 2001; Lewbel, 1997) and, in psychometrics, higher moment estimators have been proposed for SEM (Bentler, 1983) which led to the development of non-normal factor analysis (Mooijaart, 1985) and non-normal SEMs (Shimizu & Kano, 2008). Further, Mooijaart and Satorra (2012) proposed the use of higher moments to estimate latent interaction terms in SEMs. As another example, in machine learning research, non-normal SEMs motivated the development of a causal learning algorithm known as the linear non-Gaussian acyclic model (LiNGAM) (Shimizu, Hoyer, Hyvärinen, & Kerminen, 2006; see also the related Chapter 5 by Shimizu and Blöbaum, in this volume). Here, LiNGAM makes use of variable non-normality to discern causal networks from observational data alone. While model identification of LiNGAM was originally based on algorithms developed for independent component analysis (ICA) (Hyvärinen, Karhunen, & Oja, 2001), Shimizu et al. (2011) suggested using independence properties of regressors and errors (similar to DDA's independence component; see below) to identify causal networks beyond Markov-equivalent classes of models. Nonlinear extensions of LiNGAM have been proposed by, for example, Peters, Mooij, Janzing, and Schölkopf (2014).

As a parallel development, in theoretical statistics, Dodge and Rousson (2000, 2001), Muddapur (2003), and Sungur (2005) were among the first to discuss asymmetry properties of the Pearson correlation and the related linear regression line. While direction dependence in the tradition of Dodge and Rousson (2000) has been linked to the issue of homoscedasticity (Wiedermann, Artner, & von Eye, 2017) and has been extended to mediation models (Wiedermann & Li, 2019a; Wiedermann, Li, & von Eye, 2019; Wiedermann & Sebastian, ; Wiedermann & von Eye, 2015a) and log-linear models (von Eye & Wiedermann, 2018; Wiedermann & von Eye, 2020), Sungur's (2005) copula approach further led to the development of so-called copula-based directional dependence analysis (see, e.g. Kim & Kim, 2014, together with the related Chapter 3 by Sungur, in this volume). Copula-based directional dependence analysis allows one to evaluate causal structures of variables based on joint (instead of marginal) distributions of variables. Here, it is important to note that copula-based directional dependence analysis should not be conceptualized as a competitor to the DDA framework presented here. Compared to Sungur's (2005) copula-based approach, DDA (at least in part) relies on information on marginal distributions. Direction dependence in marginal distributions, however, does not imply direction of dependence in joint variable distributions (and vice versa). Therefore, both approaches are best conceptualized as complementary in nature. In Section 2.2, we review the statistical foundations of DDA, discuss methods of statistical inference, and present decision guidelines for DDA model selection.

2.2 Causation and Asymmetry of Dependence

The fundamental property of causation that distinguishes it from association is *asymmetry in the dependence structure* (Pearl, 2009). This asymmetry can manifest, for example, in distributional properties of variables (as discussed in detail below) or in the dependence structure of variables within the concept of copulas (see the related Chapter 3 by Sungur, in this volume). In general, asymmetry of cause and effect implies that changing one variable x (the cause) leads to changes in another variable y (the effect), however, changes in y do not lead to changes in x. Using probability notation, we can formalize this asymmetry as follows: Under a causal model of the form $x \rightarrow y$ (i.e. x is the cause of y) the distribution of y conditional on the intervention $do(x = x_i)$ (i.e. forcing the variable x to be constant at the value x_i, (cf. Pearl, 1995; Rubin, 1974)) is different from the marginal distribution of y, i.e. $P(y \mid do(x) = x_i) \neq P(y)$. However, since one cannot expect to see changes in x when changing y under the model $x \rightarrow y$, the distribution of x conditional on $do(y) = y_i$ will not be different from the marginal distribution of x. In other words, one obtains $P(x \mid do(y) = y_i) = P(x)$.

In the linear model, for example, we can express this fundamental asymmetry as follows: Assume that the causal model $x \rightarrow y$ takes the form $y = \beta_{yx}x + e_{yx}$ with β_{yx} being a constant and e_{yx} denoting the error term. Further, assume that, in the post-intervened model, x takes value a, i.e. $y = \beta_{yx}a + e_{yx}$ and one is interested in estimating the expected change in y when changing x from a (e.g. the control group) to b (the treatment group). In this case, the average causal effect of x on y for the two values a and b can be written as (see, e.g. Shimizu, 2019)

$$E[y \mid do(x) = a] - E[y \mid do(x) = b]$$
$$= E[\beta_{yx}a + e_{yx}] - E[\beta_{yx}b + e_{yx}]$$
$$= \beta_{yx}(a - b).$$

In other words, the average causal effect equals β_{yx} multiplied by the difference in the values a and b. In contrast, for the causally reversed model $x = \beta_{xy}y + e_{xy}$ one obtains

$$E[x \mid do(y) = a] - E[x \mid do(y) = b]$$
$$= E[\beta_{xy}y + e_{xy}] - E[\beta_{xy}y + e_{xy}]$$
$$= 0$$

because y does not affect x. Therefore, the average causal effect of y on x is zero as expected.

The inherent asymmetry of cause and effect suggests that if one is interested in making statements about the causal flow of variables (i.e. evaluating whether $x \rightarrow y$ or $y \rightarrow x$ better approximates the data-generating mechanism), one has to use dependency measures that preserve the asymmetry of cause and effect. In Section 2.3, we describe asymmetry properties of the linear model which emerge from the underlying causal mechanism when observed variables are non-normal. Based on these asymmetry properties, we describe a series of statistical measures that can be used to distinguish a causal target model from plausible alternatives (e.g. $x \rightarrow y$ versus $y \rightarrow x$).

2.3 Foundations of Direction Dependence

In general, statistical models can either be used for the purposes of *explanation* or *prediction* (Geisser, 1993; Shmueli, 2010). Both types of statistical models are crucial for theory building (Braun & Oswald, 2011). In the context of observational studies, both types of models can be characterized as association-based models. However, each model plays a unique role in the process of evaluating and refining substantive theories. In explanatory models, a priori theories carry the crucial element of causation and the goal is to match a statistical model f and an underlying mechanism $\mathcal{F}$ and use x and y as tools to estimate and validate f for the purpose of testing the causal hypothesis of interest (Shmueli, 2010). In contrast, in models used for prediction, f is considered being a tool capturing variable associations and x and y are of primary interest to build valid models for the purpose of adequately forecasting new outcome values (Shmueli, 2010).

DDA is primarily concerned with the task of evaluating statistical properties of explanatory (causal) models. Suppose that $\mathcal{X}$ is a construct that causes another construct $\mathcal{Y}$ through the causal mechanism $\mathcal{F}$, i.e. $\mathcal{Y} = \mathcal{F}(\mathcal{X})$. Let x and y be operationalizations of the constructs $\mathcal{X}$ and $\mathcal{Y}$ and let f be the statistical model to approximate the unknown causal mechanism $\mathcal{F}$, i.e. $y = f(x)$. DDA provides a set of statistical tools to evaluate asymmetry properties of the statistical model $y = f(x)$

which emerge from the hypothesized causal effect directionality implied by the causal theory $\mathcal{X} \to \mathcal{Y}$. Throughout the chapter, we assume that the linear model is sufficient to adequately represent $\mathcal{F}$. The model for the mechanism $\mathcal{X} \to \mathcal{Y}$ then takes the from (for simplicity, but without lack of generality, we assume that x and y have been standardized)

$$y = \beta_{yx}x + e_{yx} \tag{2.1}$$

with β_{yx} being the causal effect of x on y and e_{yx} denoting the corresponding error term which is assumed to be independent of x with zero mean and variance $\sigma^2_{e_{yx}}$. The causally competing model (representing the causally competing mechanism $\mathcal{Y} \to \mathcal{X}$) can be written as

$$x = \beta_{xy}y + e_{xy} \tag{2.2}$$

with β_{xy} denoting the causal effect of y on x and e_{xy} being the corresponding error term. Due to variable standardization, estimated regression coefficients in (2.1, 2.2) will be identical to the Pearson correlation coefficient $\rho_{xy} = \text{cov}(x, y)/(\sigma_x\sigma_y)$ with $\text{cov}(x, y)$ being the covariance and σ_x and σ_y being the standard deviations of x and y. Before we turn to asymmetry properties of the linear model under variable non-normality, we want to emphasize the data requirements and assumptions that are necessary to guarantee valid DDA model selection.

2.3.1 Data Requirements

Because standard association-based measures are inherently symmetric in nature (i.e. association-based measures do not depend on variable order), asymmetry properties for causally competing models are required to make statements about the causal direction of dependence. In the linear model, asymmetry properties exist under non-normality of variables. Thus, in the DDA framework considered here, the "true" predictor (e.g. x as an operationalization of the cause $\mathcal{X}$) is assumed to be a *continuous, non-normal, exogenous variable*. In other words, it is assumed that (i) x is measures at least on an interval scale, (ii) x deviates from the perfect Gaussian distribution, and (iii) the causes of x lie outside the model. Because higher moment properties of observed variables constitute the core element to test principles of direction dependence, it is assumed that distributional characteristics of x adequately reflect the underlying distributional characteristics of the construct $\mathcal{X}$. This implies that not every observed non-normal variable is suitable for the analysis of the direction of dependence. Non-normality due to, for example, outliers or ceiling/floor effects (i.e. left/right censoring) can lead to biased DDA results. Therefore, a careful evaluation of the nature of non-normality is essential to guarantee valid applications. Similarly, we assume the absence of potential selection biases. Preferential exclusion of subjects due to a non-random sample of the population can alter the distributional properties of observed variables to an extent that valid DDA model selection can be in jeopardy. In practice, selection biases can be counteracted by clearly defining the study population and the sampling procedure.

Covariates (z) can be included in DDA to address the problem of confounding and to increase the precision of causal estimates and statistical power. In general, selection of covariates is among the most difficult tasks in causal analyses. In DDA, covariates are defined as additional continuous or categorical exogenous variables (such as demographics and background variables) and need to fulfill the following criteria to be eligible for the analysis of the direction of dependence: (i) Covariates are not allowed to render the model circular (e.g. $z \to x \to y \to z$), (ii) in case of multiple covariates, cyclic relations among covariates (e.g. $z_1 \to z_2 \to z_3 \to z_1$) are not allowed (i.e. the causal flow is not allowed to go through a variable more than once), (iii) covariates are not allowed to be common causes of x and y, and (iv) it is assumed that covariates are correctly specified in the model (e.g. one properly accounts for nonlinear relations of z and y).

2.3.2 DDA Component I: Distributional Properties of Observed Variables

The first set of asymmetry properties emerges from the additive nature of the linear model, that is, the outcome variable is generated as the sum of a predictor (weighted by the causal effect) and an independent error component. For simplicity of presentation, we assume that observed variables x and y have zero means and unit variances which does not hamper generality as any variable can be trivially put in this from by way of standardization. To introduce asymmetry properties, we assume that $\mathcal{X} \to \mathcal{Y}$ corresponds to the "true" causal mechanism and $\mathcal{Y} \to \mathcal{X}$ constitutes the causally mis-specified mechanism. We use so-called higher order correlations (HOCs) to describe third and fourth moments of x and y. HOCs can be defined as higher powers of the Pearson correlation, $cor(x,y)_{ij} = cov(x,y)_{ij}/(\sigma_x^i \sigma_y^j)$ with $cov(x,y)_{ij} = E[(x - E[x])^i (y - E[y])^j]$ being the higher order covariance (in distribution theory $cov(x,y)_{ij}$ is known as the ij-th central product moment). Note that $cor(x,y)_{11} = \rho_{xy}$ refers to the standard Pearson correlation coefficient. For the third higher moments of x and y, one obtains the following HOCs:

$$cor(x,y)_{30} = \gamma_x \tag{2.3}$$

$$cor(x,y)_{21} = \rho_{xy}\gamma_x \tag{2.4}$$

$$cor(x,y)_{12} = \rho_{xy}^2 \gamma_x \tag{2.5}$$

$$cor(x,y)_{03} = \gamma_y = \rho_{xy}^3 \gamma_x + (1 - \rho_{xy}^2)^{3/2}\gamma_{e_{yx}} \tag{2.6}$$

with $\gamma_x = E[(x - E[x])^3]/\sigma_x^3$, $\gamma_y = E[(y - E[y])^3]/\sigma_y^3$, and $\gamma_{e_{yx}} = E[e_{yx}{}^3]/\sigma_{e_{yx}}^3$ denoting the skewnesses of x, y, and e_{yx}. The two measures $\text{cor}(x, y)_{30}$ and $\text{cor}(x, y)_{03}$ describe the asymmetry of marginal variable distributions (i.e. the univariate distributions of x and y) and the measures $\text{cor}(x, y)_{21}$ and $\text{cor}(x, y)_{12}$ denote co-skewnesses describing asymmetry of variables in terms of joint variable distributions (i.e. distributions that take into account the relatedness of x and y). Based on (2.3–2.6), two different DDA measures can be constructed. The first measure has been proposed by Dodge and Rousson (2000, 2001) and emerges from (2.3, 2.6). Assuming that the "true" error term e_{yx} is symmetrically distributed (i.e. $\gamma_{e_{yx}} = 0$), one obtains the following asymmetric formulation of the Pearson correlation

$$\rho_{xy}^3 = \frac{\gamma_y}{\gamma_x}. \tag{2.7}$$

Because ρ_{xy} is bounded on the interval $[-1, 1]$, the outcome variable y will always be more symmetric than the predictor (excluding $|\rho_{xy}| = 1$ due to practical irrelevance). Further, the cube of the Pearson correlation can be interpreted as the amount of skewness that is preserved by the linear model. From the perspective of direction of dependence, we can deduce the following decision rule from (2.7): Under the model $x \to y$, one obtains $|\gamma_y| < |\gamma_x|$, under the causally reversed model $y \to x$, one obtains $|\gamma_y| > |\gamma_x|$. Bias adjusted and accelerated bootstrap confidence intervals (BCa CIs) can be used to perform statistical inference on differences in skewnesses $\Delta = |\gamma_y| - |\gamma_x|$.

The second DDA measure makes use of (2.4, 2.5) and, in contrast to the measure in (2.7), does not impose any distributional assumptions on the "true" error term e_{yx}. From (2.4, 2.5), one obtains

$$\rho_{xy} = \frac{\text{cor}(x, y)_{12}}{\text{cor}(x, y)_{21}} \tag{2.8}$$

which implies that the Pearson correlation can be expressed as the ratio of co-skewnesses of x and y. Further, under the causal model $x \to y$, the co-skewness of the form $\{i = 1, j = 2\}$ will always be smaller than the co-skewness $\{i = 2, j = 1\}$. In contrast, under the causally reversed model $y \to x$, the co-skewness $\{i = 2, j = 1\}$ will always be smaller than the co-skewness $\{i = 1, j = 2\}$. Therefore, a systematic comparison of co-skewnesses can, again, inform researchers about the directionality of causal effects. When the BCa CI suggests that $|\text{cor}(x, y)_{21}| - |\text{cor}(x, y)_{12}|$ is larger than zero, one has found evidence for the causal model $x \to y$, when the BCa CI indicates $|\text{cor}(x, y)_{21}| - |\text{cor}(x, y)_{12}| < 0$, the reversed causal model $y \to x$ is more likely to reflect the underlying data generating mechanism. In general, when the symmetry assumption of the "true" error (used to construct the DDA measure based on Eq. (2.7)) holds, using differences in co-skewnesses comes at the cost of lower statistical power. This follows from the relations $\gamma_y = \rho_{xy}^3 \gamma_x$, $\text{cor}(x, y)_{12} = \rho_{yx}^2 \gamma_x$, and $\text{cor}(x, y)_{21} = \rho_{yx} \gamma_x$ which imply that $|\gamma_y| < |\text{cor}(x, y)_{12}| < |\text{cor}(x, y)_{21}| < |\gamma_x|$ and therefore $|\text{cor}(x, y)_{21}| - |\text{cor}(x, y)_{12}| < |\gamma_x| - |\gamma_y|$.

Next, we turn to fourth-higher moments of x and y. Again, two different DDA measures are available. The first fourth-moment based measure assumes that the excess kurtosis of the "true" error is zero and describes the fourth higher power of the Pearson correlation as the ratio of (excess) kurtosis values of the outcome and the predictor (Dodge & Yadegari, 2010),

$$\rho_{xy}^4 = \frac{\kappa_y - 3}{\kappa_x - 3} \tag{2.9}$$

with $\kappa_x = E[(x - E[x])^4]/\sigma_x^4$, and $\kappa_y = E[(y - E[y])^4]/\sigma_y^4$ being the kurtosis of x and y. Again, two immediate consequences emerge from the expression in (2.9): (i) the fourth power of the Pearson correlation can be interpreted as the amount of kurtosis that is preserved by the linear model, and (ii) the outcome variable will always be less kurtotic than the predictor. From the perspective of direction of dependence, one can conclude that $|\kappa_y - 3| < |\kappa_x - 3|$ under the model $x \to y$ and $|\kappa_x - 3| < |\kappa_y - 3|$ under the reversed model $y \to x$. Again, BCa CIs can be used to perform statistical inference.

The second fourth-moment based DDA measure is based on the two co-kurtosis measures obtained through defining HOCs using $i = j = \{1, 3\}$ (Hyvärinen & Smith, 2013; Wiedermann, 2018). Here, one obtains

$$\text{cor}(x, y)_{31} = \rho_{xy}\kappa_x \tag{2.10}$$

$$\text{cor}(x, y)_{13} = \rho_{xy}^3\kappa_x + 3\rho_{xy}(1 - \rho_{xy}^2) \tag{2.11}$$

and one arrives at the following interesting properties: (i) when $\kappa_x = 3$ (e.g. x is normally distributed), then $\text{cor}(x, y)_{13} = \text{cor}(x, y)_{31} = 3\rho_{xy}$ and, thus, no directional decisions can be made, (ii) when the true predictor x is leptokurtic (i.e. $\kappa_x > 3$), then $|\text{cor}(x, y)_{31}| > |\text{cor}(x, y)_{13}|$, and (iii) when x is platykurtic (i.e. $\kappa_x < 3$), then $|\text{cor}(x, y)_{31}| < |\text{cor}(x, y)_{13}|$.

In particular, the latter two properties carry valuable information to test the direction of dependence. However, to make directional decisions independent of the type of symmetric non-normal distribution, the difference measure $\Delta = |\text{cor}(x, y)_{31}| - |\text{cor}(x, y)_{13}|$ can be multiplied by the signum function $\text{sgn}(\kappa_x - 3)$ (taking value 1 if $\kappa_x - 3 > 0$ and -1 if $\kappa_x - 3 < 0$). After multiplication, when $\text{sgn}(\kappa_x - 3) = \text{sgn}(\kappa_y - 3)$ (which can be expected to hold for a broad range of applications), one expects $\Delta > 0$ when the model $x \to y$ corresponds to the data-generating mechanism and $\Delta < 0$ under $y \to x$. When $\text{sgn}(\kappa_x - 3) \neq \text{sgn}(\kappa_y - 3)$, the two quantities $\Delta_{(x)} = \{|\text{cor}(x, y)_{31}| - |\text{cor}(x, y)_{13}|\} \text{sgn}(\kappa_x - 3)$ and $\Delta_{(y)} = \{|\text{cor}(y, x)_{31}| - |\text{cor}(y, x)_{13}|\} \text{sgn}(\kappa_y - 3)$ can be computed. One then obtains $\Delta_{(x)} > 0$ under $x \to y$ and $\Delta_{(y)} > 0$ under $y \to x$. When both difference measures have the same sign (e.g. due to suboptimal sampling or model violations), no direction dependence decisions can be made.

2.3.3 DDA Component II: Distributional Properties of Errors

The next DDA component focuses on distributional properties of errors of the two causally competing models. Here, we make use of the fact that the errors of the causally mis-specified model ($y \rightarrow x$) can be expressed as a function of the predictor and errors of the "true" model, i.e. $e_{xy} = (1 - \rho_{xy}^2)x - \rho_{xy}e_{yx}$. Thus, assuming independence of x and e_{yx}, we arrive at the following third higher moments of error terms

$$\text{cor}(e_{yx}, e_{xy})_{30} = \gamma_{e_{yx}} \tag{2.12}$$

$$\text{cor}(e_{yx}, e_{xy})_{21} = -\rho_{xy}\gamma_{e_{yx}} \tag{2.13}$$

$$\text{cor}(e_{yx}, e_{xy})_{12} = \rho_{xy}^2\gamma_{e_{yx}} \tag{2.14}$$

$$\text{cor}(e_{yx}, e_{xy})_{03} = \gamma_{e_{xy}} = (1 - \rho_{xy}^2)^{3/2}\gamma_x - \rho_{xy}^3\gamma_{e_{yx}} \tag{2.15}$$

with $\gamma_{e_{xy}} = E[e_{xy}^3]/\sigma_{e_{xy}}^3$ being the skewness of the mis-specified error term and $\text{cor}(e_{yx}, e_{xy})_{21}$ and $\text{cor}(e_{yx}, e_{xy})_{12}$ denoting the two co-skewnesses of e_{yx} and e_{xy}. Again, two DDA measures can be constructed from the four equations. First, under symmetry of the "true" error term, one obtains $\text{cor}(e_{yx}, e_{xy})_{30} = \text{cor}(e_{yx}, e_{xy})_{21} = \text{cor}(e_{yx}, e_{xy})_{12} = 0$ and $\gamma_{e_{xy}} = (1 - \rho_{xy}^2)^{3/2}\gamma_x$. In this special case, the mis-specified error term will always be more skewed than the "true" error tem and one can use the following simple decision rule: If $\Delta = | \gamma_{e_{xy}} | - | \gamma_{e_{yx}} | > 0$ then $x \rightarrow y$ is preferred, if $\Delta < 0$ then the reversed model $y \rightarrow x$ is more likely to approximate the data generating mechanism. Second, when the "true" error is asymmetrically distributed ($\gamma_{e_{yx}} \neq 0$), the Pearson correlation coefficient can be re-written as

$$\rho_{xy} = -\frac{\text{cor}(e_{yx}, e_{xy})_{12}}{\text{cor}(e_{yx}, e_{xy})_{21}} \tag{2.16}$$

which implies that $|\text{cor}(e_{yx}, e_{xy})_{12}| < | \text{cor}(e_{yx}, e_{xy})_{21}|$ under $x \rightarrow y$ and $|\text{cor}(e_{yx}, e_{xy})_{12}| > | \text{cor}(e_{yx}, e_{xy})_{21}|$ under $y \rightarrow x$. Further, under non-normal "true" errors, a Gaussianization effect can be observed for the error distribution of the mis-specified model (Wiedermann & von Eye, 2015a). That is, whenever $|\gamma_x| \leq | \gamma_{e_{yx}} |$, the errors of the mis-specified model are now closer to the normal distribution and one can expect $| \gamma_{e_{yx}} | > | \gamma_{e_{xy}} |$ under $x \rightarrow y$ and $| \gamma_{e_{yx}} | < | \gamma_{e_{xy}} |$ under $y \rightarrow x$. Similar results were presented by Wiedermann and Hagmann (2016) focusing on the special case of a symmetrically distributed predictor ($\gamma_x = 0$) and a skewed "true" error ($\gamma_{e_{yx}} \neq 0$) and by Hernandez-Lobato, Morales-Mombiela, Lopez-Paz, and Suarez (2016) assuming that x and y have the same distribution (which can be achieved by transforming one variable using the probability integral transform). Model selection can be based on BCa CIs of differences in skewnesses ($\Delta = |\gamma_{e_{xy}}| - |\gamma_{e_{yx}}|$) and co-skewnesses ($\Delta = |\text{cor}(e_{yx}, e_{xy})_{21}| - |\text{cor}(e_{yx}, e_{xy})_{12}|$).

When variables are non-normal but symmetrically distributed, again fourth higher moment properties of regression errors can be used to perform DDA model selection. For example, assuming that errors associated with the correctly specified model are mesokurtic (i.e. $(\kappa_{e_{yx}} - 3) = 0$), the kurtosis of the mis-specified errors is given by $(\kappa_{e_{xy}} - 3) = (1 - \rho_{xy}^2)^2(\kappa_x - 3)$ which implies that $\Delta = |\kappa_{e_{xy}}| - |\kappa_{e_{yx}}|$ is greater than zero under $x \to y$ and smaller than zero under $y \to x$. For further details see Wiedermann (2015).

2.3.4 DDA Component III: Independence Properties

The last DDA component focuses on properties of independence of the predictor and the error term. Again, under non-normality of variables, systematic asymmetry exists between the two causally competing models, which opens the door to perform model selection with respect to the causal effect directionality. In the "true" model, the predictor (x) is assumed to be independent of the corresponding errors (e_{yx}). In other words, predictor scores and errors are not related to each other in any way. In contrast, non-independence is observed when predictor scores and errors are systematically related (e.g. higher predictor score may be associated with systematically higher errors). In the context of DDA, two sources of predictor-error non-independence exist: First, predictors and errors can be non-independent due to the fact that the model is directionally mis-specified (i.e. in the case of a reverse-causation bias). Second, non-independence can occur due to the presence of an unconsidered confounder (u; i.e. in the case of a confounding bias). As noted above, in the presence of a reverse causation bias, the error term of the mis-specified model is given by $e_{xy} = (1 - \rho_{xy}^2)x - \rho_{xy}e_{yx}$. Thus, together with the "true" model $y = \beta_{yx}x + e_{yx}$, one would expect that y and e_{xy} are stochastically related because y and e_{xy} are both linear functions of the "true" predictor x and the "true" error term e_{yx}.

While this can serve as an intuitive explanation, non-independence can be rigorously proven using the Darmois–Skitovich theorem (Darmois, 1953; Shimizu et al., 2011; Skitovich, 1953; Wiedermann et al., 2017). The theorem states that if two linear functions of independent random variables (z_j; $j = 1, ..., k$), $w = \sum_{j=1}^{k} \alpha_j z_j$ and $v = \sum_{j=1}^{k} \beta_j z_j$, $\alpha_j \beta_j \neq 0$ (with α_j and β_j being population constants) are stochastically independent, then all random variates z_j follow a Gaussian distribution. Thus, it follows that, given that there exists at least one common non-normal random variate z_j, for which $\alpha_j \beta_j \neq 0$ holds, the variables w and v will be stochastically non-independent.

Applying this line of argument in the present regression context, one concludes that, if the "true" predictor x deviates from the Gaussian distribution, y and e_{xy}

will be stochastically non-independent. Because independence will hold in any confounder-free "true" model (here $x \to y$), comparing independence properties of rival models can be informative when testing the plausibility of a model in terms of its effect directionality. When independence holds in the model $x \to y$ and non-independence occurs in the competing model $y \to x$, then $x \to y$ better reflects the data-generating mechanism. In contrast, when $y \to x$ constitutes the "true" model, then the reversed pattern is likely to occur.

Model selection can be based on any independence measure that is able to pick up dependence structures beyond first order correlations (recall that OLS residuals and predictors will always be uncorrelated by construct). Such measures include nonlinear correlation tests (Hyvärinen et al., 2001), homoscedasticity tests (Wiedermann et al., 2017), and omnibus independence tests such as the Hilbert–Schmidt Independence Criterion (HSIC) (cf. Gretton et al., 2008) and Brownian distance correlation (dCor) (Székely, Rizzo, & Bakirov, 2007). The HSIC is provably omnibus in detecting any form of dependence in the large sample limit (for a discussion of the HSIC in the context of testing independence in linear models see Sen & Sen, 2014). The dCor test provides a natural extension of the Pearson correlation and is consistent against all forms of dependent alternatives with finite second moments. Further, these omnibus measures can also be used to construct difference measures such as $\Delta = \text{HSIC}(y \to x) - \text{HSIC}(x \to y)$ and $\Delta = |\text{dCor}(y \to x)| - |\text{dCor}(x \to y)|$. Here the model $x \to y$ is preferred over $y \to x$ when $\Delta > 0$ and $y \to x$ should be selected when $\Delta < 0$.

2.3.5 Presence of Confounding

So far, we assumed the absence of confounding when testing the superiority of one causal model (e.g. $x \to y$) over the causally reversed alternative ($y \to x$). In the presence of an unconsidered confounder the model $x \to y$, for example, extends to

$$x = \beta_{xu}u + e_{xu}$$
$$y = \beta_{yx}x + \beta_{yu}u + e_{yx.}$$

(2.17)

where u is assumed to be a continuous exogenous variable and e_{xu} and e_{yx} denote error terms (with zero means and variances $\sigma_{e_{xu}}^2$ and $\sigma_{e_{yx}}^2$) that are assumed to be independent of the predictors and of each other. Here, $\beta_{yx} \neq 0$ refers to the case in which x causes y and, at the same time, confounding is present (the model above also covers $\beta_{yx} = 0$ as a special case where the x–y relation is entirely attributable to the confounder u). The estimate of the causal effect β_{yx} is known to be biased when $\beta_{xu}\beta_{yu} \neq 0$ and one fails to account for u. In contrast, when regressing y on both, x and u (i.e. "controlling" for u), one obtains an unbiased estimate for the causal effect β_{yx}.

From the perspective of direction of dependence, it is important to realize that the presence of a confounder (u) also affects the distribution of the predictor and the outcome. In other words, in addition to potential biases in the causal effect estimate, one cannot rule out biases in distribution-based DDA measures when confounding is present. Here, distributional biases depend on both, the degree of non-normality of u and the magnitude of confounding effects β_{xu} and β_{yu}. The reason for this is that higher moments of x and y can be re-expressed as a function of higher moments of the confounder u. Skewness and kurtosis values of x and y can be written as

$$\gamma_x = \rho_{xu}^3 \gamma_u$$
$$\gamma_y = \rho_{yu}^3 \gamma_u$$

$$(2.18)$$

and

$$(\kappa_x - 3) = \rho_{xu}^4 (\kappa_u - 3)$$
$$(\kappa_y - 3) = \rho_{yu}^4 (\kappa_u - 3).$$

$$(2.19)$$

with ρ_{xu} and ρ_{yu} being the Pearson correlation coefficients of x and u, and y and u. Because $0 \leq |\rho_{xu}| \leq 1$ and $0 \leq |\rho_{yu}| \leq 1$, it follows that the observed variables x and y will always be closer to normality than the confounder u. No biases in DDA model selection are expected when the confounder has a larger effect on the predictor than on the outcome. In this case, one obtains $|\rho_{yu}| < |\rho_{xu}|$ and one is more likely to observe $|\gamma_y| < |\gamma_x|$ and $|\kappa_y| < |\kappa_x|$, which suggests the model $x \rightarrow y$. In contrast, biases in direction dependence decisions are more likely to occur when $|\rho_{yu}| > |\rho_{xu}|$ which leads to $|\gamma_y| > |\gamma_x|$ and $|\kappa_y| > |\kappa_x|$ and, thus, indicates the causally mis-specified model $y \rightarrow x$.

Higher cross-moment based DDA measures are biased in a similar fashion by the presence of confounding. Under the model $x \leftarrow u \rightarrow y$, the co-skewnesses of x and y, for example, are given by $\text{cor}(x,y)_{12} = \rho_{xu}\rho_{yu}^2 \gamma_u$ and $\text{cor}(m,y)_{21} = \rho_{xu}^2 \rho_{yu} \gamma_u$ (with γ_u being the skewness of the confounder u) which implies that the co-skewness ratio no longer approximates the Pearson correlation of x and y but the ratio of pairwise correlations of x, y, and u (Wiedermann et al., 2019), that is,

$$\frac{\text{cor}(x,y)_{12}}{\text{cor}(x,y)_{21}} = \frac{\rho_{yu}}{\rho_{xu}}.$$

$$(2.20)$$

In other words, the outcome of the co-skewness-based DDA test depends on the magnitude of the ratio of two correlations ρ_{yu} and ρ_{mu}. When ρ_{yu}^2 is smaller than ρ_{xu}^2, it follows that $\Delta > 0$ and one is more likely to select the model $x \rightarrow y$. In contrast, when ρ_{yu}^2 is larger than ρ_{mu}^2, then $\Delta < 0$ which suggests at the reversed model $y \rightarrow x$ should be preferred. Thus, the HOC test assumes unconfoundedness of the "true" model and should be applied with caution whenever confounders are likely to be present. In the presence of confounding, however, the ratio of co-skewnesses can be useful to learn more about the nature of confounding. When confounding is present and $\Delta > 0$, one can conclude that the confounder has a larger impact on x than on y. $\Delta < 0$ indicates that the confounder has a larger effect on y than on x.

The distribution of the confounder and the magnitude of the confounding effects can also be expected to bias directional decisions based on asymmetry properties of error distributions (cf. Wiedermann & Sebastian, in press-a). This can be seen by analyzing properties of error terms obtained from the two directionally competing models that erroneously ignore the common cause u,

$$
\begin{aligned}
y &= \beta'_{yx}x + e'_{yx} \\
x &= \beta'_{xy}y + e'_{xy}.
\end{aligned}
\tag{2.21}
$$

In this case, higher moments of e'_{yx} and e'_{xy} can be characterized by

$$
\begin{aligned}
\gamma_{e'_{yx}} &= \rho^3_{y(u|x)}\gamma_u \\
(\kappa_{e'_{yx}} - 3) &= \rho^4_{y(u|x)}(\kappa_u - 3),
\end{aligned}
\tag{2.22}
$$

with $\rho_{y(u|x)} = (\rho_{yu} - \rho_{xy}\rho_{xu})/\sqrt{1 - \rho^2_{xy}}$ representing the semi-partial correlation of y and u given x, and

$$
\begin{aligned}
\gamma_{e'_{xy}} &= \rho^3_{x(u|y)}\gamma_u \\
(\kappa_{e'_{xy}} - 3) &= \rho^4_{x(u|y)}(\kappa_u - 3)
\end{aligned}
\tag{2.23}
$$

with $\rho_{x(u|y)} = (\rho_{xu} - \rho_{xy}\rho_{yu})/\sqrt{1 - \rho^2_{xy}}$ describing the semi-partial correlation between x and u given y. Assuming that the confounder deviates from the Gaussian distribution, unbiased DDA decisions are still possible when $|\rho_{y(u|x)}| < |\rho_{x(u|y)}|$ because $|\gamma_{e'_{xy}}| > |\gamma_{e'_{yx}}|$ and $|\kappa_{e'_{xy}}| > |\kappa_{e'_{yx}}|$ still holds suggesting the model $x \to y$. In contrast, biased DDA decisions (i.e. erroneously selecting $y \to x$ instead of $x \to y$) are likely to occur when $|\rho_{y(u|x)}| > |\rho_{x(u|y)}|$ because one then obtains $|\gamma_{e'_{xy}}| < |\gamma_{e'_{yx}}|$ and $|\kappa_{e'_{xy}}| < |\kappa_{e'_{yx}}|$. Higher cross moments of e'_{yx} and e'_{xy} can be expected to be equally affected by the magnitude of confounding effects and non-normality of the confounder.

So far, we can conclude that distribution-based direction dependence decisions are prone to biases when confounding is present. We now show that the third DDA component (asymmetry properties of the independence assumption) allows one to detect the presence of confounding, that is, one is able to distinguish the model $x \to y$ (or $y \to x$) and $x \leftarrow u \to y$. The reason is that in the presence of confounding, predictors and errors of *both* models will be non-independent. We can re-express the error terms of the two mis-specified models ($y = \beta'_{yx}x + e'_{yx}$ and $x = \beta'_{xy}y + e'_{xy}$) as a function of the true confounder model which results in

$$
\begin{aligned}
e'_{yx} &= [\beta_{yu} + (\beta_{yx} - \beta'_{yx})\beta_{xu}]u + (\beta_{yx} - \beta'_{yx})e_{xu} + e_{yu} \\
e'_{xy} &= [\beta_{xu} - \beta'_{xy}(\beta_{yu} + \beta_{yx}\beta_{xu})]u + (1 - \beta'_{xy}\beta_{yx})e_{xu} + \beta'_{xy}e_{yu}
\end{aligned}
\tag{2.24}
$$

which implies that x and e'_{yx} share the confounder u and the error term e_{xu} and y and e'_{xy} share u and e_{yu} as common independent components. Thus, under non-normality of u or the errors e_{xu} and e_{yu}, non-independence holds in both models, $x \to y$ and $y \to x$. Because none of the three components can be observed in practice, non-normality of the three components can indirectly be tested through inspecting distributional features of x and y. For example, non-normality of x implies that u, e_{xu}, or both have to deviate from the Gaussian distribution. In either case, independence is likely to be violated in both causally competing models.

2.3.6 An Integrated Framework

The DDA approach integrates all asymmetry properties of observed variables and regression errors into a unified framework (see also Wiedermann & Li, 2018; Wiedermann & Sebastian, ; Wiedermann & von Eye, 2015c). As a result, DDA patterns exists that can be used to uniquely identify the three candidate models $x \to y$, $y \to x$, and $x \gets u \to y$. Detailed patterns for each model are summarized in Figure 2.1. For example, a target model, (i.e. the causal model that represents the substantive causal theory of interest), finds empirical support over the alternative model (i.e. the model which reflects the causally competing substantive theory) when

1) the corresponding outcome variable is closer to normality than the predictor (skewness and kurtosis values can be used when one assumes normality of the "true" errors, co-skewness and co-kurtosis values can be used under non-normal "true" errors),
2) the corresponding model errors are closer to normality than the errors of the alternative model (again, skewness and kurtosis values can be used under normal "true" errors, co-skewness and co-kurtosis values are preferable under non-normality of "true" errors), and
3) predictors and errors are independent in the target model and, at the same time, independence is violated in the alternative model. When the independence assumption is *either violated or satisfied* in both models (note that confounders can render the distributions of x and y too close to normality to detect dependence structures) one has to conclude that unconsidered confounders are present and no clear-cut decisions about the direction of effect can be made.

Each DDA component offers a series of approaches for statistical inference. Distributional properties of observed variables and regression residuals can be analyzed using BCa CIs, non-independence properties can be analyzes using nonlinear correlation tests, homoscedasticity tests, and omnibus independence procedures (such as the HSIC and the dCor test). In the analysis of real-world data, it may rarely be the case that all significance procedures of all three DDA components come to the same causal conclusions. Thus, the question concerning

Model	Causal diagram	Direction dependence patterns
I.	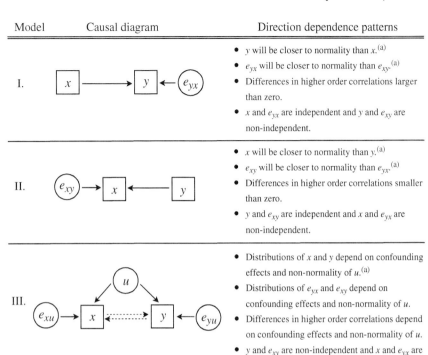	• y will be closer to normality than x.[a] • e_{yx} will be closer to normality than e_{xy}.[a] • Differences in higher order correlations larger than zero. • x and e_{yx} are independent and y and e_{xy} are non-independent.
II.		• x will be closer to normality than y.[a] • e_{xy} will be closer to normality than e_{yx}.[a] • Differences in higher order correlations smaller than zero. • y and e_{xy} are independent and x and e_{yx} are non-independent.
III.		• Distributions of x and y depend on confounding effects and non-normality of u.[a] • Distributions of e_{yx} and e_{xy} depend on confounding effects and non-normality of u. • Differences in higher order correlations depend on confounding effects and non-normality of u. • y and e_{xy} are non-independent and x and e_{yx} are non-independent.[b]

(a) This DDA criterion assumes normality of the "true" error term.
(b) Because confounders can render variable distributions being too close to normality, Model III should also be selected when independence is rejected or retained in both causally competing models.

Figure 2.1 Patterns of direction of dependence of three explanatory causal models. Rectangles represent observed variables and circles represent unobserved variates. Model I: x causes y, Model II: y causes x, Model III: an unconsidered confounder u is responsible for the $x-y$ relation.

a minimum set of DDA procedures for model selection naturally arises. Here, the DDA independence component is the most important criterion to confirm that no influential confounding is present (or, at least, that the influence of confounders is minimal). Thus, independence tests are crucial DDA procedures. Here, the HSIC and the dCor test have the advantage that both are omnibus under the large sample limit. In other words, any form of dependence can be detected with these tests. However, large sample sizes ($n \geq 800$) are usually needed to guarantee sufficient power with moderately skewed variable distributions ($\gamma \approx 1$) (Wiedermann & Li, 2019a). In contrast, nonlinear correlation tests and homoscedasticity procedures are sufficiently powerful for smaller sample sizes (e.g. $n = 200$) but may have the disadvantage of not being omnibus in detecting any form of dependence (Wiedermann et al., 2017). When independence tests indicate that confounding is likely to be present, results of the remaining distribution-based DDA procedures should be interpreted with caution. Thus, we recommend that, before proceeding with DDA, one needs to resolve the issue of confounding through proper covariate adjustment.

When preliminary model selection is possible based on independence tests, the other two DDA components can be used to evaluate distributional properties of observed variables and errors. Here, as shown above, two different sets of DDA decision rules exist for the one DDA component that focuses on asymmetry properties of errors. When the "true" error term follows a normal distribution, the errors associated with the causally mis-specified model will always be more non-normal than the "true" errors. In contrast, when the "true" errors are non-normal, a Gaussianization effect can be observed for the errors of the mis-specified model, that is, the errors of the mis-specified model will tend to be closer to normality. Because of these contradicting decision rules, we suggest starting with DDA's independence component to obtain a tentative target model. If normality of residuals holds for this tentative target model, one can proceed with applying decision rules for normal "true" errors. If normality is violated for the residuals of the target model, one needs to apply decision rules under "true" error non-normality. For all distribution-based DDA measures it holds that skewness-based procedures are more powerful than kurtosis-based approaches (Dodge & Rousson, 2016). The reason for this is that kurtosis-based measures require larger sample sizes to guarantee sufficient precision in parameter estimation. Further, separate moments tests provide stricter DDA model selection criteria (i.e. normality tests of the putative outcome and the corresponding residual term must be non-significant to allow clear-cut decisions) and can be more powerful than moment-difference procedures (Pornprasertmanit & Little, 2012; Wiedermann & von Eye, 2015b).

Finally, we want to re-emphasize that application of the presented DDA framework requires a number of crucial assumptions being fulfilled: (i) non-normality of variables, (ii) linearity of variable relations, (iii) causal effect homogeneity, (iv) absence of outliers/highly influential data points, (v) additive errors, (vi) absence of selection biases, and (vii) no measurement error in the predictor. The first four of these assumptions can be tested, the last three are routine in linear regression modeling. Apparently, the majority of assumptions listed above are not new in the context of linear regression modeling. In fact, a careful evaluation of these assumption should always be part of linear regression modeling (with or without evaluating the causal effect directionality). A summary of these assumptions together with their consequences in case of violations, methods of testing the assumptions, and potential remedies are given in the upper panel of Table 2.1. In the following sections, we discuss applications of DDA in the context of mediation and moderation models.

Assumption	Definition	Methods of evaluation	Consequences of assumption violations	Potential remedy
Assumptions of standard direction dependence analysis				
Non-normality of observed variables	Observed variables are assumed to deviate from the Gaussian distribution. Non-normality is assumed to be an inherent characteristic of the variables	Testing distributions using skewness, kurtosis, or omnibus normality tests	DDA cannot be applied under the normal distribution	None
No outliers and influential observations[a]	Absence of highly influential observations and outliers	Outlier detection and influence measures (e.g. multivariate Mahalanobis distances, Studentized residuals, leverage, and Cook's distances)	Distorted higher-moment estimates may bias DDA results	Temporarily remove or impute conspicuous observations and re-run DDA. Evaluate stability of DDA results using bootstrapping
No selection bias[a]	Preferential exclusion of subjects due to a non-random sample of the population	In case of a truncated outcome variable, a Heckman two-stage correction (Heckman, 1979) can be applied	Causal effect estimates and DDA results may be biased	Study population needs to be clearly defined; if applicable, correct for selection biases using Heckman-type two-stage estimation
Linearity of relationship[a]	The association between the focal predictor and outcome can be approximated by a linear model	Visual diagnostics (e.g. LOWESS smoothed lines); testing significance of R^2 change when adding high-order polynomials	Regression model does not describe the relationship of the tentative predictor and outcome	Evaluate direction of dependence in polynomial regression models
Additive errors[a]	The error term of the target model is assumed to be additive	Linearize regression function using proper variable transformation and evaluate resulting residual patterns	Regression model does not adequately describe the underlying causal mechanism which may bias DDA results	Evaluate direction of dependence in linearized regression models

(continued)

Table 2.1 (Continued)

Assumption	Definition	Methods of evaluation	Consequences of assumption violations	Potential remedy
Reliability of measurement[a]	The tentative predictor is assumed to be measured without measurement error	Reliability analysis; methods of moment estimation (see, e.g. Wiedermann et al., 2018)	Distributional biases may occur in case of low reliability (for details see Wiedermann et al., 2018)	Use two-step factor score regression approach (Devlieger & Rosseel, 2017) and perform DDA on factor scores
Causal effect homogeneity[a]	The causal effect is assumed to be constant across subjects	Testing the significance of interaction terms	DDA result may be biased when causal effect directionality varies as a function of a moderator	Use conditional Direction Dependence Analysis (CDDA; Li & Wiedermann, in press)
Additional assumptions specific to conditional direction dependence analysis				
Exogeneity of moderator	The moderator is assumed to be an exogenous variable	Selection of moderator has to be guided by a priori theoretical considerations	Conditioning on an endogenous moderator (e.g. a common cause of the predictor and the outcome) can lead to non-independent errors in the target model	None
Causal effect homogeneity conditional on moderator	The causal effect is assumed to be homogeneous conditional on the observed moderator, i.e. CDDA reduces but does not eliminate the causal effect heterogeneity assumption	Detection of potentially hidden moderators has to be guided by a priori theoretical considerations	CDDA results can vary as a function of hidden moderators	Account for additional moderators when residualizing the focal predictor and the focal outcome
Linearity of moderation	The predictor-outcome relation is affected by a linear function of the moderator	Evaluate the presence of curvilinear interaction terms	Model mis-specifications can lead to non-independent errors in the target model	Include curvilinear interaction terms in the proposed two-step auxiliary approach (cf. Eqs. (2.22)–(2.25)

a) Assumption is routinely made in conventional linear regression applications.

2.4   Direction Dependence in Mediation

Mediation analysis (MacKinnon, 2008) is routinely used to decompose the observed total effect of x on y into direct and indirect effect components. The indirect effect component serves as an explanation why a particular causal effect occurs through additionally considering one or more intervening (mediating) variables. The simple linear mediation model for a predictor x, a mediator m, and an outcome variable y, can be expressed as (again assuming intercepts being fixed to zero)

$$m = \beta_{mx}x + e_{m(x)}$$
$$y = \beta_{yx}x + \beta_{ym}m + e_{y(mx)},$$

(2.25)

where β_{yx} quantifies the direct effect of x on y and the product term $\beta_{mx}\beta_{ym}$ quantifies the indirect effect of x on y via m. Covariates can be incorporated in both regression models when needed. The errors $e_{m(x)}$, and $e_{y(mx)}$ are assumed to be independent of the model predictors and of each other. Resampling-based CIs (MacKinnon, Lockwood, & Williams, 2004) are routinely used to evaluate whether the indirect effect statistically differs from zero. If this is the case, one has found empirical evidence for the presence of a mediational mechanism. It is important to realize that indirect effects can only be interpreted as causal when both the hypothesized model (Wiedermann & von Eye, 2015a, 2016) and unconfoundedness assumptions hold (Imai, Keele, & Tingley, 2010; Imai, Keele, & Yamamoto, 2010; Pearl, 2012). Figure 2.2 summarizes the six possible (acyclic) causal specifications for a simple mediation model together with the corresponding indirect effect expressed as a function of pairwise correlation coefficients of standardized variables (i.e. x, m, and y are assumed to have zero means and unit variances).

Comparing indirect effect estimates of alternative mediation models is a common strategy to evaluate competing causal theories in cross-sectional data (Hayes, 2013; Iacobucci, Saldanha, & Deng, 2007; Kim et al., 2018). However, nothing new can be learned from these model comparisons, which can be explained by the fact that any indirect effect estimate obtained from x, m, and y only depends on the corresponding pairwise correlations (see right panel of Figure 2.2 (cf. Thoemmes, 2015; Wiedermann & von Eye, 2015a)). Thus, no new information about the underlying data generating mechanism is obtained when estimating competing indirect effects. Consider, for example, the case in which one is interested in comparing a mediation model in which x is the predictor, m is the mediator, and y is the outcome (all variables are assumed to be standardized). For this model, the indirect effect (IE) can be calculated as $IE_{(a)} = [\rho_{xm}(\rho_{ym} - \rho_{xy}\rho_{xm})]/(1 - \rho_{xm}^2)$. Now, consider the causally mis-specified mediation model where y serves as the predictor, m is the mediator, and x is the outcome. Here, the indirect effect can be expressed as $IE_{(b)} = [\rho_{ym}(\rho_{xm} - \rho_{xy}\rho_{ym})]/(1 - \rho_{ym}^2)$, i.e. the indirect effect is a function of the partial correlation of x and m given y). Considering the difference between the two indirect effects, $IE_{(a)} - IE_{(b)} = [(\rho_{xm}\rho_{ym} - \rho_{xy})(\rho_{xm}^2 - \rho_{ym}^2)]/[(1 - \rho_{xm}^2)(1 - \rho_{ym}^2)]$ it becomes evident that $IE_{(a)}$ and $IE_{(b)}$ will be the same only if $\rho_{xm} = \rho_{ym}$ or $\rho_{xm}\rho_{ym} = \rho_{xy}$ (which is highly unlikely in real-world data). Furthermore, the indirect effect of the causally mis-specified model is larger than the "true" indirect effect whenever $\rho_{xm}\rho_{ym} > \rho_{xy}$ and $\rho_{xm} < \rho_{ym}$ or $\rho_{xm}\rho_{ym} < \rho_{xy}$ and $\rho_{xm} > \rho_{ym}$. Thus, one can conclude that no additional information is gained through comparing indirect effect estimates of alternative mediation models which would enable one to make statements about the plausibility of a mediation model. In addition, from a SEM perspective, all possible mediation models in Figure 2.2 are class-equivalent (Lee & Hershberger, 1990) and, thus, have exactly the same support by the data in terms of model fit.

In practice, some of the six candidate models in Figure 2.2 may be excluded because of logical inconsistency. However, to demonstrate that one mediation theory is superior over an alternative theory, evaluating causal effect directionality of a target mediation model against at least a subset of competing models is needed.

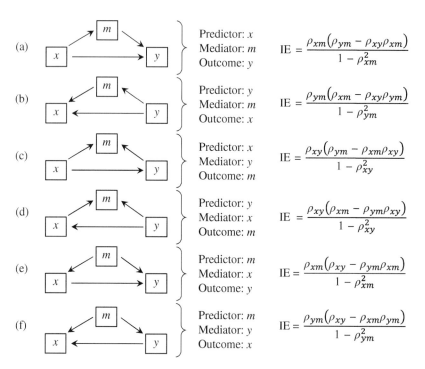

(a) Predictor: x Mediator: m Outcome: y $\quad IE = \dfrac{\rho_{xm}(\rho_{ym} - \rho_{xy}\rho_{xm})}{1 - \rho_{xm}^2}$

(b) Predictor: y Mediator: m Outcome: x $\quad IE = \dfrac{\rho_{ym}(\rho_{xm} - \rho_{xy}\rho_{ym})}{1 - \rho_{ym}^2}$

(c) Predictor: x Mediator: y Outcome: m $\quad IE = \dfrac{\rho_{xy}(\rho_{ym} - \rho_{xm}\rho_{xy})}{1 - \rho_{xy}^2}$

(d) Predictor: y Mediator: x Outcome: m $\quad IE = \dfrac{\rho_{xy}(\rho_{xm} - \rho_{ym}\rho_{xy})}{1 - \rho_{xy}^2}$

(e) Predictor: m Mediator: x Outcome: y $\quad IE = \dfrac{\rho_{xm}(\rho_{xy} - \rho_{ym}\rho_{xm})}{1 - \rho_{xm}^2}$

(f) Predictor: m Mediator: y Outcome: x $\quad IE = \dfrac{\rho_{ym}(\rho_{xy} - \rho_{xm}\rho_{ym})}{1 - \rho_{ym}^2}$

Figure 2.2 Six alternative mediation models together with the corresponding indirect effects (IE) expressed as a function of pairwise correlations of standardized variables x, m, and y.

Because a simple mediation model essentially consists of two separate regression models, application of DDA is straightforward. In the following paragraph, we focus on the comparison of the two mediation models in which x and y are exchanged in their roles as predictors and outcomes while m is the mediator, i.e. one tests $x \rightarrow m \rightarrow y$ against the alternative $y \rightarrow m \rightarrow x$ (this is line with recommendations by Iacobucci et al. (2007) that among all possible alternative mediation models, at least the one that exchanges x and y should be inspected). The causally mis-specified mediation model can be written as

$$m = \beta_{my}y + e_{m(y)}$$
$$x = \beta_{xy}y + \beta_{xm}m + e_{x(ym)},$$
(2.26)

with β_{xy} being the direct effect of y on x and $\beta_{my}\beta_{xm}$ estimating the indirect effect of y on x via m. Similar to the standard bivariate regression case (see above), the distribution of errors associated with the causally mis-specified mediation model again depend on the distribution of the "true" predictor. For example, under symmetry of the "true" errors ($e_{x(m)}$ and $e_{y(xm)}$), the skewness of $e_{m(y)}$ and $e_{x(ym)}$ can be re-expressed as

$$\gamma_{e_{m(y)}} = (1 - \rho_{my}^2)^{\frac{3}{2}} \rho_{xm}^3 \, \gamma_x$$

$$\gamma_{e_{x(ym)}} = \left(\frac{(1 - \rho_{xy}^2) - \frac{(\rho_{xm} - \rho_{xy}\rho_{my})^2}{(1-\rho_{my}^2)}}{\sigma_{e_{x(my)}}} \right)^3 \gamma_x,$$
(2.27)

from which one can conclude that $|\gamma_{e_{m(y)}}|$ and $|\gamma_{e_{x(ym)}}|$ increase with $|\gamma_x|$. In addition, the degree of non-independence of predictors and errors in the mis-specified model (the focus of DDA's third component) depends on the magnitude of non-normality of the "true" predictor x and all three pairwise correlations of x, m, and y. This can be seen by re-writing nonlinear covariances of $e_{x(ym)}$ and y as

$$cov(e_{x(ym)}, y^2) = \rho_{xy}^2 \left(1 - \rho_{xy}^2 - \frac{(\rho_{xm} - \rho_{xy}\rho_{ym})^2}{1 - \rho_{ym}^2} \right) \gamma_x$$

$$cov(e_{x(ym)}^2, y) = \rho_{xy} \left(1 - \rho_{xy}^2 - \frac{(\rho_{xm} - \rho_{xy}\rho_{ym})^2}{1 - \rho_{ym}^2} \right)^2 \gamma_x.$$
(2.28)

Similarly, nonlinear covariances of the two error terms $e_{m(y)}$ and $e_{x(ym)}$ are given by

$$cov(e_{m(y)}, e_{x(ym)}^2) = (\rho_{xm} - \rho_{xy}\rho_{ym}) \left(1 - \rho_{xy}^2 - \frac{(\rho_{xm} - \rho_{xy}\rho_{ym})^2}{1 - \rho_{ym}^2} \right)^2 \gamma_x$$

$$cov(e_{m(y)}^2, e_{x(ym)}) = (\rho_{xm} - \rho_{xy}\rho_{ym})^2 \left(1 - \rho_{xy}^2 - \frac{(\rho_{xm} - \rho_{xy}\rho_{ym})^2}{1 - \rho_{ym}^2} \right) \gamma_x$$
(2.29)

which implies that, in addition to testing the independence of model predictors and errors, evaluation of the independence of the two errors in a mediation model can also inform researchers about the causal direction of the underlying mediation mechanism.

In general, to carry out DDA in the mediation context one makes use of the Frisch–Waugh–Lovell theorem (Frisch & Waugh, 1933; Lovell, 1963). This theorem states that a partial regression coefficient of a multiple linear regression model is identical to the simple regression slope of the residualized outcome and predictor variables. Specifically, to test direction dependence properties of the direct effect of the mediation model in (2.25), one proceeds as follows: First, one starts with regressing x and y on the mediator (covariates can be included as well) and extracts the corresponding regression residuals as "purified" predictor and outcome measures, i.e.

$$
\begin{aligned}
r_x &= x - \beta_{xm} m \\
r_y &= y - \beta_{ym} m.
\end{aligned}
\tag{2.30}
$$

with r_x and r_y being the raw residuals of the two bivariate models $m \to x$ and $m \to y$. Regressing r_y on r_x then leads to the same direct effect estimate for the causal effect β_{yx} as one would obtain from the multiple regression model $\{x, m\} \to y$. In addition, the error terms of $r_x \to r_y$ and $\{x, m\} \to y$ will be identical as well (Lovell, 2008). Therefore, direction dependence decisions about the direct effect can be based on the bivariate model involving "purified" x and y variates.

Evaluating direction dependence of the predictor–mediator and the mediator–outcome paths can be performed in a similar fashion. To test the causal direction of effect of the predictor–mediator path, one compares the models $x \to m$ and $m \to x$, to test effect directionality of the mediator–outcome path, one proceeds with exchanging the roles of m and y while adjusting for x, i.e. $\{x, m\} \to y$ versus $\{x, y\} \to m$. Overall, the initial target mediation model given in (2.25) is supported when $x \to m$ is selected in the first model comparison and $\{x, m\} \to y$ is selected in the two remaining model comparisons (for details see Wiedermann et al., 2019; Wiedermann & Sebastian, ; Wiedermann & von Eye, 2015a).

2.5 Direction Dependence in Moderation

So far, we have assumed that the causal effect of the correctly specified model is constant across all subjects. In the following paragraph, we relax this assumption and discuss direction dependence properties of regression models with conditional causal effects (i.e. in the presence of interaction terms). Moderation analysis (Aiken & West, 1991) is commonly used to examine the conditional effect of a predictor (x) on an outcome (y) at values of a moderator (z). Let x be the "true" predictor, y is the "true" outcome, and z is a moderator that is known to be on the explanatory side (see lower panel of Table 2.1 for details). The "true" moderator model ($x \mid z \to y$) and the causally mis-specified model ($y \mid z \to x$) can be written as

$$\begin{aligned} y &= \beta_0 + (\beta_1 + \beta_3 z)x + \beta_2 z + e_y \\ &= \beta_0 + \beta_1 x + \beta_2 z + \beta_3 xz + e_y. \end{aligned} \tag{2.31}$$

$$\begin{aligned} x &= \beta_0' + (\beta_1' + \beta_3' z)y + \beta_2' z + e_x \\ &= \beta_0' + \beta_1' y + \beta_2' z + \beta_3' yz + e_x \end{aligned} \tag{2.32}$$

where β_0 is the population intercept, β_1, β_2, and β_3 are (population) regression weights, and e_y is the error term (assumed to be independent of the predictors) of the correctly specified model, and β_0' is the population intercept, β_1', β_2', and β_3' are population regression weights, and e_x is the error term of the mis-specified model. Using Aiken and West's (1991) procedure, significance tests of β_3 can be used to confirm the presence of moderation. Further, significance tests of ($\beta_1 + \beta_3 z$) are routinely used to test the simple (conditional) slope of y on x at different values z_i. Tests of simple slopes are usually performed by applying the linear transformation $z' = z - z_i$ so that the slope of the focal predictor (e.g. β_1 in Eq. (2.31)) estimates the main effect for the case of $z' = 0$ (i.e. $z = z_i$). Here, a pre-specified moderator value, z_i, such as fixed population values or estimated sample values can be used (for a systematic comparison of different simple slope approaches see Liu, West, Levy, and Aiken (2017)).

To test direction dependence properties of competing moderator models, one again makes use of the Frisch–Waugh–Lovell theorem. That is, one partials out the main effect of z together with the interaction effect of the focal predictor and the outcome (potential covariates can be incorporated as well), and, then, performs DDA using the simple slope model of the "purified" x and y variables, i.e. using DDA after purification using different z' values. However, in the moderation context, two separate pairs of candidate models need to be estimated because different product terms are used in residualization of the causally competing models. The product term xz is used to isolate the x–y effect of the conditional model $x \mid z \to y$ and the product term yz is used to isolate the y–x effect in the causally competing model $y \mid z \to x$. Specifically, to residualize y and x on z in the context of the model $x \mid z \to y$ one estimates two auxiliary regression models where x and y are regressed on z and the product term xz,

$$y = \beta_0 + \beta_2 z + \beta_3 xz + r_y^{(x \to y)} \tag{2.33}$$

$$x = \beta_0' + \beta_2' z + \beta_3' xz + r_x^{(x \to y)} \tag{2.34}$$

and uses the extracted residuals

$$r_y^{(x \to y)} = y - (\beta_0 + \beta_2 z + \beta_3 xz) \tag{2.35}$$

$$r_x^{(x \to y)} = x - (\beta_0' + \beta_2' z + \beta_3' xz) \tag{2.36}$$

in the bivariate model

$$r_y^{(x \to y)} = \beta_1 r_x^{(x \to y)} + e_y \tag{2.37}$$

where the regression weight and the residual term of $r_x^{(x \to y)} \to r_y^{(x \to y)}$ equal the esti- mated multiple regression coefficient and the residual term of the model $x \mid z \to y$. In a similar fashion, one can isolate the y–x effect (β_1') together with the corre- sponding error (e_x) of the model $y \mid z \to x$. Here, one starts with estimating the two auxiliary models in which y and x are regressed on z and the product term yz,

$$y = \beta_0 + \beta_2 z + \beta_3 yz + r_y^{(y \to x)} \tag{2.38}$$

$$x = \beta_0' + \beta_2' z + \beta_3' yz + r_x^{(y \to x)} \tag{2.39}$$

and the extracted residuals

$$r_y^{(y \to x)} = y - (\beta_0 + \beta_2 z + \beta_3 yz) \tag{2.40}$$

$$r_x^{(y \to x)} = x - (\beta_0' + \beta_2' z + \beta_3' yz) \tag{2.41}$$

are, again, used in the bivariate model

$$r_x^{(y \to x)} = \beta_1' r_y^{(y \to x)} + e_x. \tag{2.42}$$

The two purified models $r_x^{(x \to y)} \to r_y^{(x \to y)}$ and $r_y^{(y \to x)} \to r_x^{(y \to x)}$ can be used as a basis to make directional statements about the conditional x–y relation at values of the moderator. To evaluate the causal direction of effect at a specific value z_i, one transforms the moderator and the interaction term based on different z_i val- ues before performing DDA using the purified x and y variables. This stepwise approach allows one to address several research questions: First, through insert- ing a broad range of moderator values (z_i) one is able to identify moderator regions for which a causal model of the form $x \to y$ can be retained. Second, the approach is able to detect cut-offs for which the causal direction of effect changes (e.g. x causally affects y when $z < z_0$ and the causally reversed pattern holds for $z \geq z_0$). Third, one can use the presented approach to identify moderator regions for which latent confounders become influential (e.g. the impact of unconsidered con- founders is small for $z < z_0$ and substantial for $z \geq z_0$).

2.6 Some Applications and Software Implementations

Many previous studies made use of principles of direction dependence to test causally competing theoretical models. Dodge and Rousson (2000), for example, used Crile and Quiring's (1940) data on brain and body weights of different species to test Jerison's (1973) theory that reptiles, fish, and amphibians can be classified as "lower" vertebrates with respect to brain evolution, i.e. selection pressures did not require fundamental changes in the brain–body relationship. Thus, "lower" vertebrates may be described by a single brain:body ratio. In contrast, mammals and birds are classified as "higher" vertebrates, i.e. the brain size relative to the body size increased to increase information processing capacities as a consequence of selection pressures. Using direction dependence principles of observed variable distributions, Dodge and Rousson (2000) showed that the body weight influences the brain weight in "lower" vertebrates (reptiles and fish) and a reversed picture was observed for "higher" vertebrates which is in perfect agreement with Jerison's (1973) theory of brain development.

In developmental sciences, for example, von Eye and Deshon (2012) used direction dependence properties of observed variable distributions to test whether blood lead content can be considered being a cause of attention-deficit/hyperactivity disorder (ADHD) or whether children with ADHD behave in a way that their blood lead level is increased (see also Nigg et al., 2008). Their analysis suggested that a model of the form *blood lead → ADHD* was better able to describe the observed association (this result was later confirmed by Kim and Kim (2014), using copula-based directional dependence methods). In another example, Wiedermann and von Eye (2015b) as well as Wiedermann, Artner, and von Eye (2017) tested the hierarchical nature of the triple code model of numerical cognition (Dehaene & Cohen, 1998; von Aster & Shalev, 2007). The triple code model posits that numbers are represented by three different codes to serve different functions; the analog magnitude code (AMC) (numerical quantities are represented on a mental number line and are used for magnitude comparisons), the auditory verbal code (AVC) (numbers are presented in syntactically organized word sequences necessary for counting and retrieving arithmetic facts), and the visual Arabic code (VAC) (represents numbers in Arabic format necessary for multidigit operations and parity judgments). Because AMC is assumed to be an inherited core system that is necessary to further develop the AVC and VAC, principles of DDA were used to confirm the causal model AMC → AVC (the model AVC → VAC found empirical support in a separate study by Wiedermann and von Eye (2016)).

In education, Wiedermann and Sebastian (in press-a) used the presented DDA framework to evaluate an organizational model proposing a mediational relation between school leadership and student achievement via school safety. Here, the authors found empirical support for the mediator–outcome path of the proposed model, that is, *school safety → achievement* better approximated the underlying causal mechanism than the reversed model. However, influential confounders were present for the predictor (school leadership)–mediator (school safety) path of the proposed model. More recently, Wiedermann, Reinke, and Herman (Under

Review) evaluated the causal structure of the relationship between student behavior (prosocial behavior and emotional dysregulation) and academic competence. While DDA confirmed that *prosocial behavior → academic competence* is better able to explain the variable relation, substantial confounding was found for the relation between emotional dysregulation and academic competence.

Finally, in health sciences, Wiedermann et al. (2019) used DDA to evaluate whether headache severity mediated the relation between acupuncture and quality of life (measured as perceived health change and perceived level of energy/fatigue). Data for this re-analysis came from a study which had found that acupuncture led to improvements in quality of life (Vickers et al., 2004). Mediation analysis suggested that acupuncture lowered headache severity, which, in turn, increased quality of life. Further, DDA provided evidence that the observed indirect effect can be interpreted as causal only when focusing on perceived energy/fatigue. In contrast, for perceived health change, DDA suggested that confounding biased the mediator (headache severity)–outcome (health change) relation.

To make methods of DDA accessible to a broad audience of researchers, several software implementation have been provided. Wiedermann and Li (2018), for example, provide SPSS macros that implement significance tests compatible with the three DDA components. An SPSS implementation using custom dialogs is introduced in a related Chapter 14 by Li and Wiedermann, in this volume. An R implementation of the presented DDA framework is given by Wiedermann and Li (2019b). All source codes and introductory material are freely available from http://www.ddaproject.com.

2.7 Conclusions and Future Directions

Within the last two decades, research on the direction of dependence in linear models has made tremendous progress. Fundamental asymmetry properties of linear models under variable non-normality have been identified, significance tests compatible with direction dependence have been proposed, and decision guidelines to perform model selection have been developed. Further, direction dependence principles have been extended to various statistical models such as mediation and moderation models (extensions of the direction dependence principles to categorical data are discussed in a related Chapter 8 by von Eye and Wiedermann, in this volume).

The presented DDA framework and the related work on copula-based directional dependence are important first steps in providing researchers with statistical tools to critically evaluate competing causal theories. However, DDA needs to be applied with caution. The reason for this is that DDA operates on third and fourth higher moments which are known to be prone to biases due to influential observations and outliers. While various regression diagnostic tools and outlier detection methods are available for the normal case, DDA requires that non-normality of variables is (at least to some extent) preserved when outliers are imputed or temporarily removed from the sample (for a discussion or regression diagnostics in the context of direction dependence methods see the Chapter 6 by Wood, in this volume).

Further, sensitivity analyses can be informative when performing model selection. Wiedermann and Sebastian (in press-b), for example, used a nonparametric bootstrap approach to test whether DDA model decision depends on random differences in sample composition (see also Rosenström et al., 2012). An algorithm to test causal model decision against additional (latent) confounding is discussed in Chapter 11 by Rosenström and García-Velázquez (2020, in this volume). A related Monte-Carlo based approach has been discussed in Wiedermann and Sebastian (in press-b).

Future research in the area of direction of dependence can go in a number of different directions. For example, the presented DDA components assume that the relationship between the tentative predictor and the tentative outcome can be adequately described by a linear model. While the DDA framework can straightforwardly be applied in linearizable regression models (i.e. cases where the linearity assumption is met after proper variable transformation), extensions of DDA principles to nonlinear cases are still underdeveloped. However, in machine learning research, previous studies were able to successfully extend the basic LiNGAM to nonlinear variable relations (Hernandez-Lobato et al., 2016; Mooij, Peters, Janzing, Zscheischler, & Schölkopf, 2016; Jonas Peters et al., 2014). These extensions may provide a useful basis to describe direction dependence principles in curvilinear variable relations. As another example, the present DDA framework assumes that the tentative predictor and the tentative outcome are continuous (i.e. measured at least on an interval scale). Although previous studies discussed direction dependence methods and causal learning algorithms when predictors and outcomes are categorical in nature (Inazumi et al., 2012; Peters, Janzing, & Scholkopf, 2011; von Eye & Wiedermann, 2017; Wiedermann & von Eye, 2020), extending direction dependence principles to the generalized linear modeling framework (McCullagh & Nelder, 1989) would be most promising for evaluating cause–effect relations among categorical, count, and bounded continuous variables. First attempts of learning the causal structure of mixed data are given by, for example, Li and Shimizu (2018) and Yamayoshi, Tsuchida, and Yadohisa (2020).

References

Aiken, L. S., & West, S. G. (1991). *Multiple regression: Testing and interpreting interactions.* Thousand Oaks, CA: Sage.

Bartlett, M. S. (1935). The effect of non-normality on the t distribution. *Mathematical Proceedings of the Cambridge Philosophical Society, 31*(2), 223–231. doi:10.1017/S0305004100013311

Bellemare, M. F., Masaki, T., & Pepinsky, T. B. (2017). Lagged explanatory variables and the estimation of causal effect. *Journal of Politics, 79*, 949–963. doi:10.2139/ssrn.2568724

Bentler, P. M. (1983). Some contributions to efficient statistics in structural models: Specification and estimation of moment structures. *Psychometrika, 48*(4), 493–517. doi:10.1007/BF02293875

Blanca, M. J., Arnau, J., López-Montiel, D., Bono, R., & Bendayan, R. (2013). Skewness and kurtosis in real data samples. *Methodology European Journal of Research Methods for the Behavioral and Social Sciences, 9*(2), 78–84. doi:10.1027/1614-2241/a000057

Boneau, C. A. (1960). The effects of violations of assumptions underlying the *t* test. *Psychological Bulletin, 57*(1), 49–64. doi:10.1037/h0041412

Box, G. E., & Watson, G. S. (1962). Robustness to non-normality of regression tests. *Biometrika, 49*(1–2), 93–106. doi:10.1093/biomet/49.1-2.93

Box, G. E. P. (1953). Non-normality and tests on variances. *Biometrika, 40*(3/4), 318–335. doi:10.2307/2333350

Braun, M. T., & Oswald, F. L. (2011). Exploratory regression analysis: A tool for selecting models and determining predictor importance. *Behavior Research Methods, 43*(2), 331–339. doi:10.3758/s13428-010-0046-8

Cain, M. K., Zhang, Z., & Yuan, K.-H. (2017). Univariate and multivariate skewness and kurtosis for measuring nonnormality: Prevalence, influence and estimation. *Behavior Research Methods, 49*(5), 1716–1735. doi:10.3758/s13428-016-0814-1

Cook, D. L. (1959). A replication of Lord's study on skewness and kurtosis of observed test-score distributions. *Educational and Psychological Measurement, 19*(1), 81–87. doi:10.1177/001316445901900109

Cragg, J. G. (1997). Using higher moments to estimate the simple errors-in-variables model. *The Rand Journal of Economics, 28*, S71–S91. doi:10.2307/3087456

Crile, G., & Quiring, D. P. (1940). A record of the body weight and certain organ and gland weights of 3690 animals. *Ohio Journal of Science, 40*(5), 219–259.

Dagenais, M. G., & Dagenais, D. L. (1997). Higher moment estimators for linear regression models with errors in the variables. *Journal of Econometrics, 76*(1–2), 193–221. doi:10.1016/0304-4076(95)01789-5

Darmois, G. (1953). Analyse générale des liaisons stochastiques: Etude particulière de l'analyse factorielle linéaire [General analysis of stochastic links]. *Revue de l'Institut*

International de Statistique/Review of the International Statistical Institute, 21(1/2), 2–8. doi:10.2307/1401511

Dehaene, S., & Cohen, L. (1998). Levels of representation in number processing. In B. Stemmer & H. A. Whitaker (Eds.), *Handbook of neurolinguistics* (pp. 331–431). San Diego, CA: Academic Press.

Devlieger, I., & Rosseel, Y. (2017). Factor score path analysis. *Methodology, 13*, 31–38. doi:10.1027/1614-2241/a000130

Dodge, Y., & Rousson, V. (2000). Direction dependence in a regression line. *Communications in Statistics—Theory and Methods, 29*(9–10), 1957–1972. doi:10.1080/03610920008832589

Dodge, Y., & Rousson, V. (2001). On asymmetric properties of the correlation coefficient in the regression setting. *The American Statistician, 55*(1), 51–54. doi:10.1198/000313001300339932

Dodge, Y., & Rousson, V. (2016). Statistical inference for direction of dependence in linear models. In W. Wiedermann & A. von Eye (Eds.), *Statistics and causality: Methods for applied empirical research* (pp. 45–62). Hoboken, NJ: Wiley.

Dodge, Y., & Yadegari, I. (2010). On direction of dependence. *Metrika, 72*(1), 139–150. doi:10.1007/s00184-009-0273-0

Drion, E. F. (1951). Estimation of the parameters of a straight line and of the variances of the variables, if they are both subject to error. *Indagationes Mathematicae, 54*, 256–260. doi:10.1016/S1385-7258(51)50036-7

Durbin, J. (1954). Errors in variables. *Review of the International Statistical Institute, 22*, 23–32. 10.2307/1401917

Erickson, T. (2001). Constructing instruments for regressions with measurement error when no additional data are available: Comment. *Econometrica, 69*(1), 221–222. 10.1111/1468-0262.00185

Frisch, R., & Waugh, F. V. (1933). Partial time regressions as compared with individual trends. *Econometrica, 1*, 387–401. doi:10.2307/1907330

Geary, R. C. (1949). Determination of linear relations between systematic parts of variables with errors of observation the variances of which are unknown. *Econometrica, 17*(1), 30–58. doi:10.2307/1912132

Geisser, S. (1993). *Predictive inference.* doi:10.1201/9780203742310

Gillard, J. (2014). Method of moments estimation in linear regression with errors in both variables. *Communications in Statistics—Theory and Methods, 43*(15), 3208–3222. doi:10.1080/03610926.2012.698785

Greenland, S., Pearl, J., & Robins, J. M. (1999). Causal diagrams for epidemiologic research. *Epidemiology, 10*(1), 37–48. doi:10.1097/00001648-199901000-00008

Greenland, S., & Robins, J. M. (1986). Identifiability, exchangeability, and epidemiological confounding. *International Journal of Epidemiology, 15*(3), 413–419. doi:10.1093/ije/15.3.413

Gretton, A., Fukumizu, K., Teo, C. H., Song, L., Schölkopf, B., & Smola, A. J. (2008). A kernel statistical test of independence. *Advances in Neural Information Processing Systems, 20*, 585–592.

Gunderson, E. A., Park, D., Maloney, E. A., Beilock, S. L., & Levine, S. C. (2018). Reciprocal relations among motivational frameworks, math anxiety, and math achievement in early elementary school. *Journal of Cognition and Development, 19*(1), 21–46. doi:10.1080/15248372.2017.1421538

Hayes, A. F. (2013). *Introduction to mediation, moderation, and conditional process analysis: A regression-based approach.* New York, NY: Guilford Press.

Heckman, J. J. (1979). Sample selection bias as a specification error. *Econometrica, 47*, 153–161. doi:10.2307/1912352

Hernandez-Lobato, D., Morales-Mombiela, P., Lopez-Paz, D., & Suarez, A. (2016). Non-linear causal inference using gaussianity measures. *Journal of Machine Learning Research, 17*, 1–39.

Ho, A. D., & Yu, C. C. (2015). Descriptive statistics for modern test score distributions: Skewness, kurtosis, discreteness, and ceiling effects. *Educational and Psychological Measurement, 75*(3), 365–388. doi:10.1177/0013164414548576

Holland, P. W. (1986). Statistics and causal inference. *Journal of the American Statistical Association, 81*(396), 945–960. doi:10.1080/01621459.1986.10478354

Hyvärinen, A., Karhunen, J., & Oja, E. (2001). *Independent component analysis.* New York, NY: Wiley & Sons.

Hyvärinen, A., & Smith, S. M. (2013). Pairwise likelihood ratios for estimation of non-Gaussian structural equation models. *Journal of Machine Learning Research, 14*, 111–152.

Iacobucci, D., Saldanha, N., & Deng, X. (2007). A meditation on mediation: Evidence that structural equations models perform better than regressions. *Journal of Consumer Psychology, 17*(2), 139–153. doi:10.1016/S1057-7408(07)70020-7

Imai, K., Keele, L., & Tingley, D. (2010). A general approach to causal mediation analysis. *Psychological Methods, 15*(4), 309–334. doi:10.1037/a0020761

Imai, K., Keele, L., & Yamamoto, T. (2010). Identification, inference and sensitivity analysis for causal mediation effects. *Statistical Science, 25*, 51–71. doi:10.1214/10-sts321

Inazumi, T., Washio, T., Shimizu, S., Suzuki, J., Yamamoto, A., & Kawahara, Y. (2012). Discovering causal structures in binary exclusive-or skew acyclic models. ArXiv:1202.3736 [Cs, Stat]. Retrieved from http://arxiv.org/abs/1202.3736

Jerison, H. J. (1973). *Evolution of the brain and intelligence.* New York, NY: Academic Press.

Kim, D., & Kim, J.-M. (2014). Analysis of directional dependence using asymmetric copula-based regression models. *Journal of Statistical Computation and Simulation, 84*(9), 1990–2010. doi:10.1080/00949655.2013.779696

Kim, J., Hwang, E., Phillips, M., Jang, S., Kim, J.-E., Spence, M. T., & Park, J. (2018). Mediation analysis revisited: Practical suggestions for addressing common deficiencies. *Australasian Marketing Journal*, *26*(1), 59–64. doi:10.1016/j.ausmj.2018.03.002

Lee, S., & Hershberger, S. (1990). A simple rule for generating equivalent models in structural equation modeling. *Multivariate Behavioral Research*, *25*(3), 313–334. doi:10.1207/s15327906mbr2503-4

Lewbel, A. (1997). Constructing instruments for regressions with measurement error when no additional data are available, with an application to patents and R&D. *Econometrica*, *65*(5), 1201–1213. doi:10.2307/2171884

Li, C., & Shimizu, S. (2018). Combining linear non-Gaussian acyclic model with logistic regression model for estimating causal structure from mixed continuous and discrete data. ArXiv:1802.05889 [Cs, Stat]. Retrieved from http://arxiv.org/abs/1802.05889

Li, X., & Wiedermann, W. (in press). Conditional direction dependence analysis: Testing the causal direction of linear models with interaction terms. *Multivariate Behavioral Research*. doi:10.1080/00273171.2019.1687276

Liu, Y., West, S. G., Levy, R., & Aiken, L. S. (2017). Tests of simple slopes in multiple regression models with an interaction: Comparison of four approaches. *Multivariate Behavioral Research*, *52*(4), 445–464. doi:10.1080/00273171.2017.1309261

Lord, F. M. (1955). A survey of observed test-score distributions with respect to skewness and kurtosis. *Educational and Psychological Measurement*, *15*(4), 383–389. doi:10.1177/001316445501500406

Lovell, M. C. (1963). Seasonal adjustment of economic time series and multiple regression analysis. *Journal of the American Statistical Association*, *58*(304), 993–1010. doi:10.1080/01621459.1963.10480682

Lovell, M. C. (2008). A simple proof of the FWL theorem. *Journal of Economic Education*, *39*(1), 88–91. doi:10.3200/JECE.39.1.88-91

Ma, X., & Xu, J. (2004). The causal ordering of mathematics anxiety and mathematics achievement: A longitudinal panel analysis. *Journal of Adolescence*, *27*(2), 165–179. doi:10.1016/j.adolescence.2003.11.003

MacKinnon, D. P. (2008). *Introduction to statistical mediation analysis*. New York, NY: Lawrence Erlbaum Associates.

MacKinnon, D. P., Lockwood, C. M., & Williams, J. (2004). Confidence limits for the indirect effect: Distribution of the product and resampling methods. *Multivariate Behavioral Research*, *39*(1), 99–128. doi:10.1207/s15327906mbr3901-4

McCullagh, P., & Nelder, J. A. (1989). *Generalized linear models* (2nd ed.). London, UK: Chapman & Hall.

Micceri, T. (1989). The unicorn, the normal curve, and other improbable creatures. *Psychological Bulletin*, *105*(1), 156–166. doi:10.1037/0033-2909.105.1.156

Mooij, J. M., Peters, J., Janzing, D., Zscheischler, J., & Schölkopf, B. (2016). Distinguishing cause from effect using observational data: Methods and benchmarks. *Journal of Machine Learning Research, 17*(32), 1–102.

Mooijaart, A. (1985). Factor analysis for non-normal variables. *Psychometrika, 50*(3), 323–342. doi:10.1007/BF02294108

Mooijaart, A., & Satorra, A. (2012). Moment testing for interaction terms in structural equation modeling. *Psychometrika, 77*(1), 65–84. doi:10.1007/s11336-011-9232-6

Muddapur, M. (2003). On directional dependence in a regression line. *Communications in Statistics—Theory and Methods, 32*(10), 2053–2057. doi:10.1081/STA-120023266

Neyman, J. (1923). Sur les applications de la thar des probabilities aux experiences Agaricales: Essay des principle. Excerpts reprinted (1990) in English. In: Dabrowska D, Speed T, translators. *Statistical Science, 5*, 463–472.

Nigg, J. T., Knottnerus, G. M., Martel, M. M., Nikolas, M., Cavanagh, K., Karmaus, W., & Rappley, M. D. (2008). Low blood lead levels associated with clinically diagnosed attention-deficit/hyperactivity disorder and mediated by weak cognitive control. *Biological Psychiatry, 63*(3), 325–331. doi:10.1016/j.biopsych.2007.07.013

Pal, M. (1980). Consistent moment estimators of regression coefficients in the presence of errors in variables. *Journal of Econometrics, 14*(3), 349–364. doi:10.1016/0304-4076(80)90032-9

Pearl, J. (1995). Causal diagrams for empirical research. *Biometrika, 82*(4), 669–688. doi:10.1093/biomet/82.4.669

Pearl, J. (2009). *Causality: Models, reasoning, and inference* (2nd ed.). New York, NY: Cambridge University Press.

Pearl, J. (2012). The causal mediation formula—A guide to the assessment of pathways and mechanisms. *Prevention Science, 13*(4), 426–436. doi:10.1007/s11121-011-0270-1

Peters, J., Janzing, D., & Scholkopf, B. (2011). Causal inference on discrete data using additive noise models. *IEEE Transactions on Pattern Analysis and Machine Intelligence, 33*(12), 2436–2450. doi:10.1109/TPAMI.2011.71

Peters, J., Mooij, D., Janzing, D., & Schölkopf, B. (2014). Causal discovery with continuous additive noise models. *Journal of Machine Learning Research, 15*, 2009–2053.

Pornprasertmanit, S., & Little, T. D. (2012). Determining directional dependency in causal associations. *International Journal of Behavioral Development, 36*(4), 313–322. doi:10.1177/0165025412448944

Posten, H. O. (1978). The robustness of the two-sample t-test over the Pearson system. *Journal of Statistical Computation and Simulation, 6*(3–4), 295–311. doi:10.1080/00949657808810197

Posten, H. O. (1982). Two-sample Wilcoxon power over the Pearson system and comparison with the t-test. *Journal of Statistical Computation and Simulation, 16*(1), 1–18. doi:10.1080/00949658208810602

Ramirez, G., Chang, H., Maloney, E. A., Levine, S. C., & Beilock, S. L. (2016). On the relationship between math anxiety and math achievement in early elementary school: The role of problem solving strategies. *Journal of Experimental Child Psychology, 141*, 83–100. doi:10.1016/j.jecp.2015.07.014

Rasch, D., & Guiard, V. (2004). The robustness of parametric statistical methods. *Psychology Science, 46*(2), 175–208.

Reiersøl, O. (1950). Identifiability of a linear relation between variables which are subject to error. *Econometrica, 18*(4), 375–389. doi:10.2307/1907835

Rosenström, T., & García-Velázquez, R. (2020). Distribution-based causal inference: A review and practical guidance for epidemiologists. In W. Wiedermann, D. Kim, E. A. Sungur, & A. von Eye (Eds.), *Direction dependence in statistical modeling: Methods of analysis*. Hoboken, NJ: Wiley & Sons. this volume

Rosenström, T., Jokela, M., Puttonen, S., Hintsanen, M., Pulkki-Råback, L., Viikari, J. S., ... Keltikangas-Järvinen, L. (2012). Pairwise measures of causal direction in the epidemiology of sleep problems and depression. *PLoS ONE, 7*(11), e50841. doi:10.1371/journal.pone.0050841

Rubin, D. B. (1974). Estimating causal effects of treatments in randomized and nonrandomized studies. *Journal of Educational Psychology, 66*(5), 688–701. doi:10.1037/h0037350

Scott, E. L. (1950). Note on consistent estimates of the linear structural relation between two variables. *The Annals of Mathematical Statistics, 21*(2), 284–288. doi:10.1214/aoms/1177729846

Sen, A., & Sen, B. (2014). Testing independence and goodness-of-fit in linear models. *Biometrika, 101*(4), 927–942. doi:10.1093/biomet/asu026

Shimizu, S. (2019). Non-Gaussian methods for causal structure learning. *Prevention Science, 20*(3), 431–441. doi:10.1007/s11121-018-0901-x

Shimizu, S., Hoyer, P. O., Hyvärinen, A., & Kerminen, A. (2006). A linear non-Gaussian acyclic model for causal discovery. *The Journal of Machine Learning Research, 7*, 2003–2030.

Shimizu, S., Inazumi, T., Sogawa, Y., Hyvärinen, A., Kawahara, Y., Washio, T., ... Bollen, K. (2011). DirectLiNGAM: A direct method for learning a linear non-Gaussian structural equation model. *Journal of Machine Learning Research, 12*, 1225–1248.

Shimizu, S., & Kano, Y. (2008). Use of non-normality in structural equation modeling: Application to direction of causation. *Journal of Statistical Planning and Inference, 138*(11), 3483–3491. doi:10.1016/j.jspi.2006.01.017

Shmueli, G. (2010). To explain or to predict? *Statistical Science, 25*(3), 289–310. doi:10.1214/10-STS330

Skitovich, W. P. (1953). On a property of the normal distribution. *Doklady Akademii Nauk SSSR [Reports of the Academy of Sciences USSR], 89*, 217–219.

Spiegelman, C. (1979). On estimating the slope of a straight line when both variables are subject to error. *The Annals of Statistics, 7*(1), 201–206. doi:10.1214/aos/1176344565

Stigler, S. M. (2010). The changing history of robustness. *The American Statistician, 64*(4), 277–281. doi:10.1198/tast.2010.10159

Sungur, E. A. (2005). A note on directional dependence in regression setting. *Communications in Statistics—Theory and Methods, 34*(9–10), 1957–1965. doi:10.1080/03610920500201228

Székely, G. J., Rizzo, M. L., & Bakirov, N. K. (2007). Measuring and testing dependence by correlation of distances. *Annals of Statistics, 35*(6), 2769–2794. doi:10.1214/009053607000000505

Thoemmes, F. (2015). Reversing arrows in mediation models does not distinguish plausible models. *Basic and Applied Social Psychology, 37*(4), 226–234. doi:10.1080/01973533.2015.1049351

Vickers, A. J., Rees, R. W., Zollman, C. E., McCarney, R., Smith, C. M., Ellis, N., … Van Haselen, R. (2004). Acupuncture for chronic headache in primary care: Large, pragmatic, randomised trial. *British Medical Journal, 328*(7442), 744. doi:10.1136/bmj.38029.421863.eb

von Aster, M. G., & Shalev, R. S. (2007). Number development and developmental dyscalculia. *Developmental Medicine and Child Neurology, 49*(11), 868–873. doi:10.1111/j.1469-8749.2007.00868.x

von Eye, A., & DeShon, R. P. (2012). Directional dependence in developmental research. *International Journal of Behavioral Development, 36*(4), 303–312. doi:10.1177/0165025412439968

von Eye, A., & Wiedermann, W. (2017). Direction of effects in categorical variables: Looking inside the table. *Journal for Person-Oriented Research, 3*(1), 11–27. doi:10.17505/jpor.2017.02

von Eye, A., & Wiedermann, W. (2018). Locating event-based causal effects: A configural perspective. *Integrative Psychological and Behavioral Science, 52*(2), 307–330. doi:10.1007/s12124-018-9423-0

Wiedermann, W. (2015). Decisions concerning the direction of effects in linear regression models using fourth central moments. In M. Stemmler, A. von Eye, & W. Wiedermann (Eds.), *Dependent data in social sciences research: Forms, issues, and methods of analysis* (Vol. 145, pp. 149–169). doi:10.1007/978-3-319-20585-4_7

Wiedermann, W. (2018). A note on fourth moment-based direction dependence measures when regression errors are non normal. *Communications in Statistics—Theory and Methods, 47*(21), 5255–5264. doi:10.1080/03610926.2017.1388403

Wiedermann, W., & Alexandowicz, R. (2007). A plea for more general tests than those for location only. *Psychology Science, 49*(1), 2–12.

Wiedermann, W., & Alexandrowicz, R. W. (2011). A modified normal scores test for paired data. *Methodology, 7*(1), 25–38. doi:10.1027/1614-2241/a000020

Wiedermann, W., Artner, R., & von Eye, A. (2017). Heteroscedasticity as a basis of direction dependence in reversible linear regression models. *Multivariate Behavioral Research, 52*(2), 222–241. doi:10.1080/00273171.2016.1275498

Wiedermann, W., & Hagmann, M. (2016). Asymmetric properties of the Pearson correlation coefficient: Correlation as the negative association between linear regression residuals. *Communications in Statistics: Theory and Methods, 45*(21), 6263–6283. doi:10.1080/03610926.2014.960582

Wiedermann, W., & Li, X. (2018). Direction dependence analysis: A framework to test the direction of effects in linear models with an implementation in SPSS. *Behavior Research Methods, 50*(4), 1581–1601. doi:10.3758/s13428-018-1031-x

Wiedermann, W., & Li, X. (2019a). Confounder detection in linear mediation models: Performance of kernel-based tests of independence. *Behavior Research and Methods, 52*, 342–359. doi:10.3758/s13428-019-01230-4

Wiedermann, W., & Li, X. (2019b). *Direction Dependence Analysis in R.* http://www .ddaproject.com.

Wiedermann, W., Li, X., & von Eye, A. (2019). Testing the causal direction of mediation effects in randomized intervention studies. *Prevention Science, 20*(3), 419–430. doi:10.1007/s11121-018-0900-y

Wiedermann, W., Merkle, E. C., & Eye, A. (2018). Direction of dependence in measurement error models. *British Journal of Mathematical and Statistical Psychology, 71*(1), 117–145. doi:10.1111/bmsp.12111

Wiedermann, W., & Sebastian, J. (in press-a). Direction dependence analysis in the presence of confounders: Applications to linear mediation models. *Multivariate Behavioral Research.*

Wiedermann, W., & Sebastian, J. (in press-b). Sensitivity analysis and extensions of testing the causal direction of dependence: A rejoinder to Thoemmes (2019). *Multivariate Behavioral Research.* doi:10.1080/00273171.2019.1659127

Wiedermann, W., & von Eye, A. (2015a). Direction of effects in mediation analysis. *Psychological Methods, 20*(2), 221–244. doi:10.1037/met0000027

Wiedermann, W., & von Eye, A. (2015b). Direction of effects in multiple linear regression models. *Multivariate Behavioral Research, 50*(1), 23–40. doi:10.1080/00273171.2014.958429

Wiedermann, W., & von Eye, A. (2015c). Direction-dependence analysis: A confirmatory approach for testing directional theories. *International Journal of Behavioral Development, 39*(6), 570–580. doi:10.1177/0165025415582056

Wiedermann, W., & von Eye, A. (2016). Directionality of effects in causal mediation analysis. In W. Wiedermann & A. von Eye (Eds.), *Statistics and causality: Methods for applied empirical research* (pp. 63–106). doi:10.1002/9781118947074.ch4

Wiedermann, W., & von Eye, A. (2020). Log-linear models to evaluate direction of effect in binary variables. *Statistical Papers, 61*(1), 317–346. doi:10.1007/s00362-017-0936-2

Yamayoshi, M., Tsuchida, J., & Yadohisa, H. (2020). An estimation of causal structure based on latent LiNGAM for mixed data. *Behaviormetrika, 47*(1), 105–121. doi:10.1007/s41237-019-00095-3

3

The Use of Copulas for Directional Dependence Modeling

Engin A. Sungur

Statistics Discipline, University of Minnesota, Morris, MN, USA

3.1 Introduction and Definitions

Copula is simply a p-dimensional distribution function with uniform univariate marginals on the interval 0 to 1. Let $(U_1, U_2, \ldots, U_p)$ be a random vector where U_i, $i = 1, 2, \ldots, p$ has a uniform distribution on the interval 0 to 1. Then,

$$P(U_1 \leq u_1, U_2 \leq u_2, \ldots, U_p \leq u_p) = F(u_1, u_2, \ldots, u_p) = C(u_1, u_2, \ldots, u_p)$$

(3.1)

where $F()$ and $C()$ represent distribution function and copula of the random vector, respectively.

The power of the copulas on dependence modeling comes from the result that is proven by Sklar (1959). Starting with a p-dimensional random vector $(X_1, X_2, \ldots, X_p)$ with continuous marginal distribution functions $F_{X_1}(x_1), F_{X_2}(x_2), \ldots, F_{X_p}(x_p)$, the joint distribution function of them, F, can be written uniquely as a function of the marginal distribution functions through its copula C. That is

$$F(x_1, x_2, \ldots, x_p) = C(F_{X_1}(x_1), F_{X_2}(x_2), \ldots, F_{X_p}(x_p)).$$

(3.2)

When $F_{X_1}(x_1), F_{X_2}(x_2), \ldots, F_{X_p}(x_p)$ are not continuous C is uniquely determined on Range $F_{X_1} \times$ Range $F_{X_2} \times \cdots \times$ Range F_{X_p}.

As a historical note, the main idea behind the copulas came from Hoeffding (1994) who studied distributions with $[-1/2, 1/2]^2$ support and uniform marginal. Fréchet (1951) independently rediscovered many of the same results such as Fréchet–Hoeffding bounds. Access to the concept of copulas is provided by Nelsen (2006) who presented a comprehensive and integrated overview of the concept of copulas.

Direction Dependence in Statistical Modeling: Methods of Analysis, First Edition.
Edited by Wolfgang Wiedermann, Daeyoung Kim, Engin A. Sungur, and Alexander von Eye.
© 2021 John Wiley & Sons, Inc. Published 2021 by John Wiley & Sons, Inc.

3.1.1 Why Copulas?

Figure 3.1 provides a summary of the basic concepts behind copulas. The important part from dependence modeling point of view is the fact that the joint behavior of a random vector reflected on the joint distribution function can be decomposed into two parts: the marginal behavior, and the dependence structure. It is clear that marginal distributions of individual variables do not provide any information on the dependence structure. To clarify this statement, let X and Y be continuous random variables with a copula C. If g and h are strictly increasing on the range of X and Y, respectively, then the copula of $(g(X), h(Y))$ is the same as the copula of (X, Y). See Nelsen (2006) for the proof of this result. Therefore, even if the strictly increasing transformation of X, Y, or both leads to a different distribution functions than the distribution function of (X, Y), the dependence structure given by the copula will stay the same. For example, transforming the random variables to make the conditional expectation of Y given $X = x$ linear does not change the dependence mechanism between the variables. When such transformations have been done with copulas all lead to the same copula regression functions under strictly increasing transformations. Copulas provide a clear look at dependence structure.

3.1.2 Defining Directional Dependence

Modeling and measuring directional dependence is of interest for many researchers, both applied and theoretical. In most of these cases, the objective is

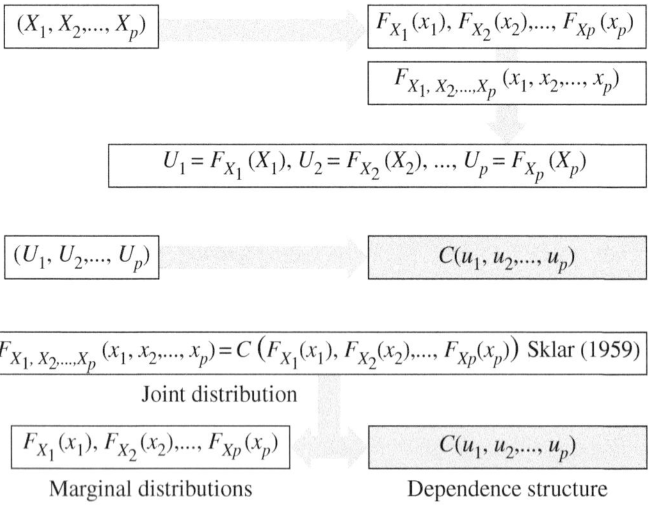

$$(X_1, X_2,..., X_p)$$

$$F_{X_1}(x_1), F_{X_2}(x_2),..., F_{Xp}(x_p)$$

$$F_{X_1, X_2,...,X_p}(x_1, x_2,..., x_p)$$

$$U_1 = F_{X_1}(X_1),\ U_2 = F_{X_2}(X_2),\ ...,\ U_p = F_{X_p}(X_p)$$

$$(U_1, U_2,..., U_p)$$

$$C(u_1, u_2,..., u_p)$$

$$F_{X_1, X_2,...,X_p}(x_1, x_2,..., x_p) = C\left(F_{X_1}(x_1), F_{X_2}(x_2),..., F_{Xp}(x_p)\right) \text{ Sklar (1959)}$$

Joint distribution

$$F_{X_1}(x_1), F_{X_2}(x_2),..., F_{Xp}(x_p)$$

Marginal distributions

$$C(u_1, u_2,..., u_p)$$

Dependence structure

Figure 3.1 Construction of a copula and its use.

to be able to set up causal relationships that can only be set through experiments. In this section, directional dependence will be defined and differences between *direction of dependence* and *causal dependence* will be explained.

First, let us look at the *direction of dependence* for the two-dimensional copula. The *Fréchet–Hoeffing bounds* state that any copula C and for all $\{(u, v); 0 \le u \le 1, 0 \le v \le 1\}$,

$$C^-(u, v) = \max\{0, u + v - 1\} \le C(u, v) \le C^+(u, v) = \min\{u, v\}. \tag{3.3}$$

Considering that $C^0(u, v) = uv$ is the copula in case of independence, one may say that direction of dependence is positive if and only if $C^0(u, v) \le C(u, v) < C^+(u, v)$, and negative if and only if $C^-(u, v) \le C(u, v) < C^0(u, v)$, for all $\{(u, v); 0 \le u \le 1, 0 \le v \le 1\}$. If a measure of dependence or association is used, one expects that the measure will be positive if the direction of dependence is positive and negative if the direction of dependence is negative. On the other hand, in the analysis of *directional dependence* we look at changes in the dependence structure when we "move" from U to V and V to U. It is clear that the directional dependence does not exist if the copula C is symmetric, that is $C(u, v) = C(v, u)$ for all $\{(u, v); 0 \le u \le 1, 0 \le v \le 1\}$. By using a different terminology, this case occurs when (U, V) are exchangeable, that is the distribution of the random pair (U, V) is the same as (V, U).

Let (X, Y) be a random pair with a distribution function F_{XY}, copula C_{XY}, and marginals F_X and F_Y. To be able to set up a directional dependence between them the random pair should not be exchangeable. If they are not, three cases will occur:

I. *Directional Dependence in Marginal Behavior*: Different marginals but same copulas for (X, Y) and (Y, X).

II. *Directional Dependence in Joint Behavior*: Same marginals but different copulas for (X, Y) and (Y, X).

III. *Directional Dependence in Both Marginal and Joint Behavior*: Different marginals and different copulas for (X, Y) and (Y, X).

In all of these three cases as we have emphasized in the previous section, cases II and III are the ones directly related with the dependence structure. In both of these cases, copula of (X, Y) random pair needs to be asymmetric. We can now define the directional dependence for a random pair as follows.

Definition 3.1 Let (X, Y) be a random pair with marginal distribution functions F_X, F_Y, and a copula C. Also, let $U = F_X(X)$, and $V = F_Y(Y)$. The pair (X, Y) is directionally dependent in the joint behavior if and only if the conditional distributions of the U given $V = v$, and the V given $U = u$ differ.

The conditional distribution functions of U given $V = v$, and V given $U = u$ are known as the conditional copulas and they will be represented by $C_{U|V}$ and $C_{V|U}$, respectively, i.e.

$$C_{U|V}(u;v) = P(U \le u \mid V = v) = \frac{\partial C(u,v)}{\partial v}, \tag{3.4}$$

$$C_{V|U}(v;u) = P(V \le v \mid U = u) = \frac{\partial C(u,v)}{\partial u}. \tag{3.5}$$

Therefore, $C_{U|V}(w; z) \neq C_{V|U}(w; z)$. In the regression setting, the existence of directional dependence implies that the form of the regression functions for $U \mid V = v$ and $V \mid U = u$ differ.

Definition 3.2 Let (U, V) be a random pair with uniform marginal distributions over the interval $[0, 1]$. The expected values of U given $V = v$, and V given $U = u$ are known as the copula regression functions denoted as $r_{U|V}$ and $r_{V|U}$, respectively, i.e.

$$r_{U|V}(v) = E[U \mid V = v], \tag{3.6}$$

$$r_{V|U}(u) = E[V \mid U = u]. \tag{3.7}$$

If the conditional distribution functions of X given $Y = y$ and Y given $X = x$ differ, but the corresponding conditional copulas are the same, then the random pair (X, Y) is directionally dependent only in marginals. This case has been studied by Dodge and Rousson (2000), Wiedermann and Li (2018), and Wiedermann, Li, and von Eye (2019).

To understand the differences between directional dependence in the joint behavior and marginals, consider a random pair (X, Y), where X has a uniform distribution on the interval $[0, 1]$, and Y has an exponential distribution with parameter $\mu = 2$. Consider two alternatives for the copula of the random pair.

$$C^M(u,v) = uv[1 + \theta(1 - u)(1 - v)], \quad \theta \in [0, 1] \tag{3.8}$$

$$C^D(u,v) = uv[1 + \theta(1 - u)(1 - v^2)], \quad \theta \in [0, 1/2] \tag{3.9}$$

Corresponding joint distribution functions, regression functions, and the copula regression functions for these two cases are:

$$F^M(x,y) = xe^{-2\mu y}(1 - e^{\mu y})(\theta x - \theta - e^{\mu y}),$$

$$F^D(x,y) = xe^{-3\mu y}(1 - e^{\mu y})(\theta x - \theta - e^{2\mu y}),$$

$$r^M_{X|Y}(y) = \frac{3+\theta}{6} - \frac{\theta}{3}e^{-\mu y}, \quad r^M_{Y|X}(x) = \frac{2-\theta}{2\mu} + \frac{\theta}{\mu}x,$$

$$r^D_{X|Y}(y) = \frac{3+2\theta}{6} + \frac{1}{2}\theta e^{-2\mu y} - \theta e^{\mu y}, \quad r^D_{Y|X}(x) = \frac{6-5\theta}{6\mu} + \frac{5\theta}{3\mu}x,$$

$$r^M_{U|V}(v) = \frac{3-\theta}{6} + \frac{\theta}{3}v, \quad r^M_{V|U}(u) = \frac{3-\theta}{6} + \frac{\theta}{3}u,$$

$$r^D_{U|V}(v) = \frac{3-\theta}{6} + \frac{\theta}{2}v^2, \quad r^D_{V|U}(u) = \frac{2-\theta}{4} + \frac{\theta}{2}u.$$

Since the copula C^M is symmetric, we end up with the case of the random pair being directionally dependent in marginal behavior. Even if the regression functions differ, the copula regression functions turn out to be the same. On the other, the copula C^D leads to the case that the random pair is directionally dependent in joint behavior. Note that the form of both regression and copula regression functions differ.

Another case is when the random pair is directionally dependent in joint behavior but not in marginal. That is, the copula regression functions differ but not the regression functions. As it is given in the following example, such a case can be easily constructed by starting with copula regression functions such that $r_{U|V}(w) \neq r_{V|U}(w)$, and taking marginals F_X and F_Y such that $F_Y^{-1}(r_{V|U}(F_X(z))) = F_X^{-1}(r_{U|V}(F_Y(z)))$. Therefore, it is clear that directional dependence in joint behavior does not imply directional dependence in the marginals, and vice versa.

Example 3.1   Let (X, Y) be a random pair with copula C^D given in Eq. (3.9), and marginals $F_X(x) = x, x \in [0, 1]$, and $F_Y(y) = \frac{\sqrt{-6+\sqrt{15}+12y}}{3^{3/4}5^{1/4}}, y \in \left[\frac{1}{12}(6 - \sqrt{15}), \frac{1}{6}(3 + \sqrt{15}) \right]$. Therefore, the regression functions are

$$r_{X|Y}^D(z) = \frac{1}{2} + \frac{\theta(-1 + 2z)}{\sqrt{15}} = r_{Y|X}^D(z). \tag{3.10}$$

On the other hand, the copula regression functions are

$$r_{U|V}^D(w) = \frac{3 - \theta}{6} + \frac{\theta}{2}w^2 \neq r_{V|U}^D(w) = \frac{2 - \theta}{4} + \frac{\theta}{2}w. \tag{3.11}$$

As it can be seen from this example, directional dependence may disappear when we move from the joint behavior to marginal behavior for certain marginal distributions.

3.2 Directional Dependence Between Two Numerical Variables

In this section, we will take a closer look at directional dependence for the bivariate case where the marginal distributions are continuous. Since the directional dependence in joint behavior requires asymmetric copulas, first we will present how to construct copula models with directional dependence property. Then, the results in directional dependence in regression setting will be given, leading to a discussion on how to measure the directional dependence.

3.2.1 Asymmetric Copulas

In this section, various ways of constructing copulas with directional dependence property will be discussed. Such copulas need to be asymmetric to have this property. There are basically two approaches to construct classes of copula models with directional dependence properties.

For the first approach, that will be referred as the functional equations approach, the starting point is one of the existing class of symmetric copulas. Consider the copula that has the form $C(u, v) = C^*(u, v) + k(u, v)$, where C^* is a symmetric copula and k is an asymmetric two-place real function. Some special cases of this type of copulas have been extensively studied by Rodríguez-Lallena and Flores (2004), Bairamov and Kotz (2002), and Lai and Xie (2000). For the Rodríguez-Lallena and Úbeda Flores family copula has the general form of

$$C(u, v) = uv + f(u)g(v), \tag{3.12}$$

where f and g are two real functions defined on the unit interval.

Rodríguez-Lallena and Flores (2004) provide the conditions that these two functions need to satisfy that is given in Theorem 3.1.

Theorem 3.1 Let f and g be two non-zero real functions defined on $I = [0, 1]$. Let $C : I^2 \rightarrow R$ be the function defined by Eq. (3.12). Then C is a copula, if and only if

1. $f(0) = f(1) = g(0) = g(1) = 0$,
2. f and g are absolutely continuous and
3. $\min\{\alpha\delta, \beta\gamma\} \geq -1$, where $\alpha = \inf\{f'(u) : u \in A\} < 0$, and $\beta = \sup\{f'(u) : u \in A\} > 0$, $\gamma = \inf\{g'(v) : v \in B\} < 0$, and $\delta = \sup\{g'(v) : v \in B\} > 0$, with $A = \{u \in I : f'(u)$ exists$\}$ and $B = \{v \in I : g'(v)$ exists$\}$. Furthermore, in such a case, C is absolutely continuous.

In this case, the conditional copulas and copula regression functions will have the following forms:

$$C_{U|V}(u; v) = u + f(u)g'(v), \tag{3.13}$$

$$C_{V|U}(v; u) = v + f'(u)g(v), \tag{3.14}$$

$$r_{U|V}(v) = \frac{1}{2} - \rho_C g'(v) \left[12 \int_0^1 g(v) dv \right]^{-1}, \tag{3.15}$$

$$r_{V|U}(u) = \frac{1}{2} - \rho_C f'(u) \left[12 \int_0^1 f(u) du \right]^{-1}, \tag{3.16}$$

where ρ_C is the Pearson correlation. When f and g are different, the copula will not be symmetric and the form of the regression functions for V and U will differ. Another example of the asymmetric class of copulas that has been generated by using an exchangeable/symmetric bivariate copula C is proposed by Frees and Valdez (1998). In their construction, the form of the copula is

$$C_{\kappa,\lambda}(u,v) = u^{1-\kappa}v^{1-\lambda}C(u^{\kappa},v^{\lambda}), \quad 0 < \kappa, \quad \lambda < 1, \tag{3.17}$$

where C is any exchangeable/symmetric bivariate copula.

The second approach directly uses probabilistic arguments to generate a class of copulas with the directional dependence property. Sungur and Celebioglu (2011) started with looking at the copulas of linear combinations of independently distributed uniform random variables U and V. They showed that the copula of the two linear combinations, $X = \alpha_1 U + \beta_1 V$ and $Y = \alpha_2 U + \beta_2 V$, have the form

$$C(u,v) = \int_0^1 D[k(u,w), l(v,w)]dw, \tag{3.18}$$

where, $D(u,v) = \min\{u,v\}$, $k(u,w) = \frac{F_X^{-1}(u) - \beta_1 w}{\alpha_1}$, and $l(v,w) = \frac{F_Y^{-1}(v) - \beta_2 w}{\alpha_2}$ for $\alpha_1 > \beta_1$, $\alpha_2 \le -\beta_2$, $\alpha_1, \beta_1 > 0$, $\alpha_2 \ge 0$, and $\beta_2 \le 0$.

Note that as long as the functions k and l are non-decreasing in their first arguments, the resulting function will be 2-increasing. In addition, the structure of the class allows to produce asymmetric copulas. A special case for a class of copulas that has this form can be constructed as follows:

Consider a random triple (U, V, W) with uniform marginals. W will be called the direction variable. Then,

$$C(u,v) = P(U \le u, V \le v) = \int_0^1 P(U \le u, V \le v \mid W = w)dw. \tag{3.19}$$

Therefore,

$$C(u,v) = \int_0^1 C_{U,V|W}[C_{U|W}(u;w), C_{V|W}(v;w)]dw, \tag{3.20}$$

where, $C_{U,V|W}$, $C_{U|W}$, and $C_{V|W}$ are the conditional copulas of (U, V), U, and V given $W = w$, respectively. It is trivial to show that the resulting two place function is grounded, 2-increasing and $C(1, v) = v$, and $C(u, 1) = u$. For the random triple (U, V, W), the direction variable W need not to be continuous. Sungur and Celebioglu (2011) consider the case that W has a Bernoulli distribution and look at the class of copulas of the form

$$C(u,v) = \sum_{w=0}^1 C_{U,V|W=w}[P(U \le u \mid W = w), P(V \le v \mid W = w)]P(W = w). \tag{3.21}$$

Taking the copula model one step further the conditional distribution of the direction variable can be modeled by using various link functions such as logit, probit, and complementary log–log. For the logistic regression set up let

$$\pi(u, \alpha_1, \beta_1) = P(W = 1 \mid U = u) = \frac{1}{1 + \exp[-\alpha_1 - \beta_1 u]},$$

$$\pi(v, \alpha_2, \beta_2) = P(W = 1 \mid V = v) = \frac{1}{1 + \exp[-\alpha_2 - \beta_2 v]}.$$

Therefore,

$$C(u, v) = C_{U,V|W=0}\left[\frac{u - \Pi(u, \alpha_1, \beta_1)}{P(W = 0)}, \frac{v - \Pi(v, \alpha_2, \beta_2)}{P(W = 0)}\right] P(W = 0)$$

$$+ C_{U,V|W=1}\left[\frac{\Pi(u, \alpha_1, \beta_1)}{P(W = 1)}, \frac{\Pi(v, \alpha_2, \beta_2)}{P(W = 1)}\right] P(W = 1), \tag{3.22}$$

where

$$\Pi(u, \alpha_1, \beta_1) = \frac{1}{\beta_1} \log\left[\frac{1 + e^{\alpha_1 + \beta_1 u}}{1 + e^{\alpha_1}}\right], \tag{3.23}$$

$$\Pi(v, \alpha_2, \beta_2) = \frac{1}{\beta_2} \log\left[\frac{1 + e^{\alpha_2 + \beta_2 v}}{1 + e^{\alpha_2}}\right]. \tag{3.24}$$

To complete the construction of the copula model two issues need to be addressed. The first one is the selection of the $C_{U,V|W=w}$, and the second one is the restriction on the model parameters. The restriction on the parameters is due to the fact that

$$P(W = 1) = \Pi(1, \alpha_1, \beta_1) = \Pi(1, \alpha_2, \beta_2). \tag{3.25}$$

To get around of this problem, we will fix the success probability of the W at p, and determine the values of α_i in terms of β_i, $i = 1, 2$, i.e.

$$\alpha_i = \log\left[\frac{1 - \exp[\beta_i p]}{\exp[\beta_i p] - \exp[\beta_i]}\right].$$

Under this restriction,

$$\Pi(u, \alpha_1, \beta_1) = \Pi_r(u, \beta_1, p) = \frac{1}{\beta_1} \log\left[\frac{e^{\beta_1 p} - e^{\beta_1} + e^{\beta_1 u} - e^{\beta_1 p} e^{\beta_1 u}}{1 - e^{\beta_1}}\right], \tag{3.26}$$

$$\Pi(v, \alpha_2, \beta_2) = \Pi_r(v, \beta_2, p) = \frac{1}{\beta_2} \log\left[\frac{e^{\beta_2 p} - e^{\beta_2} + e^{\beta_2 v} - e^{\beta_2 p} e^{\beta_2 v}}{1 - e^{\beta_2}}\right]. \tag{3.27}$$

There are many options for the selection of the conditional copula of the (U, V) given $W = w$, the simplest one being the conditional independence, i.e.

$$P(U \le u, V \le v \mid W = w) = P(U \le u \mid W = w)P(V \le v \mid W = w), \tag{3.28}$$

$$C_{U,V|W=w}(u, v; w) = uv. \tag{3.29}$$

Sungur and Celebioglu (2011) discuss the cases that the conditional copula is a member of the Farlie–Gumbel–Morgenstern class and Fréchet class, i.e.

$$C_{U,V|W=w}(u, v; w) = uv[1 + \theta(1 - u)(1 - v)], \tag{3.30}$$

$$C_{U,V|W=w}(u,v;w) = \alpha \min\{u,v\} + (1-\alpha)\max\{u+v-1,0\}. \tag{3.31}$$

The class of copula which is constructed through the logistic regression model given in Eq. (3.22) is very rich. It allows one to work with both symmetric and asymmetric dependence structures and the conditional copula $C_{U,V|W=w}$ could be a member of any class of copulas. The parameters of this class of copula can be grouped into three: (i) directional dependence parameters, β_1 and β_2, that are originated from the logistic regression, (ii) dependence parameter, θ, that is the parameter for the conditional copula $C_{U,V|W=w}$, and (iii) mixture parameter, $p = P(W = 1)$, the parameter of the Bernoulli direction variable. The value of p can be selected based on prior knowledge, predetermined through the design of experiment, or estimated together with the other parameters of the copula model. The symmetry behavior of the class is determined by the model parameters. Symmetry has been achieved when $\beta_1 = \beta_2$. Therefore, these parameters can provide information on the directional dependence. The direction variable W can be defined in many different ways in application. As an example, one may want to study the relationship between wife and husband's ages at death and consider a Bernoulli direction variable that identifies who died first. In medical research, the dependence between two vital variables during a surgery can be modeled by considering outcome of the surgery as direction variable. In the example below, we present a special case of such copula models.

Example 3.2   Let C be a member of the class of copula given in Eq. (3.22), under the conditional independence, i.e.

$$P(U \le u, V \le v \mid W = w) = P(U \le u \mid W = w)P(U \le u \mid W = w).$$

The copula will have the form

$$C(u,v) = \frac{(u - \Pi_r(u,\beta_1,p))(v - \Pi_r(v,\beta_2,p))}{1-p} + \frac{\Pi_r(u,\beta_1,p)\Pi_r(v,\beta_2,p)}{p} \tag{3.32}$$

where, $\Pi_r(u,\beta_1,p)$ and $\Pi_r(v,\beta_2,p)$ are defined in Eqs. (3.26) and (3.27), respectively. To understand properties of this copula we need the following lemma.

Lemma 3.1   Consider $\Pi_r(u,\beta_1,p)$ and $\Pi_r(v,\beta_2,p)$ that are defined in Eqs. (3.26) and (3.27), respectively. [label=]

1. $\Pi_r(u,\beta_1,1) = u$ and $\Pi_r(v,\beta_2,1) = v$.
2. $\Pi_r(u,\beta_1,0) = \Pi_r(v,\beta_2,0) = 0$.
3. $\Pi_r(1,\beta_1,p) = \Pi_r(1,\beta_2,p) = p$.
4. $\Pi_r(0,\beta_1,p) = \Pi_r(0,\beta_2,p) = 0$.

Theorem 3.2 Let C be a copula defined in Eq. (3.32), then

i. $C(u, v) = uv$ if $p = 0$, 1 or $\beta_1 = \beta_2 = 0$
ii. Copula regression functions are

$$r_{U|V}(v) = 1 + \frac{\int_0^1 \Pi_r(u, \beta_1, p)du - \frac{1}{2}}{1 - p} - \frac{\int_0^1 \Pi_r(u, \beta_1, p)du - \frac{p}{2}}{p(1 - p)} \pi_r(v, \beta_2, p),$$

$$r_{V|U}(u) = 1 + \frac{\int_0^1 \Pi_r(v, \beta_2, p)dv - \frac{1}{2}}{1 - p} - \frac{\int_0^1 \Pi_r(v, \beta_2, p)dv - \frac{p}{2}}{p(1 - p)} \pi_r(u, \beta_1, p),$$

where

$$\pi_r(v, \beta_2, p) = \frac{\partial \Pi_r(v, \beta_2, p)}{\partial v} = \frac{e^{\beta_2 v}(1 - e^{\beta_2 p})}{e^{\beta_2 v}(1 - e^{\beta_2 p}) + e^{\beta_2 p} - e^{\beta_2}},$$

$$\pi_r(u, \beta_1, p) = \frac{e^{\beta_1 u}(1 - e^{\beta_1 p})}{e^{\beta_1 u}(1 - e^{\beta_1 p}) + e^{\beta_1 p} - e^{\beta_1}}.$$

Proof.

i. Directly follows from the Lemma 3.1.
ii. By using the facts that

$$r_{U|V}(v) = 1 - \int_0^1 \frac{\partial C(u, v)}{\partial v} du,$$

$$r_{V|U}(u) = 1 - \int_0^1 \frac{\partial C(u, v)}{\partial u} dv,$$

the results follow. □

Figures 3.2–3.4 give the plots of the two copula regression functions for selected values of β_1, β_2, and p. Figure 3.5 demonstrates how the difference between two copula functions changes as a function of the value of the conditioning variable. All of these plots clearly indicate existence of the directional dependence. This example will be revisited in the following section when we discuss the directional dependence in the regression setting.

We will end this section by looking at directional dependence copula models generated by using the random triplet (U, V, W), where direction covariate W has a uniform distribution on a unit interval. Note that the selection of functions triple $(C_{U,V|W}, C_{U|W}, C_{V|W})$ is restricted with the compatibility conditions. Since,

$$C(u, v) = \int_0^1 C_{U,V|W}[C_{U|W}(u; w), C_{V|W}(v; w)]dw, \qquad (3.33)$$

directional dependence property of the copula could be initiated from various sources; (i) directional dependence on the conditional copulas $C_{U|W}$ and $C_{V|W}$,

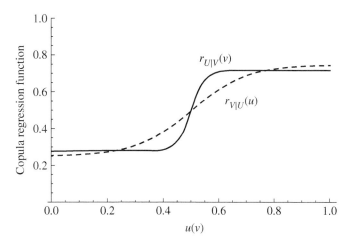

Figure 3.2 Plot of the copula regression functions for $\beta_1 = 10$, $\beta_2 = 40$, and $p = 0.5$.

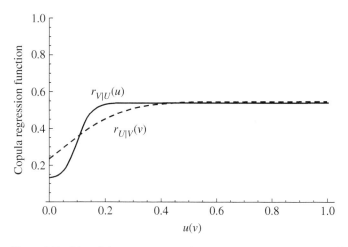

Figure 3.3 Plot of the copula regression functions for $\beta_1 = 10$, $\beta_2 = 40$, and $p = 0.9$.

(ii) structure of the copula connecting these two conditional copulas, i.e. $C_{U, V|W}$, (iii) or both. Note that for the source (ii), the directional dependence will exists if and only if $C_{U, V|W}$ is asymmetric. The following example provides a way of producing a simple directional dependence copula model by using a uniformly distributed direction covariate.

Example 3.3 Suppose that the uniformly distributed direction covariate W is independent of U and dependent on V through the Farlie–Gumbel–Morgenstern copula. Furthermore, assume that conditioned on W, U, and V are perfectly

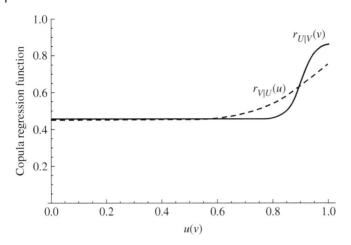

Figure 3.4 Plot of the copula regression functions for $\beta_1 = 10$, $\beta_2 = 40$, and $p = 0.1$.

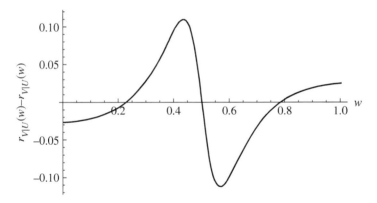

Figure 3.5 Plot of the difference between the copula regression functions for $\beta_1 = 10$, $\beta_2 = 40$, and $p = 0.5$.

dependent. The copula of the random pair (U, V) is

$$C(u, v) = \int_0^1 P(U \leq \min\{u, v(1 + \theta - \theta v) - 2\theta wv(1 - v)\})dw \tag{3.34}$$

$$C(u, v) = \begin{cases} \frac{1}{2}\left(1 - \frac{u-v}{B_\theta(v)}\right) & \text{if } v - B_\theta(v) \leq u \leq v + B_\theta(v) \\ \left\{u - v - \frac{B_\theta(v)}{2}\left(1 + \frac{u-v}{B_\theta(v)}\right)\right\}, & \\ u, & \text{if } u \leq v - B_\theta(v) \\ v, & \text{if } u \geq v + B_\theta(v) \end{cases}$$

$$\tag{3.35}$$

where, $B_\theta(v) = \theta v(1 - v)$. It is trivial to show that the two place function defined above is in fact a copula. Pearson's correlation for this copula is $1 - \frac{\theta^2}{15}$, and the copula regression functions are

$$r_{U|V}(v) = 1 - \int_0^1 C_{U|V}(u; v)du = \frac{1}{3v(3 + \theta^2(1 - 3v + 2v^2))}, \tag{3.36}$$

$$r_{V|U}(u) = 1 - \left[\frac{v^U - v^L}{2} + \frac{1}{2\theta}\log\left[\left(\frac{v^U - 1}{v^L - 1}\right)^{u-1} \Big/ \left(\frac{v^U}{v^L}\right)^u\right] + v^L\right], \tag{3.37}$$

where

$$v^L = \frac{1 + \theta - \sqrt{(1 + \theta)^2 - 4\theta u}}{2\theta},$$

$$v^U = \frac{-1 + \theta - \sqrt{(1 - \theta)^2 + 4\theta u}}{2\theta}.$$

In summary, existence of directional dependence requires an asymmetric copula. Such copulas can be constructed by simply taking a symmetric copula form and modifying it or by making probabilistic arguments. The classes of the copulas introduced in this section possess directional dependence properties and use direction covariates in discrete and continuous forms giving researchers a chance to move away from the symmetric case and create more general dependence models. The dependence structure could change based on a third variable which might be discrete or continuous. Selection of a discrete or continuous direction variable depends on the researcher's interest. In the discrete setup, the Bernoulli direction variable has an advantage of using developed inferential tools for the logistic regression. Increasing the dimension of directional dependence covariates, W and the responses, U and V, is a promising research area.

3.2.2 Regression Setting

Directional dependence can be defined and studied from different perspectives. Dodge and Rousson (2000), and Muddapur (2003) studied directional dependence in a regression line. In their approach, the objective is to decide about the direction of regression line, that is whether Y is dependent on X or X is dependent on Y through a regression line. They show that in a linear regression setting under the assumption that the error term is symmetric, the Pearson correlation, ρ_{XY}, satisfies the following relationship:

$$\rho_{XY}^3 = \frac{\gamma_Y}{\gamma_X}, \quad \gamma_X \neq 0 \tag{3.38}$$

where γ_X and γ_Y are the coefficients of skewness of the variables X and Y, respectively. The coefficients of the skewness are defined as

$$\gamma_X = \frac{E[X - E(X)]^3}{\sigma_X^3},$$

$$\gamma_Y = \frac{E[Y - E(Y)]^3}{\sigma_Y^3},$$

where σ_X^2 and σ_Y^2 are the variances of the corresponding variables. They argue that $(\rho_{XY}^2)^3 = \frac{\gamma_Y^2}{\gamma_X^2}$ can be used to determine the direction of dependence as follows: Since the left hand side of the equation is always less than or equal to 1, $\gamma_Y^2 \le \gamma_X^2$. Thus, Y is linearly dependent on X with the error term being symmetric. They provide a similar argument for the linear regression dependence of X on Y.

Since the directional dependence can arise from marginals or joint behavior or both Sungur (2005) considers the following as a general measure of the directional dependence in the regression setting:

$$\rho_{X \to Y}^{(k)} = \frac{E\{[E(Y \mid X) - E(Y)]^k\}}{E\{[Y - E(Y)]^k\}} = \frac{E\{[r_{Y|X}(X) - E(Y)]^k\}}{\mu_k(Y)}, \quad \mu_k(Y) \ne 0,$$

(3.39)

$$\rho_{Y \to X}^{(k)} = \frac{E\{[r_{X|Y}(Y) - E(X)]^k\}}{\mu_k(X)}, \quad \mu_k(X) \ne 0.$$

(3.40)

Positive version of $\rho_{X \to Y}^{(k)}$ can be interpreted as the proportion of kth central moment of Y that can be explained by the regression of Y on X. Therefore, $\rho_{X \to Y}^{(2)}$ is the proportion of total variation of the variable that is explained by the regression on the paired variable. Similarly, $\mid \rho_{X \to Y}^{(3)} \mid$ is the proportion of skewness of the variable that is preserved by the regression.

For the directional dependence in joint behavior, first the random pair (X, Y) can be transformed to $(U, V) = (F_X(X), F_Y(Y))$, and then $\rho_{U \to V}^{(k)}$ and $\rho_{V \to U}^{(k)}$ can be defined as above.

The following result provides the basic properties of this measure.

Theorem 3.3
1. $\rho_{X \to Y}^{(1)} = \rho_{Y \to X}^{(1)} = 0.$
2. Suppose that $r_{Y|X}(x)$ is linear. Then

$$\rho_{X \to Y}^{(2)} = \rho_{XY}^2 [\rho_{Y \to X}^{(2)} = \rho_{XY}^2], \quad \text{and}$$

(3.41)

$$\rho_{X \to Y}^{(3)} = \rho_{XY}^3 \frac{\frac{\mu_3(X)}{\sigma_X^3}}{\frac{\mu_3(Y)}{\sigma_Y^3}} = \rho_{XY}^3 \frac{\gamma_X}{\gamma_Y} \left[\rho_{Y \to X}^{(3)} = \rho_{XY}^3 \frac{\gamma_Y}{\gamma_X} \right].$$

(3.42)

3. Suppose that both of the copula regression functions are linear. Then the random pair $(U, V) = (F_X(X), F_Y(Y))$ cannot be directionally dependent, i.e. $r_{V|U}(w) = r_{U|V}(w)$.

4.

$$\rho_{U \to V}^{(2)} = \frac{\mathrm{Var}(r_{V|U}(U))}{\mathrm{Var}(V)} = 12\mathrm{Var}(r_{V|U}(U)) = 12E\left[\left(r_{V|U}(U) - \frac{1}{2}\right)^2\right].$$

This result shows that even one can make decisions about the directional dependence in marginals when the regression functions are linear, this is not the case for the directional dependence in joint behavior. In addition, the last part of the theorem shows that $\rho_{U \to V}^{(2)}$ is the proportion of variation that is explained by using the copula regression function in the direction U to V. Since the copula regression function is one half for the independence case, $\mathrm{Var}(r_{V|U}(U))$ could be seen as the expected squared distance of the copula regression function from independence in the direction of U to V. Therefore, $\mathrm{Var}(r_{V|U}(U)) > \mathrm{Var}(r_{U|V}(V))$ implies that the predictive power of the regression in the direction U to V is higher than V to U.

Example 3.2 *(continued)* In this example we will revisit the Example 3.2, where C is a member of the class of copula given in Eq. (3.22), under conditional independence. In this case, copula regression functions are:

$$r_{U|V}(v) = \frac{1}{2} - p\psi(\beta_1, p) + \psi(\beta_1, p)\pi_r(v, \beta_2, p),$$

$$r_{V|U}(u) = \frac{1}{2} - p\psi(\beta_2, p) + \psi(\beta_2, p)\pi_r(u, \beta_1, p),$$

where,

$$\Pi_r(w, \beta, p) = \frac{1}{\beta} \log\left[\frac{e^{\beta p} - e^{\beta} + e^{\beta w} - e^{\beta p}e^{\beta w}}{1 - e^{\beta}}\right],$$

$$\pi_r(w, \beta, p) = \frac{\partial \Pi_r(w, \beta, p)}{\partial w},$$

$$\psi(\beta, p) = \frac{\frac{p}{2} - \int_0^1 \Pi_r(w, \beta, p)dw}{p(1 - p)}.$$

Therefore,

$$\rho_{U \to V}^{(2)} = 12E\left[\left(r_{U|V}(U) - \frac{1}{2}\right)^2\right] = 12\psi^2(\beta_1, p)\int_0^1 (p - \pi_r(v, \beta_2, p))^2 dv. \tag{3.43}$$

Similarly,

$$\rho_{V \to U}^{(2)} = 12\psi^2(\beta_2, p)\int_0^1 (p - \pi_r(u, \beta_1, p))^2 du. \tag{3.44}$$

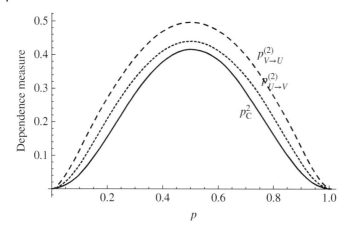

Figure 3.6 Plots of directional dependence measures $\rho^{(2)}_{U \to V}$ and $\rho^{(2)}_{V \to U}$ together with the squared Pearson correlation, ρ^2_C, as a function of p for $\beta_1 = 10$ and $\beta_2 = 30$.

Note that for this copula the Pearson correlation is

$$\rho_C = 12 \int_0^1 \int_0^1 [C(u,v) - uv] \mathrm{d}u \mathrm{d}v = 12p(1-p)\psi(\beta_1,p)\psi(\beta_2,p). \qquad (3.45)$$

Theorem 3.4 Let C be a copula defined in Eq. (3.32), then

$$\rho^2_C = \frac{\rho^{(2)}_{U \to V} \rho^{(2)}_{V \to U}}{\int_0^1 (p - \pi_r(v,\beta_2,p))^2 \mathrm{d}v \int_0^1 (p - \pi_r(u,\beta_1,p))^2 \mathrm{d}u}.$$

Proof.
The result directly follows from the Eqs. (3.43)–(3.45). □

The behavior of the directional dependence measures $\rho^{(2)}_{U \to V}$ and $\rho^{(2)}_{V \to U}$ are graphically described in the Figure 3.6 for $\beta_1 = 10$ and $\beta_2 = 30$ as a function of p. Since $\beta_2 > \beta_1$, $\rho^{(2)}_{V \to U} > \rho^{(2)}_{U \to V}$ for all $p \in (0,1)$, indicating that the dominant direction of dependence is from V to U. Figure 3.6 also gives the plot of the squared Pearson correlation as a function of p. Both of the directional dependence measures are higher than the squared Pearson correlation, ρ^2_C. This result is expected since ρ^2_C measures the predictive power of the linear regression for the case that regression functions in both directions are not linear.

3.2.3 An Alternative Approach to Directional Dependence

Causation can be treated as the summary of the behavior under intervention, see Pearl (2009). In this approach, setting up cause- and-effect relationships require

carefully designed experiments. For example, in experiments we physically manipulate say X by setting $X = x$ and observe the effect of this manipulation on Y for
different values of x. Then, X is a cause of Y if one can change Y by physically
manipulating X, but changing Y does not change X. One can adopt this idea for the
concept of directional dependence. Since there is no experimental control, instead
of physical manipulation one may consider a distributional/probabilistic manipulation. The main ingredients are directional dependence operator, conditional
operator, and impact function. The *directional dependence operator* can be identity, order or record statistics, subsetting/partitioning, truncation, or model-based
such as regression. The *conditional operator* depends on the operator. For example,
if the operator used on X is the ordering, then the conditional operator could be
the Y concomitants of the order statistics. The impact of the distributional intervention through the operator defined can be quantified by looking at the changes
on Y after the application of the conditional operator from the original distributional behavior of Y without the intervention of X. The *impact function* can be
any distance function that involves intervened and not intervened behavior of Y.
Some of the examples of the directional impact function are correlation, Pearson's
distance, cosine distance, proportional distance, sum of squared distance, etc. The
values of the impact functions can be compared in two directions to determine the
"dominant" direction of dependence. The following diagram summarizes these
ideas and compares the causal and directional dependence.

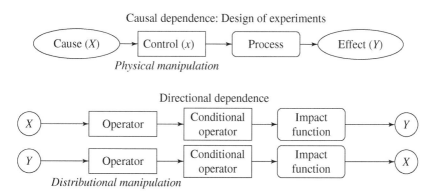

Let us consider ordering as the directional dependence operator. Let (X, Y) be an
absolutely continuous random vector with joint cumulative distribution function,
cdf, $F(x, y)$, and marginal cdfs $F_X(x)$ and $F_Y(y)$. Consider a random sample of n
observations from the distribution of X and Y, and let $X_{(1:n)} \leq X_{(2:n)} \leq \cdots \leq X_{(n:n)}$
be the order statistics of X. The value of Y associated with the $X_{(i:n)}$ is called the
concomitant of the ith ordered statistics of the X and will be represented by $Y_{[i:n]}$.
Similarly, $X_{[i:n]}$ is the X-concomitant of the $Y_{(i:n)}$. A function of $Y_{(i:n)}$ and $Y_{[i:n]}$

$(X_{(i:n)}$ and $X_{[i:n]})$ can be considered as an estimator of the population correlation. In this direction Schezhtman and Yitzhaki (1987) have proposed a sample Gini correlation coefficient for the bivariate normal case. Barnett, Green, and Robinson (1976) studied the estimation of the correlation coefficient in a bivariate normal distribution using only the information contained in the concomitants of the ordered observations of one of the variables. The joint distribution of $(Y_{(i:n)}, Y_{[j:n]})$ is obtained by Tsukibayashi (1998) for the case $i = j = n$, and by He and Nagaraja (2009a) for all $i, j = 1, 2, \ldots, n$. Also, He and Nagaraja (2009b) have proposed a family of estimators of correlation between X and Y using Y-concomitants of X-order statistics and Y-order statistics from bivariate normal samples. We will look at $Y_{[i:n]}(X_{[i:n]})$, ordering on $Y(X)$ introduced by ordering in $X(Y)$, and compare it with the $Y_{(i:n)}(X_{(i:n)})$, marginal ordering on $Y(X)$.

Each one of these random vectors provides invaluable information on the underlying dependence structure. For example, "the $Y_{[j:n]}$ are not necessarily ordered, but can be expected to reflect the association between tending to lead to the values of $Y_{[j:n]}$ in roughly ascending order, and, similarly negative association to $Y_{[i:n]}$ in descending order" (see Barnett et al. (1976)). Figure 3.7 gives six of the key variables and their roles in understanding the distribution of the random pair (X, Y). The pairwise relationships between these variables provide various information such as marginal trend, equality of the marginal distributions, (unidirectional) dependence, and directional dependence. Stronger dependence on original data on $X(Y)$ and order statistics of it will indicate an upward or downward trend in the data. Studying original data pairs on X and Y, and order statistics and their concomitants will inform us about the strength of dependence but not the directional dependence. Also, the equality of the two marginal distributions will reflect on the strength of dependence between associated order statistics.

If the directional dependence from X to Y (Y to X) is strong then $X(Y)$ will initiate a "substantial" change in the ordering of $Y(X)$ reflected on the concomitants leading to a weaker dependence between concomitant and the marginal ordering of $Y(X)$. The strength of change imposed (weakness of dependence) may be used to determine the dominant directional dependence (see Figure 3.8).

The magnitude of change in ordering of one variable from its marginal ordering that is imposed by the other variable can be measured in various ways that we will discuss in the next section.

Any measure of dependence such as Pearson's, Spearman's correlation, and Kendall's tau can be used to measure the dependence between the six key variables that we have introduced above. For understanding the order-based directional dependence, other than measures of dependence one can use the simple regression of order statistics on the concomitant. In addition, since the objective is to measure amount of change in the marginal ordering of one variable from the ordering imposed by the other variable, one can employ Mann-Kendall

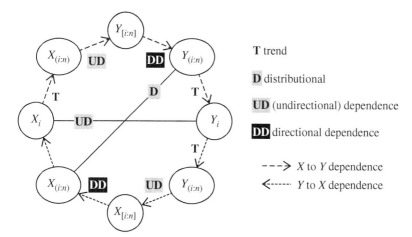

Figure 3.7 Distributional cycle. The key random variables and their roles on understanding marginal and joint behaviors of (X, Y).

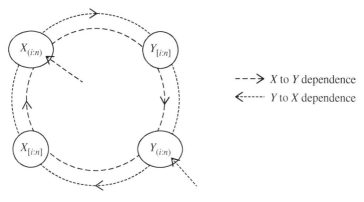

Figure 3.8 Dependence cycle. The key random variables and their roles on understanding dependence structure of (X, Y).

test statistics for monotonic trend (Gibbons & Chakraborti, 2003) on the vectors of concomitants.

We can state our objective in simple terms as measuring the change in the natural ordering of Y induced by the ordering of X through its concomitant. It can be argued if X is affecting Y through ordering it will lead to a weaker relationship between the order statistics of Y and the Y concomitant of the X. Under this objective the key components are

$$r_{XY} = \frac{1}{(n-1)s_X s_Y} \sum_{i=1}^{n} (x_i - \bar{x})(y_i - \bar{y}) \tag{3.46}$$

$$r_{X \to Y} = r_{Y_{:n}Y_{[:n]}} = \frac{1}{(n-1)s_Y^2} \sum_{i=1}^{n} (y_{(i:n)} - \bar{y})(y_{[i:n]} - \bar{y}) \tag{3.47}$$

$$r_{Y \to X} = r_{X_{:n}X_{[:n]}} = \frac{1}{(n-1)s_X^2} \sum_{i=1}^{n} (x_{(i:n)} - \bar{x})(x_{[i:n]} - \bar{x}) \tag{3.48}$$

where, s_X^2 and s_Y^2 are the sample variances for X and Y, respectively. In addition, we will consider the case that marginal distributions of X and Y are known.

$$R_{XY} = \frac{1}{n\sigma_X\sigma_Y} \sum_{i=1}^{n} (X_i - \mu_X)(Y_i - \mu_Y) \tag{3.49}$$

$$R_{X \to Y} = R_{Y_{:n}Y_{[:n]}} = \frac{1}{n\sigma_Y^2} \sum_{i=1}^{n} (Y_{(i:n)} - \mu_Y)(Y_{[i:n]} - \mu_Y) \tag{3.50}$$

$$R_{Y \to X} = R_{X_{:n}X_{[:n]}} = \frac{1}{n\sigma_X^2} \sum_{i=1}^{n} (X_{(i:n)} - \mu_X)(X_{[i:n]} - \mu_X) \tag{3.51}$$

where, μ_X and σ_X^2 (μ_Y and σ_Y^2) are the population mean and variance for $X(Y)$, respectively. Note that, $E[R_{XY}] = \rho_{XY}$, is the population correlation between X and Y.

He and Nagaraja (2009b) suggested $r_{Y_{:n}Y_{[:n]}}$ as an estimator of ρ_{XY}. Goel and Hall (1994) used $SS_{Y_{:n}Y_{[:n]}} = \sum_{i=1}^{n} (y_{(i:n)} - y_{[i:n]})^2$ as a cost function. They define it as the cost of mismatching when the bivariate sample information is broken so that one can only observe the X's and Y's separately, but not as pairs. Figure 3.9 shows the values of pairwise correlations by using the distributional circle for various data sets with particular characteristics.

Therefore, one can use the following criteria for determining the dominant direction of dependence in terms of ordering:

Definition 3.3   We will say that X to Y is the dominant direction of dependence in terms of ordering, if and only if $| E[r_{Y_{:n}Y_{[:n]}}] | < | E[r_{X_{:n}X_{[:n]}}] |$ ($| E[R_{Y_{:n}Y_{[:n]}}] | < | E[R_{X_{:n}X_{[:n]}}] |$) for a fixed n.

To be able to understand the implications of the criteria that we have identified, we need the following results.

Theorem 3.5   Let $X_n(Y_n)$, $X_{(+:n)}(Y_{(+:n)})$, and $X_{[+:n]}(Y_{[+:n]})$ be the random vectors of observations, order statistics, and concomitants, respectively. Then

i.

$$r_{XY} = r_{X_{:n}Y_{[:n]}} = r_{Y_{:n}X_{[:n]}} \quad \text{and} \quad R_{XY} = R_{X_{:n}Y_{[:n]}} = R_{Y_{:n}X_{[:n]}}$$

ii.

$$r_{Y_{:n}Y_{[:n]}} = 1 - \frac{\sum_{i=1}^{n} (y_{(i:n)} - y_{[i:n]})^2}{2 \sum_{i=1}^{n} (y_i - \bar{y})^2} = 1 - \frac{\sum_{i=1}^{n} (y_{[i:n]} - y_{(i:n)})(y_{[i:n]} - \bar{y})}{\sum_{i=1}^{n} (y_i - \bar{y})^2}$$

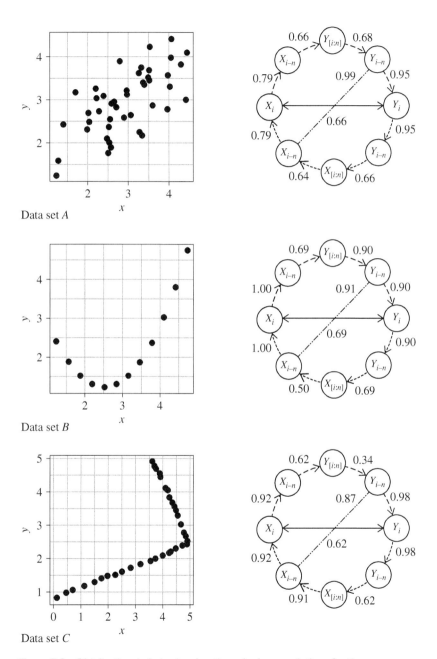

Figure 3.9 Distributional circle showing the pairwise correlations for three cases.

iii.

$$r_{X_{:n}Y_{[:n]}.Y_{:n}} = \frac{r_{XY} - r_{X_{:n}Y_{(:n)}} r_{Y_{:n}Y_{[:n]}}}{\sqrt{(1 - r^2_{X_{:n}Y_{(:n)}})(1 - r^2_{Y_{:n}Y_{[:n]}})}}$$

Based on Theorem 3.5 (ii), alternative order-based directional dependence measures are

$$d_{X \to Y} = \frac{\sum_{i=1}^{n}(y_{[i:n]} - y_{(i:n)})(y_{[i:n]} - \bar{y})}{\sum_{i=1}^{n}(y_i - \bar{y})^2} \tag{3.52}$$

$$d_{Y \to X} = \frac{\sum_{i=1}^{n}(x_{[i:n]} - x_{(i:n)})(x_{[i:n]} - \bar{x})}{\sum_{i=1}^{n}(x_i - \bar{x})^2} \tag{3.53}$$

For positively correlated variables $d_{X \to Y}$ can be interpreted as the proportion of variation in Y that can be explained by the order-based directional impact of X on Y. Therefore, $d_{X \to Y}$ and $d_{Y \to X}$ provide evidence on stronger dependence from X to Y and vice versa. To preserve this interpretation for the negative correlations we will simply multiply one of the variables under consideration by -1, since directional dependence conceptually is unrelated with the sign of dependence.

Another interpretational tool can be created by using conditional/partial correlation as it is introduced in Theorem 3.5 (iii). We will borrow the terminology from statistical mediation analysis (see MacKinnon (2009)). First note that $r_{XY} = r_{X_{:n}Y_{[:n]}} = r_{Y_{:n}X_{[:n]}}$ by Theorem 3.5 (i). Based on the dependence cycle given in Figure 3.8, one can consider various cases that will shape the structure of the order-based directional dependence. The dependence from $X_{(i:n)}$ to $Y_{[i:n]}$, therefore X_i to Y_i, can be decomposed into two paths. One of these paths links the $X_{(i:n)}$ to $Y_{[i:n]}$ directly, and the other links the $X_{(i:n)}$ to $Y_{[i:n]}$ through the $Y_{(i:n)}$. Similarly the dependence from $Y_{(i:n)}$ to $X_{[i:n]}$ can be decomposed by considering the $X_{(i:n)}$. In one case, $X_{(i:n)}$ or $Y_{(i:n)}$ or both may be mediators. That is, if $Y_{(i:n)}$ is a mediating variable on order-based dependence, then $r_{XY} = r_{X_{:n}Y_{[:n]}}$ will be significantly different than zero, but $r_{X_{:n}Y_{[:n]}.Y_{:n}}$ will be zero. Since the correlation between two variables is only explained by their order statistics, we will not have a directional effect of X to Y. An example of such a case is given in Figure 3.10. In addition, Table 3.1 gives the values of the all of the measures that we have introduced so far.

Theorem 3.6 Let (X, Y) be an exchangeable pair of random variables, i.e. $F(x, y) = F(y, x)$ for all $x, y \in \mathbb{R}$ where F is the joint cumulative distribution function. Then, $E[r_{Y \to X}] = E[r_{X \to Y}]$, $E[R_{XY}] = E[R_{Y \to X}] = E[R_{X \to Y}]$, $E[d_{X \to Y}] = E[d_{Y \to X}]$, and $E[r_{X_{:n}Y_{[:n]}.Y_{:n}}] = E[r_{Y_{:n}X_{[:n]}.X_{:n}}]$.

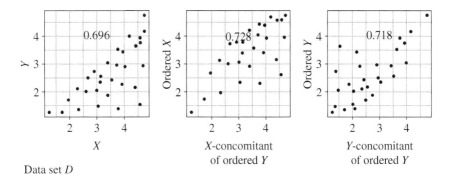

Data set D

Figure 3.10 The plots of a data set where order statistics of X and Y are both order-based directional dependence mediating variables.

Table 3.1 The values of directional dependence measures for the data sets A–D.

Measure	Data set			
	A	B	C	D
r_{XY}	0.663	0.693	0.624	0.696
$r_{Y \to X} = r_{X_{:n} X_{[:n]}}$	0.644	0.505	0.913	0.718
$r_{X \to Y} = r_{Y_{:n} Y_{[:n]}}$	0.680	0.903	0.342	0.728
$d_{X \to Y}$	0.320	0.097	0.658	0.282
$d_{Y \to X}$	0.356	0.495	0.087	0.272
$r_{X_{:n} Y_{[:n]} \cdot Y_{:n}}$	−0.140	−0.749	0.705	0.020
$r_{Y_{:n} X_{[:n]} \cdot X_{:n}}$	0.266	0.660	−0.845	0.064

Theorem 3.6 clarifies that for exchangeable pairs directional dependence does not exists. In fact, this was the case for the data set A.

It is well known that for the case that (X, Y) has a bivariate Normal distribution with mean $\mu = (\mu_X, \mu_Y)^{\mathrm{T}}$, and variance-covariance matrix

$$\Sigma = \begin{pmatrix} \sigma_X^2 & \rho_{XY} \sigma_X \sigma_Y \\ \rho_{XY} \sigma_X \sigma_Y & \sigma_Y^2 \end{pmatrix},$$

$$Y_{[i:n]} = \mu_Y + \rho_{XY} \frac{\sigma_Y}{\sigma_X} (X_{(i:n)} - \mu_X) + \varepsilon_{[i]}, \tag{3.54}$$

where $\varepsilon_{[i]}$ is independent of $X_{(i:n)}$ and has $N(0, \sigma_Y^2(1 - \rho^2))$, (see David and Nagaraja (1998)). Therefore,

$$R_{X \to Y} = R_{Y_{:n} Y_{[:n]}} = \frac{1}{n \sigma_Y^2} \sum_{i=1}^{n} (Y_{(i:n)} - \mu_Y)(Y_{[i:n]} - \mu_Y) = \rho_{XY} R_{X_{:n} Y_{:n}} + E_n$$

$$\tag{3.55}$$

where $E_n = \frac{1}{n\sigma_Y^2}\sum_{i=1}^n (Y_{(i:n)} - \mu_Y)\varepsilon_{[i]}$. Therefore, based on the distributional and dependence cycles that we have proposed, the directional dependence can be explained completely by using the undirectional dependence, ρ_{XY}, and marginal distributional characteristics, $R_{Y_{:n}Y_{[:n]}}$. In fact, this result is true for any random pair linked by the linear relationship

$$Y_i = \mu_Y + \rho_{XY}\frac{\sigma_Y}{\sigma_X}(X_i - \mu_X) + \varepsilon_i, \tag{3.56}$$

special case being the bivariate Normal distribution. This observation explains why we are observing a directional dependence in data sets B and C but not in A.

3.3 Directional Association Between Two Categorical Variables

In this section, we will look at a way of modeling and understanding directional association between categorical variables following the core idea behind the copula approach. To keep the discussion simple, binary variables will be considered. Sungur and Orth (2017) use dummy coding on the p binary categorical variables and define a random vector $\mathbf{X} = (X_1, X_2, ..., X_p)^T$ that is composed of Bernoulli random variables with $P(X_i = 1) = p_i = 1 - P(X_i = 2)$, $i = 1, ..., p$. Let us represent the joint distribution of the $\mathbf{X}$ by the table $\mathbf{P_X} = \{p_{x_1 x_2...x_p}, x_1, x_2, ..., x_p = 1, 2\}$. Throughout this section, we will refer to $\mathbf{P_X}$ as $\mathbf{P}$ - type table. If $P(X_i = 1) = P(X_i = 2) = \frac{1}{2}$, $i = 1, ..., p$, the resulting $\mathbf{P}$ - type table will be called a $\mathbf{Q}$ - type table. This characterization of a $\mathbf{P}$ - type table can be seen similar to the copula representation of the joint distribution of a continuous random pair. For the case where $p = 2$, the joint distribution will be represented by the table $\mathbf{P}_{XY} = \{p_{xy}; x, y = 1, 2\}$, and $P(X = 1) = p_{1+}$, $P(X = 2) = p_{2+}$, $P(Y = 1) = p_{+1}$, $P(Y = 2) = p_{+2}$. The classes of copulas have parameter(s) that are connected with a measure of the strength of dependence/association between the variables. In our case, we will use the odds ratio, **OR**, the ratio of the odds of $Y = 1$ when $X = 1$ to the $Y = 1$ when $X = 2$. In the medical literature, **OR** is reported as a measure of association between exposure (X) and outcome (Y), and interpreted as the ratio of the odds of outcome in the exposed to the odds of outcome in the non exposed. Note that the way that the odds ratio has been defined, it can also be interpreted as the ratio of the odds of $X = 1$ when $Y = 1$ to the $X = 1$ when $Y = 2$. Another measure of association commonly used in medical and epidemiological studies is the ratio of the conditional probability of outcome given the exposed to the conditional probability of outcome given the non exposed. This measure is called risk ratio. A modified form of risk ratio will be used to measure the direction of association and referred as conditional ratio.

The characterization of $\mathbf{P}$ - type table by using $\mathbf{Q}$ - type, just like the copulas, will eliminate marginal behavior of Bernoulli variables and emphasizes association structures between them more clearly. For example, all of the $\mathbf{P}$ - type tables under independence will have a one unique representation making interpretation and inference directly possible from the cross-table.

Following Sungur and Orth (2017), $\mathbf{P}$ - type for $p = 2$, can be easily characterized as follows. Since, for the $\mathbf{Q}$ - type, $q_{1+} = q_{2+} = q_{+1} = q_{+2} = 1/2$, q_{11} will uniquely determine the table. Therefore, $\mathbf{Q}$ - type can be constructed by letting

$$q_{11} = \frac{\sqrt{\mathbf{OR_P}}}{2(1 + \sqrt{\mathbf{OR_P}})} \tag{3.57}$$

where, $\mathbf{OR_P} = \frac{p_{11}p_{22}}{p_{12}p_{21}}$, that is, the odds ratio of the $\mathbf{P}$ - type table.

Under this characterization, $\mathbf{OR_P} = \mathbf{OR_Q} = \mathbf{OR}$, and given the $\mathbf{Q}$ - type table and marginals one can uniquely determine the $\mathbf{P}$ - type table.

For the directional association/dependence, we need to define the conditional $\mathbf{P}$ and $\mathbf{Q}$ - type tables as follows:

$$\mathbf{P}_{X|Y} = \left\{ \frac{p_{ij}}{p_{+j}}; i, j = 1, 2 \right\},$$

$$\mathbf{P}_{Y|X} = \left\{ \frac{p_{ij}}{p_{i+}}; i, j = 1, 2 \right\},$$

$$\mathbf{Q}_{X|Y} = \mathbf{Q}_{Y|X} = \{2q_{ij}; i, j = 1, 2\}.$$

The relationship between the risk ratios for the $\mathbf{P}$ and $\mathbf{Q}$ - type tables is

$$\mathbf{RR_P} = (\mathbf{RR_Q})^2 \frac{(p_{1+} - p_{11})(1 - p_{1+})}{p_{1+}(1 - p_{1+} - p_{+1} + p_{11})}. \tag{3.58}$$

Now, by noting this relationship, we will define the conditional ratios:

$$\mathbf{CR}_{\mathbf{P}_{X|Y}} = (\mathbf{CR_Q})^2 \frac{(p_{+1} - p_{11})(1 - p_{+1})}{p_{+1}(1 - p_{+1} - p_{1+} + p_{11})} =: (\mathbf{CR_Q})^2 \mathbf{D}_{X \to Y}, \tag{3.59}$$

$$\mathbf{CR}_{\mathbf{P}_{Y|X}} = (\mathbf{CR_Q})^2 \frac{(p_{1+} - p_{11})(1 - p_{1+})}{p_{1+}(1 - p_{1+} - p_{+1} + p_{11})} =: (\mathbf{CR_Q})^2 \mathbf{D}_{Y \to X}. \tag{3.60}$$

Note that

$$\mathbf{CR}_{\mathbf{P}_{Y|X}} = \mathbf{OR_P} \mathbf{D}_{X \to Y} = \mathbf{OR_Q} \mathbf{D}_{X \to Y},$$

$$\mathbf{CR}_{\mathbf{P}_{X|Y}} = \mathbf{OR_P} \mathbf{D}_{Y \to X} = \mathbf{OR_Q} \mathbf{D}_{Y \to X}.$$

In this representation, observe that conditional ratios have two components. The first component, $\mathbf{OR_Q}$, does not depend on the marginal behavior of two variables, but the second component, $\mathbf{D}_{X \to Y}(\mathbf{D}_{Y \to X})$ does. The second component depends on conditioning variable and, therefore, reflects a direction.

We can view the conditional ratio for the Y given X, $\mathbf{CR}_{\mathbf{P}_{Y|X}}$, and X given Y, $\mathbf{CR}_{\mathbf{P}_{X|Y}}$, in three different ways. The first is when all of the effects of two marginals have been eliminated ($\mathbf{CR}_Q = \sqrt{\mathbf{OR}_Q} = \sqrt{\mathbf{OR}_P}$). The second is when only the effect of row variable X has been eliminated ($\mathbf{CR}_{\mathbf{P}_{Y|X}} = \mathbf{OR}_P\mathbf{D}_{X \to Y}$). The third is when only the effect of column variable Y eliminated ($\mathbf{CR}_{\mathbf{P}_{X|Y}} = \mathbf{OR}_P\mathbf{D}_{Y \to X}$). As it is illustrated in the following example, elimination of the marginal effects one by one helps researchers to identify directional association as well as the underlying factors such as sampling design and rarity easily.

Example 3.4   Now we will illustrate the usefulness of such decompositions on understanding directional association with an example given in LaMorte (2015). In this example, the risk of wound infections is given when an incidental appendectomy was done during a staging laparotomy for Hodgkin's disease. The variables are X = Occurrence of incidental appendectomy (Yes = 1, No = 2) and Y = Wound infection (Yes = 1, No = 2). Assume that P-type table is

$$P_{XY} = \begin{bmatrix} 7/210 & 124/210 \\ 1/210 & 78/210 \end{bmatrix}. \tag{3.61}$$

In this case, $\mathbf{CR}_{\mathbf{P}_{Y|X}} = 4.22$, $\mathbf{OR}_P = \mathbf{OR}_Q = 4.40$, $\mathbf{D}_{X \to Y} = 0.95$, $\mathbf{CR}_{\mathbf{P}_{X|Y}} = 1.4$, and $\mathbf{D}_{Y \to X} = 0.32$. The odds ratio can be interpreted in two different ways because of its "symmetric" nature: (i) the odds of wound infection when the incidental appendectomy occurred is 4.4 times larger than the odds for non-occurrence group being infected, (ii) the odds of occurrence of incidental appendectomy when wound infection occurred is 4.4 times larger than the odds for not infected group showing occurrence of incidental appendectomy. The odds ratio does not provide any information on direction of association reflected on the interpretation unless it is assumed. On the other hand, the conditional ratios can be interpreted as follows: "Patients who undergo incidental appendectomy have 4.22 times the risk of post-operative wound infection compared to patients who do not undergo incidental appendectomy" and "Patients who have post-operative wound infection are 1.4 times more likely undergo incidental appendectomy compared to patients who do not have post-operative wound infection." If we eliminated the effect of marginals completely, $\mathbf{D}_Q = \mathbf{D}_{Y \to X} = \mathbf{D}_{X \to Y} = (\mathbf{OR}_P)^{-1/2} = 0.48$. Note that when the direction of association is from incidental appendectomy to wound infections (risk factor) 0.95 times of the association will be preserved. On the other hand, when we move in the other direction, that is, from wound infection to incidental appendectomy, 0.32 will be preserved. Also note that after elimination of marginal effects of both factors $\mathbf{CR}_Q = 2.10$.

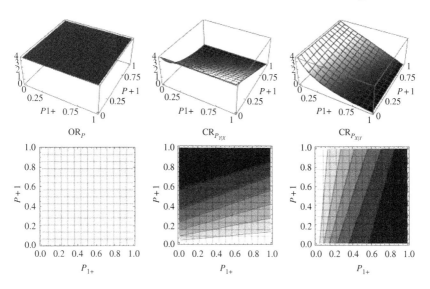

Figure 3.11 Surface and contour plots of odds ratio and conditional ratios for the Example 3.2.

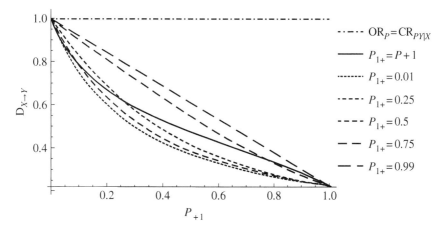

Figure 3.12 The behavior of $D_{X \to Y}$ as a function of p_{+1} for various values of p_{1+}.

Figures 3.11 and 3.12 display surfaces of odds ratio and two conditional ratios as a function of marginal behaviors of X and Y. As it can be seen from these plots, the conditional ratio for Y given X approaches to odds ratio as probability of wound infection approaches to 0, that is, as it becomes rare. Similarly, conditional ratio for X given Y approaches odds ratio as the occurrence of incidental appendectomy becomes more rare. Also, the directional contribution from risk factor to risk gets higher when the probability of occurrence of risk factor goes up.

3.4 Concluding Remarks and Future Directions

Understanding and modeling dependence structures even for the bivariate case is challenging, but extremely useful in application. It becomes more challenging when one introduces the direction of dependence. We need to emphasize again that directional dependence analysis is not the same as causal analysis. The causality can only be set up through carefully designed experiments.

Since dependence modeling involves the joint behavior between the variables not the marginals, the copulas turns out to be a crucial tool. The directional dependence exists when copula of the variables involved are asymmetric. The classes of copulas that possess directional dependence property and use directional variable in discrete and continuous forms gives researchers a chance to move away from the symmetric case and create more general dependence models. The dependence structure could change based on a third variable which might be discrete or continuous. Selection of a discrete or continuous direction variable heavily depends on the researcher's interest. In discrete setup the Bernoulli direction variable has an advantage of using developed inferential tools for the logistic regression. Increasing the dimension of dependence covariates and the responses is a promising new research area.

The reflection of the directional dependence in regression functions that we presented can be generalized to multiple and multivariate regression set-ups. This generalization will require the use of tools from graph theory.

Directional dependence modeling is closely related with other commonly used models and methods used in statistics. For example, the copula class that is based on the linear combinations of independent uniform variables originates from ideas in principal component analysis and logistic regression. The research on directional dependence by using copulas can introduce a fresh look to these techniques.

An alternative approach to directional dependence modeling is to create an impact on the random observations taken from one variable through an operator applied on the other. Then the impact can be compared and the largest impact determines the dominant directional dependence. Therefore directional dependence is associated and based on the operator. We identified the operator as the ordering of the observations from one variable, obtain the ordering imposed on the other variable, i.e. concomitants, and consider the changes from the natural/univariate ordering of the impacted variable. There are different ways of determining the directional dependence operator and measuring the impact.

We propose using correlation for order-based directional dependence. Other operators and measures of impact can be developed and studied.

Similar to the continuous case, directional association for the categorical variables can be defined and studied. We introduce an approach on modeling the dependence structure of a Bernoulli random vector. Main results provided are for the case of two Bernoulli variables. Complete and partial elimination of univariate marginals are useful in simplifying and understanding dependence structures. We attempted to eliminate the effect of marginals by transforming them to discrete uniform distributions. The resulting distribution in the form of a table is called $\mathbf{Q}$ - type table. This characterization of 2×2 tables has a potential on the analysis of two-way tables. We aim to generalize our results to $i \times j$ tables. This study is limited to probabilistic model building and does not include inferential implications based on data. The inferential component of the approach needs to be studied. For example it is easy to see that testing on odds ratio can be easily done by using q_{11} element of the $\mathbf{Q}$ - type table. In this chapter we introduce an approach on modeling the dependence structure of a Bernoulli random vector.

References

Bairamov, I., & Kotz, S. (2002). Dependence structure and symmetry of Huang–Kotz FGM distributions and their extensions. *Metrika*, *56*(1), 55–72. doi:10.1007/s001840100158

Barnett, V., Green, P. J., & Robinson, A. (1976). Concomitants and correlation estimates. *Biometrika*, *63*(2), 323–328. Retrieved from http://www.jstor.org/stable/2335626

David, H. A., & Nagaraja, H. N. (1998). 18 Concomitants of order statistics. In *Order statistics: Theory & methods, handbook of statistics* (Vol. *16*, pp. 487–513). New York, NY: Elsevier. doi:10.1016/S0169-7161(98)16020-0 Retrieved from http://www.sciencedirect.com/science/article/pii/S0169716198160200

Dodge, Y., & Rousson, V. (2000). Direction dependence in a regression line. *Communications in Statistics - Theory and Methods*, *29*(9–10), 1957–1972. doi:10.1080/03610920008832589

Fréchet, M. (1951). Sur les tableaux de corrélation dont les marges sont données. *Annales de I'Université de Lyon. Section A: Sciences mathématiques et astronomie*, *3*(14), 53–77.

Frees, E. W., & Valdez, E. A. (1998). Understanding relationships using copulas. *North American Actuarial Journal*, *2*(1), 1–25. doi:10.1080/10920277.1998.10595667

Gibbons, J. D., & Chakraborti, S. (2003). *Nonparametric statistical inference*. New York, NY: Marcel Dekker.

Goel, P. K., & Hall, P. (1994). On the average difference between concomitants and order statistics. *Annals of Probability, 22*(1), 126–144. doi:10.1214/aop/1176988851

He, Q., & Nagaraja, H. N. (2009a). Correlation estimation using concomitants of order statistics from bivariate normal samples. *Communications in Statistics - Theory and Methods, 38*(12), 2003–2015. doi:10.1080/03610920802155452

He, Q., & Nagaraja, H. N. (2009b). Distribution of concomitants of order statistics and their order statistics. *Journal of Statistical Planning and Inference, 139*(8), 2643–2655. doi:10.1016/j.jspi.2008.12.007 Retrieved from http://www.sciencedirect.com/science/article/pii/S0378375808004503

Hoeffding, W. (1994). Scale—Invariant correlation theory. In N. I. Fisher & P. K. Sen (Eds.), *The collected works of Wassily Hoeffding* (pp. 57–107). New York, NY: Springer. ISBN: 978-1-4612-0865-5

Lai, C. D., & Xie, M. (2000). A new family of positive quadrant dependent bivariate distributions. *Statistics & Probability Letters, 46*(4), 359–364. 10.1016/S0167-7152(99)00122-4. Retrieved from http://www.sciencedirect.com/science/article/pii/S0167715299001224

LaMorte, W. W.. (2015). Risk ratios and rate ratios (relative risk). Retrieved from http://sphweb.bumc.bu.edu/otlt/MPH-Modules/.

MacKinnon, D. P. (2009). *Introduction to statistical mediation analysis.* New York, NY: Erlbaum.

Muddapur, M. V. (2003). On directional dependence in a regression line. *Communications in Statistics - Theory and Methods, 32*(10), 2053–2057. doi:10.1081/STA-120023266

Nelsen, R. B. (2006). *An introduction to copulas. Springer Series in Statistics.* Berlin, Heidelberg, Germany: Springer-Verlag. ISBN: 0387286594

Pearl, J. (2009). *Causality.* Cambridge, UK: Cambridge University Press. doi:10.1017/CBO9780511803161

Rodríguez-Lallena, J. A., & Flores, M. Ú. (2004). A new class of bivariate copulas. *Statistics & Probability Letters, 66*(3), 315–325. 10.1016/j.spl.2003.09.010. Retrieved from http://www.sciencedirect.com/science/article/pii/S0167715203003584

Schezhtman, E., & Yitzhaki, S. (1987). A measure of association based on gin's mean difference. *Communications in Statistics - Theory and Methods, 16*(1), 207–231. doi:10.1080/03610928708829359

Sklar, A. (1959). Fonctions de repartitiona n dimensions et leurs marges. *Publications de l'Institut de Statistique de Universite de Paris, 8*, 229–231.

Sungur, E., & Celebioglu, S. (2011). Copulas with directional dependence property. *Gazi University Journal of Science, 24*(3), 415–424.

Sungur, E. A. (2005). A note on directional dependence in regression setting. *Communications in Statistics - Theory and Methods, 34*(9–10), 1957–1965. doi:10.1080/03610920500201228

Sungur, E. A., & Orth, J. M. (2017). A note on dependence modeling for bernoulli variables. *Communications in Statistics - Theory and Methods, 46*(16), 8217–8229. doi:10.1080/03610926.2016.1177080

Tsukibayashi, S. (1998). The joint distribution and moments of an extreme of the dependent variable and the concomitant of an extreme of the independent variable. *Communications in Statistics - Theory and Methods, 27*(7), 1639–1651. doi:10.1080/03610929808832182

Wiedermann, W., & Li, X. (2018). Direction dependence analysis: A framework to test the direction of effects in linear models with an implementation in spss. *Behavior Research Methods, 50*(4), 1581–1601. doi:10.3758/s13428-018-1031-x

Wiedermann, W., Li, X., & von Eye, A. (2019). Testing the causal direction of mediation effects in randomized intervention studies. *Prevention Science, 20*(3), 419–430. doi:10.1007/s11121-018-0900-y

Part II

Direction Dependence in Continuous Variables

4

Asymmetry Properties of the Partial Correlation Coefficient

Foundations for Covariate Adjustment in Distribution-Based Direction Dependence Analysis

Wolfgang Wiedermann

Department of Educational, School, and Counseling Psychology, College of Education, & Missouri Prevention Science Institute, University of Missouri, Columbia, MO, USA

The Pearson correlation coefficient (Bravais, 1846; Pearson, 1920) is the most widely used measure of association to quantify the linear relation between two continuous variables. Almost every introductory statistics textbook introduces the Pearson correlation as the centered and standardized sum of cross-products of two variables x and y, i.e. the ratio of the covariance and the product of standard deviations, $\rho_{xy} = \text{cov}(x, y)/(\sigma_x \sigma_y)$. However, several authors presented alternative definitions of the Pearson correlation with the aim to introduce different conceptual interpretations (Falk & Well, 1997; Nelsen, 1998; Rodgers & Nicewander, 1988; Rovine & von Eye, 1997). The majority of these alternative conceptualizations focuses on the relation between the correlation coefficient and the linear regression model and can roughly be classified as either *regression lines* – (e.g. the correlation as the slope of a standardized regression line) or *predicted values*-based formulations (e.g. correlation as the association between the response and its linear model prediction).

Most alternative conceptualizations have in common that they present the Pearson correlation in its symmetric form, i.e. because the correlation coefficient does not depend on variable order, one obtains $\rho_{xy} = \rho_{yx}$. However recently, asymmetric formulations of the Pearson correlation have been introduced (Dodge & Rousson, 2000, 2001; Dodge & Yadegari, 2010; Muddapur, 2003; Sungur, 2005; Wiedermann & Hagmann, 2016; Wiedermann, Merkle, & Eye, 2018). Asymmetry properties exist for non-normal variables and refer to data situations in which it is no longer possible to exchange variables in their roles as predictor and response without causing systematic model violations. In other words, variable information beyond second order moments (i.e. skewness and excess-kurtosis) opens the door to evaluate the causal direction of a regression line in observational data.

Direction Dependence in Statistical Modeling: Methods of Analysis, First Edition.
Edited by Wolfgang Wiedermann, Daeyoung Kim, Engin A. Sungur, and Alexander von Eye.

Evaluating distributional characteristics of variables is part of a broader statistical framework known as direction dependence analysis (DDA; Wiedermann & Li, 2018; Wiedermann & Sebastian, 2019; Wiedermann & von Eye, 2015a) which can be used to test (i) whether the model $x \to y$ or the causally reversed model $y \to x$ better approximates the underlying data-generating mechanism and (ii) whether meaningful confounders are likely to be present (details are given in the related Chapter 2 by Wiedermann, Li, and von Eye, in this volume). Because variable distribution-based DDA measures rely on the assumption of unconfoundedness, the present chapter focuses on the former task, i.e. selecting between causally competing models $x \to y$ and $y \to x$ with no confounders.

To be more specific, suppose that the continuous random variables x and y are related by the linear model

$$y = \beta_0 + \beta_{yx}x + \varepsilon_{y(x)} \tag{4.1}$$

where β_0 denotes the model intercept, β_{yx} is the causal effect of x on y (assumed to be constant across subjects), and $\varepsilon_{y(x)}$ denotes the error term exhibiting zero mean and variance $\sigma^2_{\varepsilon_{y(x)}}$. The directionally mis-specified model, i.e. the one that erroneously treats x as the response and y as the explanatory variable, is given by

$$x = \beta'_0 + \beta_{xy}y + \varepsilon_{x(y)}. \tag{4.2}$$

DDA assumes that the true predictor x is a non-normally distributed exogenous variable. When the error term of the true model in (4.1) is normally distributed and independent of x, Dodge and Rousson (2000, 2001), Muddapur (2003), as well as Dodge and Yadegari (2010) showed that higher powers of the Pearson correlation can be expressed as the ratio of higher moments of x and y (see also the related Chapter 1 by Dodge and Rousson, in this volume). That is, for the third and the fourth power of the Pearson correlation, one obtains

$$\rho^3_{xy} = \frac{\gamma_y}{\gamma_x} \tag{4.3}$$

and

$$\rho^4_{xy} = \frac{\kappa_y}{\kappa_x} \tag{4.4}$$

where $\gamma_x = \mathrm{cum}_3(x)/\sigma^3_x$ and $\gamma_y = \mathrm{cum}_3(y)/\sigma^3_y$ are the skewnesses and $\kappa_x = \mathrm{cum}_4(x)/\sigma^4_x$ and $\kappa_y = \mathrm{cum}_4(y)/\sigma^4_y$ denote the excess-kurtosis values of x and y (cum$_j$ defines the jth cumulant with $\mathrm{cum}_3(x) = E[(x - E[x])^3]$, $\mathrm{cum}_4(x) = E[(x - E[x])^4] - 3E[(x - E[x])^2]$, and E being the expected value operator; see, e.g. Kendall & Stuart, 1979). Equations (4.3) and (4.4) introduce the Pearson correlation in its asymmetric form. Because ρ_{xy} is bounded on the interval $[-1, 1]$, the (absolute value of the) skewness of y will always be smaller than the (absolute value of the) skewness of x (excluding $|\rho_{xy}| = 1$ due to practical irrelevance). Similarly, the excess-kurtosis of y will always be smaller than the excess-kurtosis of x. In other words, decisions on the direction of dependence can be based on distributional properties of observed variables. If y is closer to the normal distribution than x then the model $x \rightarrow y$ given in (4.1) is more likely to describe the underlying causal mechanism. Conversely, if x is closer to normality than y then the model $y \rightarrow x$ given in (4.4) is more likely to hold.

In some applications, the assumption of normality of the true error term may be unrealistic. In case of non-normally distributed errors $\varepsilon_{y(x)}$, distributional properties of x and y can be evaluated using higher-order correlations (HOCs)

$$\text{cor}(x, y)_{ij} = \frac{\text{cov}(x, y)_{ij}}{\sigma_x^i \sigma_y^j} \tag{4.5}$$

with $\text{cov}(x, y)_{ij} = E[(x - E[x])^i (y - E[y])^j]$ being the higher-order covariance and σ_x and σ_y being the standard deviations of x and y. Using $i, j = \{1, 2\}$ results in measures of co-skewnesses $\text{cor}(x, y)_{12} = \rho_{xy}^2 \gamma_x$ and $\text{cor}(x, y)_{21} = \rho_{xy} \gamma_x$ and one obtains

$$\rho_{xy} = \frac{\text{cor}(x, y)_{12}}{\text{cor}(x, y)_{21}}. \tag{4.6}$$

From (4.6), it follows that $|\text{cor}(x, y)_{12}| < |\text{cor}(x, y)_{21}|$ holds under the model $x \rightarrow y$ and $|\text{cor}(x, y)_{21}| < |\text{cor}(x, y)_{12}|$ is observed under the causally competing model $y \rightarrow x$ (Dodge & Rousson, 2000, 2001).

Previous studies on direction dependence properties of observed variable distributions focused on methods for statistical inference (see, e.g. von Eye & DeShon, 2012), extensions to copula regressions (Sungur, 2005), measurement error models (Wiedermann et al., 2018), and applications to causal benchmark data sets (Thoemmes, 2015). All these studies have in common that they focus on the simple bivariate case. Pornprasertmanit and Little (2012) used simulations to evaluate the robustness of direction dependence methods and Dodge and Rousson (2016) studied properties of the error terms of (4.1) and (4.4) in the presence of lurking variables. However, neither study discusses strategies to apply Dodge and Rousson's (2000, 2001) direction dependence measures in the multi-variable case. In other words, statistical foundations for covariate adjustment in variable distribution-based DDA measures have not been presented yet. The present chapter fills this gaping void. Extensions of Dodge and Rousson's (2001) direction of dependence methodology to the multiple variable setting are discussed by introducing asymmetry properties of the partial correlation coefficient. Based on these asymmetric formulations, we propose decision guidelines for model selection in the multiple variable setting. Further, we present the results of two Monte-Carlo simulation studies designed to (i) evaluate the parameter recovery of asymmetric formulas for the partial correlation, and (ii) assess the Type I error coverage and power behavior of resampling-based CIs of covariate-adjusted direction dependence measures. An empirical example is given to illustrate the application of the method. The chapter closes with recommendations for selecting proper direction dependence measures in practice, a discussion of the limitations of direction dependence measures based on semi-partial correlations, and the relation of direction dependence principles and existing causal inference methods.

4.1 Asymmetry Properties of the Partial Correlation Coefficient

In the following section, we show that the partial correlation coefficient exhibits asymmetry properties that are similar to those obtained for the simple bivariate Pearson correlation. The partial correlation can be defined as the linear association between that portion of y that is independent of a pre-defined set of independent variables ($z_j, j = 1, \ldots, k$; so-called covariates) and that portion of x that is independent of the same set z_j. Covariates can be continuous or categorical. For simplicity of presentation (but without loss of generality), we focus on the case $k = 1$, i.e. let z be a covariate that is known to be explanatory in nature (e.g. due to temporality or logical order of effects). The partial correlation coefficient ($\rho_{xy|z}$) then quantifies the magnitude of the linear correlation of x and y while adjusting for z and can be written as

$$\rho_{xy|z} = \frac{\rho_{xy} - \rho_{xz}\rho_{yz}}{\sqrt{1 - \rho_{xz}^2}\sqrt{1 - \rho_{yz}^2}} \tag{4.7}$$

where ρ_{xy}, ρ_{xz}, and ρ_{yz} denote the pairwise correlations of x, y, and z. Further, in the presence of a covariate z, the "true" model $x \rightarrow y$ given in (4.1) extends to

$$y = \beta_0 + \beta_{yx}x + \beta_{yz}z + \varepsilon_{y(xz)} \tag{4.8}$$

where $\beta_{yx} = [(\rho_{xy} - \rho_{xz}\rho_{yz})/(1 - \rho_{xz}^2)](\sigma_y/\sigma_x)$ and $\beta_{yz} = [(\rho_{yz} - \rho_{xy}\rho_{xz})/(1 - \rho_{xz}^2)](\sigma_y/\sigma_z)$ describe the partial regression slopes and $\varepsilon_{y(xz)}$ denotes the error term which is assumed to be normally distributed and independent of the predictors x and z. Similarly, the mis-specified model in (4.4) extends to

$$x = \beta_0' + \beta_{xy}y + \beta_{xz}z + \varepsilon_{x(yz)}. \tag{4.9}$$

By making use of the Frisch–Waugh–Lovell theorem (Frisch & Waugh, 1933; Lovell, 1963) alternative expressions for the partial correlation coefficient in (4.7) and β_{yx} in (4.8) are available through covariate-residualizing the variables x and y

on z. Let $\varepsilon_{x(z)}$ and $\varepsilon_{y(z)}$ be the residuals obtained from separately regressing x and y on z,

$$\varepsilon_{x(z)} = x - (\beta_{x0} + \beta'_{xz}z)$$
$$\varepsilon_{y(z)} = y - (\beta_{y0} + \beta'_{yz}z) \tag{4.10}$$

with $\beta_{x0} + \beta'_{xz}z$ and $\beta_{y0} + \beta'_{yz}z$ being the linear model predictions of $z \to x$ and $z \to y$. Then, the Pearson correlation of $\varepsilon_{x(z)}$ and $\varepsilon_{y(z)}$,

$$\rho_{\varepsilon_{x(z)}\varepsilon_{y(z)}} = \frac{\text{cov}(\varepsilon_{x(z)}, \varepsilon_{y(z)})}{\sigma_{\varepsilon_{x(z)}}\sigma_{\varepsilon_{y(z)}}}, \tag{4.11}$$

corresponds to the partial correlation coefficient defined in (4.7) and regressing $\varepsilon_{y(z)}$ on $\varepsilon_{x(z)}$, i.e.

$$\varepsilon_{y(z)} = \alpha_0 + \alpha_{yx}\varepsilon_{x(z)} + \theta_{y(x)} \tag{4.12}$$

leads to an estimate for the partial regression slope, $\beta_{yx} = \alpha_{yx}$ (Frisch & Waugh, 1933; Lovell, 1963). In addition, the error terms of the two models given in (4.8) and (4.12) are identical which can be shown through re-writing the error terms as functions of the observed variables. That is, for the model in (4.12), one obtains

$$\begin{aligned}\theta_{y(x)} &= \varepsilon_{y(z)} - (\alpha_0 + \alpha_{yx}\varepsilon_{x(z)}) \\ &= y - (\beta_{y0} - \beta_{yx}\beta_{x0}) - \beta_{yx}x - (\beta'_{yz} - \beta_{yx}\beta'_{xz})z \\ &= \varepsilon_{y(xz)}\end{aligned} \tag{4.13}$$

with $\beta_0 = \beta_{y0} - \beta_{yx}\beta_{x0}$ and $\beta_{yz} = \beta'_{yz} - \beta_{yx}\beta'_{xz}$.

Asymmetric facets of $\rho_{xy|z}$ can then be obtained through inserting the jth cumulants (cum_j) into Eq. (4.12),

$$\text{cum}_j(\varepsilon_{y(z)}) = \alpha^j_{yx}\text{cum}_j(\varepsilon_{x(z)}) + \text{cum}_j(\theta_{y(x)}). \tag{4.14}$$

Because $\varepsilon_{y(xz)}$ is assumed to be normally distributed, one obtains $\text{cum}_j(\theta_{y(x)}) = 0$ for $j > 2$ and solving (4.14) for α^j_{yx} leads to

$$\alpha^j_{yx} = \frac{\text{cum}_j(\varepsilon_{y(z)})}{\text{cum}_j(\varepsilon_{x(z)})}. \tag{4.15}$$

Further, multiplying (4.15) by $\sigma^j_{\varepsilon_{x(z)}}/\sigma^j_{\varepsilon_{y(z)}}$ and making use of $\sigma^2_{\varepsilon_{x(z)}} = 1 - \rho^2_{xz}$ and $\sigma^2_{\varepsilon_{y(z)}} = 1 - \rho^2_{yz}$, for $j = \{3, 4\}$, one arrives at

$$\rho^3_{xy|z} = \frac{\gamma_{\varepsilon_{y(z)}}}{\gamma_{\varepsilon_{x(z)}}} \tag{4.16}$$

and

$$\rho^4_{xy|z} = \frac{\kappa_{\varepsilon_{y(z)}}}{\kappa_{\varepsilon_{x(z)}}} \tag{4.17}$$

where $\gamma_{\varepsilon_{y(z)}}$ and $\gamma_{\varepsilon_{x(z)}}$ denote the skewnesses and $\kappa_{\varepsilon_{y(z)}}$ and $\kappa_{\varepsilon_{x(z)}}$ denote the excess-kurtosis values of $\varepsilon_{y(z)}$ and $\varepsilon_{x(z)}$. Equations (4.16) and (4.17) describe asymmetric definitions of the partial correlation coefficient. Because $-1 \le \rho_{xy|z} \le 1$, one concludes that $|\gamma_{\varepsilon_{y(z)}}| < |\gamma_{\varepsilon_{x(z)}}|$ and $\kappa_{\varepsilon_{y(z)}} < \kappa_{\varepsilon_{x(z)}}$ (again excluding the perfect linear case $|\rho_{xy|z}| = 1$) which enables researchers to test the direction of dependence of covariate-adjusted pairs of variables. In other words, comparing higher moments of $\varepsilon_{x(z)}$ and $\varepsilon_{y(z)}$ opens the door to test hypotheses that are compatible with direction dependence in multiple linear regression settings. Further, Eqs. (4.16) and (4.17) have natural interpretations: The cube of partial correlation coefficient represents the amount of skewness that is preserved by the linear relation $x \rightarrow y$ when adjusting for z. Similarly, the fourth power of the partial correlation describes the amount of excess-kurtosis that is preserved in the linear model $x \rightarrow y$ while adjusting for z.

4.2 Direction Dependence Measures when Errors Are Non-Normal

Normality of the error term $\varepsilon_{y(xz)}$ is required when considering both asymmetric facets in (4.16) and (4.17). Taken separately, expression (4.16) holds as long as the error term is symmetrically distributed (i.e. $\gamma_{\varepsilon_{y(xz)}} = 0$). In other words, comparing the skewness of $\varepsilon_{x(z)}$ and $\varepsilon_{y(z)}$ to derive directional decisions is valid irrespective of the excess-kurtosis of $\varepsilon_{y(xz)}$. Similarly, expression (4.17) holds as long as $\kappa_{\varepsilon_{y(xz)}} = 0$ which implies that the skewness of $\varepsilon_{y(xz)}$ is allowed to vary within the range $-\sqrt{2}$ and $\sqrt{2}$ according to the skewness-kurtosis inequality $\kappa \ge \gamma^2 - 2$ (Teuscher & Guiard, 1995). The assumption of normality of $\varepsilon_{y(xz)}$ can entirely be relaxed when focusing on HOCs of $\varepsilon_{x(z)}$ and $\varepsilon_{y(z)}$,

$$\text{cor}(\varepsilon_{x(z)}, \varepsilon_{y(z)})_{ij} = \frac{\text{cov}(\varepsilon_{x(z)}, \varepsilon_{y(z)})_{ij}}{\sigma^i_{\varepsilon_{x(z)}} \sigma^j_{\varepsilon_{y(z)}}} \tag{4.18}$$

with $\text{cov}(\varepsilon_{x(z)}, \varepsilon_{y(z)})_{ij} = E[(\varepsilon_{x(z)} - E[\varepsilon_{x(z)}])^i (\varepsilon_{y(z)} - E[\varepsilon_{y(z)}])^j]$. From this definition, various alternative expressions for $\rho_{xy|z}$ can be developed: First, the skewness of $\varepsilon_{x(z)}$ and $\varepsilon_{y(z)}$ can be written as $\gamma_{\varepsilon_{x(z)}} = \rho(\varepsilon_{x(z)}, \varepsilon_{y(z)})_{30}$ and $\gamma_{\varepsilon_{y(z)}} = \rho(\varepsilon_{x(z)}, \varepsilon_{y(z)})_{03}$ from which follows that the asymmetric facet in (4.16) can be re-expressed as (again assuming symmetry of $\varepsilon_{y(xz)}$)

$$\rho^3_{xy|z} = \frac{\rho(\varepsilon_{x(z)}, \varepsilon_{y(z)})_{03}}{\rho(\varepsilon_{x(z)}, \varepsilon_{y(z)})_{30}}. \tag{4.19}$$

Further, for $k \geq 2$, one obtains

$$
\begin{aligned}
&\mathrm{cov}(\varepsilon_{x(z)}, \varepsilon_{y(z)})_{k-1,k} \\
&= E[(\varepsilon_{x(z)} - E[\varepsilon_{x(z)}])^{k-1}(\alpha_{yx}\varepsilon_{x(z)} + \theta_{yx} - \alpha_{yx}E[\varepsilon_{x(z)}] + E[\theta_{yx}])] \\
&= \alpha_{yx}E[(\varepsilon_{x(z)} - E[\varepsilon_{x(z)}])^{k}] + E[(\varepsilon_{x(z)} - E[\varepsilon_{x(z)}])^{k-1}]E[\theta_{yx} - E[\theta_{yx}]] \\
&= \alpha_{yx}\mathrm{cov}(\varepsilon_{x(z)}, \varepsilon_{y(z)})_{k0}
\end{aligned}
\tag{4.20}
$$

from which follows that the partial correlation coefficient can also be defined as

$$
\rho_{xy|z} = \frac{\mathrm{cor}(\varepsilon_{x(z)}, \varepsilon_{y(z)})_{k-1,1}}{\mathrm{cor}(\varepsilon_{x(z)}, \varepsilon_{y(z)})_{k,0}}
\tag{4.21}
$$

provided that $\mathrm{cor}(\varepsilon_{x(z)}, \varepsilon_{y(z)})_{k,0} \neq 0$.

Inserting $k = 3$ into (4.20) leads to

$$
\rho_{xy|z} = \frac{\mathrm{cor}(\varepsilon_{x(z)}, \varepsilon_{y(z)})_{2,1}}{\gamma_{\varepsilon_{x(z)}}}
\tag{4.22}
$$

and due to $-1 \leq \rho_{xy|z} \leq 1$ one can conclude that (the absolute value of) $\mathrm{cor}(\varepsilon_{x(z)}, \varepsilon_{y(z)})_{2,1}$ will always be smaller than (the absolute value of) $\gamma_{\varepsilon_{x(z)}}$. Similarly, using

$$
\begin{aligned}
&\mathrm{cov}(\varepsilon_{x(z)}, \varepsilon_{y(z)})_{1,2} \\
&= E[(\varepsilon_{x(z)} - E[\varepsilon_{x(z)}])(\alpha_{yx}\varepsilon_{x(z)} + \theta_{yx} - \alpha_{yx}E[\varepsilon_{x(z)}] + E[\theta_{yx}])^{2}] \\
&= \alpha_{yx}^{2}E[(\varepsilon_{x(z)} - E[\varepsilon_{x(z)}])^{3}] + 2\alpha_{yx}E[(\varepsilon_{x(z)} - E[\varepsilon_{x(z)}])^{2}]E[\theta_{yx} - E[\theta_{yx}]] \\
&\quad + E[(\varepsilon_{x(z)} - E[\varepsilon_{x(z)}])]E[(\theta_{yx} - E[\theta_{yx}])^{2}]] \\
&= \alpha_{yx}^{2}\mathrm{cum}_{3}(\varepsilon_{x(z)})
\end{aligned}
\tag{4.23}
$$

one obtains

$$
\rho_{xy|z} = \frac{\mathrm{cor}(\varepsilon_{x(z)}, \varepsilon_{y(z)})_{1,2}}{\mathrm{cor}(\varepsilon_{x(z)}, \varepsilon_{y(z)})_{2,1}}
\tag{4.24}
$$

which implies that $|\mathrm{cor}(\varepsilon_{x(z)}, \varepsilon_{y(z)})_{1,2}| < |\mathrm{cor}(\varepsilon_{x(z)}, \varepsilon_{y(z)})_{2,1}|$ holds whenever the linear model $\{x, z\} \to y$ describes the data-generating mechanism. Finally, multiplying (4.22) by (4.24) gives

$$
\rho_{xy|z}^{2} = \frac{\mathrm{cor}(\varepsilon_{x(z)}, \varepsilon_{y(z)})_{1,2}}{\gamma_{\varepsilon_{x(z)}}},
\tag{4.25}
$$

Dividing (4.16) by (4.25) leads to

$$
\rho_{xy|z} = \frac{\gamma_{\varepsilon_{y(z)}}}{\mathrm{cor}(\varepsilon_{x(z)}, \varepsilon_{y(z)})_{1,2}},
\tag{4.26}
$$

and, overall, we obtain the three related asymmetric expressions for the partial correlation coefficient,

$$
\rho_{xy|z} = \frac{\mathrm{cor}(\varepsilon_{x(z)}, \varepsilon_{y(z)})_{2,1}}{\gamma_{\varepsilon_{x(z)}}} = \frac{\mathrm{cor}(\varepsilon_{x(z)}, \varepsilon_{y(z)})_{1,2}}{\mathrm{cor}(\varepsilon_{x(z)}, \varepsilon_{y(z)})_{2,1}} = \frac{\gamma_{\varepsilon_{y(z)}}}{\mathrm{cor}(\varepsilon_{x(z)}, \varepsilon_{y(z)})_{1,2}}.
\tag{4.27}
$$

Under the model $\{x, z\} \to y$ and assuming that $\rho_{xy|z} \neq 0$, $\gamma_{\varepsilon_{x(z)}} \neq 0$, and $\gamma_{\varepsilon_{y(xz)}} = 0$, the following sequence is observed,

$$| \gamma_{\varepsilon_{y(z)}} | < | \text{cor}(\varepsilon_{x(z)}, \varepsilon_{y(z)})_{1,2} | < | \text{cor}(\varepsilon_{x(z)}, \varepsilon_{y(z)})_{2,1} | < | \gamma_{\varepsilon_{x(z)}} | . \tag{4.28}$$

Because no distributional assumptions are imposed on the error term for the expression in (4.24),

$$| \text{cor}(\varepsilon_{x(z)}, \varepsilon_{y(z)})_{1,2} | < | \text{cor}(\varepsilon_{x(z)}, \varepsilon_{y(z)})_{2,1} | \tag{4.29}$$

will hold independently of the distribution of $\varepsilon_{y(xz)}$ provided that $\rho_{xy|z} \neq 0$ and $\gamma_{\varepsilon_{x(z)}} \neq 0$. Thus, Eq. (4.29) can be used to develop a HOC-based direction dependence measure that is well-suited for skewed variables (note that HOCs using $i = j = \{1, 2\}$ can be re-conceptualized as co-skewness measures of $\varepsilon_{x(z)}$ and $\varepsilon_{y(z)}$) and robust to non-normal true errors.

In a similar fashion, robust HOC-based measures for lepto-/platykurtic variables are available through considering $\text{cor}(\varepsilon_{x(z)}, \varepsilon_{y(z)})_{i,j}$ for $i = j = \{1, 3\}$ (i.e. so-called co-kurtosis values of $\varepsilon_{x(z)}$ and $\varepsilon_{y(z)}$; see Wiedermann (2018)). To simplify notation, we assume that $\varepsilon_{x(z)}$ and $\varepsilon_{y(z)}$ have been standardized. Then, under model (4.8), one obtains

$$\begin{aligned} \text{cor}(\varepsilon_{x(z)}, \varepsilon_{y(z)})_{1,3} &= E[\varepsilon_{x(z)}(\rho_{xy|z}\varepsilon_{x(z)} + \varepsilon_{y(xz)})^3] \\ &= \rho_{xy|z}^3 \kappa_{\varepsilon_{x(z)}} + 3\rho_{xy|z}(1 - \rho_{xy|z}^2) \end{aligned} \tag{4.30}$$

and

$$\begin{aligned} \text{cor}(\varepsilon_{x(z)}, \varepsilon_{y(z)})_{3,1} &= E[\varepsilon_{x(z)}^3(\rho_{xy|z}\varepsilon_{x(z)} + \varepsilon_{y(xz)})] \\ &= \rho_{xy|z}\kappa_{\varepsilon_{x(z)}}. \end{aligned} \tag{4.31}$$

Thus, when the true covariate-adjusted predictor $\varepsilon_{x(z)}$ is leptokurtic (i.e. $\kappa_{\varepsilon_{x(z)}} > 0$) then $|\text{cor}(\varepsilon_{x(z)}, \varepsilon_{y(z)})_{3,1}| > |\text{cor}(\varepsilon_{x(z)}, \varepsilon_{y(z)})_{1,3}|$ and when $\varepsilon_{x(z)}$ is platykurtic (i.e. $\kappa_{\varepsilon_{x(z)}} < 0$) then $|\text{cor}(\varepsilon_{x(z)}, \varepsilon_{y(z)})_{3,1}| < |\text{cor}(\varepsilon_{x(z)}, \varepsilon_{y(z)})_{1,3}|$. To make direction dependence properties of (4.30) and (4.31) independent of the type of symmetric non-normality, one can adjust for the sign of $\kappa_{\varepsilon_{x(z)}}$. The resulting difference measure

$$\{| \text{cor}(\varepsilon_{x(z)}, \varepsilon_{y(z)})_{3,1} | - | \text{cor}(\varepsilon_{x(z)}, \varepsilon_{y(z)})_{1,3} | \} \times sgn(\varepsilon_{x(z)}) \tag{4.32}$$

(with $sgn(\varepsilon_{x(z)}) = 1$ if $\kappa_{\varepsilon_{x(z)}} > 0$ and $sgn(\varepsilon_{x(z)}) = -1$ if $\kappa_{\varepsilon_{x(z)}} < 0$) will then always be greater than zero under model (4.8). Further, the quantity in (4.32) does not impose any distributional assumptions on the error term of the true model. Because multiplying the difference measure by the signum function requires knowledge of the tentative predictor, Wiedermann (2018) suggested using the maximum kurtosis (in absolute values) for the signum function, i.e. using $sgn(\varepsilon_{x(z)})$ if $| \kappa_{\varepsilon_{x(z)}} | > | \kappa_{\varepsilon_{y(z)}} |$ and $sgn(\varepsilon_{y(z)})$ if $| \kappa_{\varepsilon_{y(z)}} | > | \kappa_{\varepsilon_{x(z)}} |$. While the expressions given in (4.16), (4.17), (4.29), and (4.32) can readily be used for descriptive purposes when replacing population parameters with sample estimates, we next focus on statistical inference methods for performing directional model selection.

4.3 Statistical Inference on Direction Dependence

Several approaches have been suggested for the purpose of statistical inference of direction dependence. von Eye and DeShon (2012), for example, suggested using normality tests to evaluate hypotheses compatible with direction dependence. Using this approach, one has found empirical evidence for the model in (4.8) when the null hypothesis of normality is rejected for $\varepsilon_{x(z)}$ and, at the same time, retained for $\varepsilon_{y(z)}$. While this is a straightforward and computationally simple approach, one disadvantage may be that joint statistical decisions of separate significance tests are needed for model selection. As an alternative, resampling-based CIs of higher moment differences can be used instead (Pornprasertmanit & Little, 2012; Wiedermann & von Eye, 2015b). In the present study, we focus on the two higher moment-based difference measures

$$\Delta(\gamma) = \mid \gamma_{\varepsilon_{x(z)}} \mid - \mid \gamma_{\varepsilon_{y(z)}} \mid \tag{4.33}$$

$$\Delta(\kappa) = \mid \kappa_{\varepsilon_{x(z)}} \mid - \mid \kappa_{\varepsilon_{y(z)}} \mid \tag{4.34}$$

and the two HOC-based measures

$$\Delta(\gamma)^* = \mid \text{cor}(\varepsilon_{x(z)}, \varepsilon_{y(z)})_{2,1} \mid - \mid \text{cor}(\varepsilon_{x(z)}, \varepsilon_{y(z)})_{1,2} \mid \tag{4.35}$$

$$\Delta(\kappa)^* = \{\mid \text{cor}(\varepsilon_{x(z)}, \varepsilon_{y(z)})_{3,1} \mid - \mid \text{cor}(\varepsilon_{x(z)}, \varepsilon_{y(z)})_{1,3} \mid\} \times sgn(\varepsilon_{x(z)}). \tag{4.36}$$

The model $\{x, z\} \to y$ in (4.8) finds empirical support when $\Delta(\gamma)$, $\Delta(\kappa)$, $\Delta(\gamma)^*$, and/or $\Delta(\kappa)^*$ are significantly *greater than zero*. In contrast, the reverse model (x, $z) \to y$ given in (4.9) should be preferred when $\Delta(\gamma)$, $\Delta(\kappa)$, $\Delta(\gamma)^*$, and/or $\Delta(\kappa)^*$ are significantly *smaller than zero*. While moment-based measures, $\Delta(\gamma)$ and $\Delta(\kappa)$, assume that the true error term is normally distributed, HOC-based measures, $\Delta(\gamma)^*$ and $\Delta(\kappa)^*$, do not impose distributional assumptions on the true error term.

Because theoretical sampling distributions of (4.33)–(4.36) are, to the best of our knowledge, yet unknown, bootstrapping (Davison & Hinkley, 1997) can be used to approximate the distributions of the test statistics: First, pairs of $\varepsilon_{x(z)}$ and $\varepsilon_{y(z)}$ are randomly sampled with replacement from the original sample. Let Δ' be one of the difference statistics given in (4.33)–(4.36) for one of B re-samples. Percentile bootstrap CI limits are then approximated by $[\Delta'_{B\alpha/2}; \Delta'_{B(1-\alpha/2)}]$ with α being the nominal significance level. For the bivariate case, Dodge and Rousson (2016) found that the percentile bootstrap approach for differences in skewness and kurtosis tends to be overly conservative in statistical decisions. Thus, we also focus on bias-corrected and accelerated (BCa) CIs (for computational details see Efron (1987)) which correct the confidence bounds when statistics are asymmetrically distributed and, thus, may constitute an improvement compared to the percentile method.

4.4 Monte-Carlo Simulations

The following section reports the results of two Monte-Carlo simulation experiments that were conducted: (i) to evaluate the parameter recovery of asymmetric facets of the partial correlation, and (ii) to assess the performance of resampling-based CIs in terms of both, Type I error coverage and statistical power. Simulation studies were performed using the R Statistical Programming Environment (R Core Team, 2019). In all simulations, data were generated from the model $y = \beta_0 + \beta_{yx}x + \beta_{yz}z + \varepsilon_{y(xz)}$ where x represents the explanatory variable of interest, z is treated as a background covariate that is known to be explanatory in nature, and $\varepsilon_{y(xz)}$ constitutes the independent error component.

4.4.1 Study I: Parameter Recovery

To assess the parameter recovery of the partial correlation we focus on the higher moment-based expressions given in (4.16) and (4.17) and on the HOC-based expression for power values $i = j = \{1, 2\}$ given in (4.24). We do not consider HOCs for $i = j = \{1, 3\}$ because there exists no straightforward expression for $\rho_{xy|z}$. Predictors and error term of the true model were sampled from various non-normal populations. Specifically, elliptic copulas of the normal family (see, e.g. Nelsen, 1996) with gamma-distributed marginals were used to simulate correlated non-normal predictors with pre-specified skewness (excess-kurtosis) values of 0.75 (0.84), 1.5 (3.38), and 2.25 (7.59) and correlations $\rho_{xz} = 0, 0.2, 0.4$, and 0.6. The error term $\varepsilon_{y(xz)}$ was independently sampled from either the standard normal or the same gamma distributions noted above. Regression coefficients (β_{yx} and β_{yz}) were fixed at 0.14, 0.39, and 0.59 to account for small (2% of the variance), medium (13% of the variance), and large (26% of the variance) effects (cf. Cohen, 1988). For each sample, $n = 200$ observations were generated which is considered a common sample size in psychological research (Marszalek, Barber, Kohlhart, & Holmes, 2011). Fully crossed, the simulation design consisted of 3 (skewness of x) × 3 (skewness of z) × 4 (skewness of $\varepsilon_{y(xz)}$) × 3 (magnitude of β_{yx}) × (magnitude of β_{yz}) × 4 (magnitude of ρ_{xz}) = 1296 simulation conditions. Under each condition, 500 samples were generated. For each generated sample, the direction dependence measures given in (4.16), (4.17), and (4.24) were computed and retained for further analyses. Biases were calculated using $\hat{\theta} - \theta$

with $\hat{\theta}$ referring to the estimated partial correlation (using either (4.16), (4.17), and (4.24)) and θ being the true partial correlation. Percentage biases were obtained using $(\hat{\theta} - \theta)/(1 + \theta) \times 100$ (a re-scaled denominator was used to avoid denominators close to zero which occur when computing higher powers of small partial correlation coefficients). In line with previous studies, percentage biases >40% were considered practically significant (cf. Ames, 2013; Collins, Schafer, & Kam, 2001; Wiedermann et al., 2018).

4.4.1.1 Results

Table 4.1 gives median biases, median absolute deviations, and median percentage biases of direction dependence estimates as a function of β_{yx}, γ_x, and $\gamma_{\varepsilon_{y(xz)}}$ for $\rho_{xz} = 0.6$ (because biases slightly increase with ρ_{xz} the values in Table 4.1 summarize the most extreme scenarios observed in the simulation study; the remaining results for $\rho_{xz} < 0.6$ can be obtained upon request). In general, the magnitude of β_{yz} and the distribution of z had virtually no impact on the magnitude of biases which can be expected because the effect of z was partialled out prior computation of direction dependence measures.

For normally distributed errors, biases for all measures are close to zero (ranging from −7.1 to 3.7%). For non-normal errors, biases drastically increase for $\gamma_{\varepsilon_{y(z)}}/\gamma_{\varepsilon_{x(z)}}$ and $\kappa_{\varepsilon_{y(z)}}/\kappa_{\varepsilon_{x(z)}}$ (for skewness ratios biases range from 28.1 to 361.2%, for excess-kurtosis ratios biases range from 5.5 to 588.1%). In general, bias decreases with the magnitude of non-normality of x and (to a lesser extent) with the magnitude of β_{yx}. Biases for the HOC-based measure using $i = j = \{1, 2\}$ are close to zero regardless of the distribution of the error term. The only exception concerns cases of small β_{yx} effects together with less skewed true predictors x. Here, bias also slightly increases with non-normality of $\varepsilon_{y(xz)}$ (with a maximum of 26.9% when $\gamma_{\varepsilon_{y(xz)}} = 2.25$).

4.4.2 Study II: CI Coverage and Statistical Power

To study nominal coverage probabilities of resampling-based CIs, we focus on data situations in which a direction dependence procedure prefers one of the two possible models ($\{x, z\} \to y$ or $\{y, z\} \to x$) when this should not be possible. Such data situations occur when variables of the data-generating mechanism are normally distributed. Thus, to assess the Type I error coverage of direction dependence CIs, the predictors (x and z) were sampled from a standard normal distribution. To evaluate the adequacy of covariate adjustment (i.e. direction dependence CIs based on residualized x and y variables are expected to protect Type I error coverage rates), z was sampled from gamma distributions with skewness (excess-kurtosis) values of 0.75 (0.84), 1.5 (3.38), and 2.25 (7.59). Predictor correlations were again set to $\rho_{xz} = 0$, 0.2, 0.4, and 0.6. The error term $\varepsilon_{y(xz)}$ was independently sampled from

Table 4.1 Median bias and median percent bias of three direction dependence measures as a function β_{yx}, γ_x, and $\gamma_{\varepsilon_{y(xz)}}$ for $\rho_{xz} = 0.6$.

β_{yx}	γ_x	$\gamma_{\varepsilon_{y(xz)}}$	$\dfrac{\gamma_{\varepsilon_{y(z)}}}{\gamma_{\varepsilon_{x(z)}}}$		$\dfrac{\kappa_{\varepsilon_{y(z)}}}{\kappa_{\varepsilon_{x(z)}}}$		$\dfrac{\mathrm{cor}(\varepsilon_{x(z)},\varepsilon_{y(z)})_{1,2}}{\mathrm{cor}(\varepsilon_{x(z)},\varepsilon_{y(z)})_{2,1}}$	
			MD (MAD)	%bias	MD (MAD)	%bias	MD (MAD)	%bias
0.14	0.75	0.00	0.00 (0.31)	0.0	−0.05 (0.61)	−5.3	0.04 (1.10)	3.7
0.14	0.75	0.75	1.24 (0.60)	123.4	0.43 (1.29)	43.4	0.15 (1.32)	13.1
0.14	0.75	1.50	2.43 (1.05)	242.9	2.49 (3.84)	249.3	0.22 (1.77)	19.7
0.14	0.75	2.25	3.62 (1.60)	361.2	5.88 (8.04)	588.1	0.30 (2.25)	26.9
0.14	1.50	0.00	0.00 (0.15)	0.1	−0.03 (0.14)	−2.8	−0.01 (0.73)	−0.7
0.14	1.50	0.75	0.61 (0.25)	60.6	0.20 (0.32)	20.3	0.06 (0.84)	5.0
0.14	1.50	1.50	1.22 (0.44)	121.6	1.01 (0.94)	100.9	0.10 (1.19)	8.6
0.14	1.50	2.25	1.78 (0.63)	177.9	2.30 (1.92)	229.7	0.13 (1.55)	11.4
0.14	2.25	0.00	0.00 (0.10)	0.0	−0.01 (0.06)	−1.2	−0.02 (0.50)	−2.0
0.14	2.25	0.75	0.41 (0.17)	40.8	0.10 (0.14)	9.6	0.02 (0.60)	2.1
0.14	2.25	1.50	0.82 (0.28)	81.4	0.47 (0.41)	47.1	0.04 (0.81)	3.1
0.14	2.25	2.25	1.20 (0.42)	119.4	1.05 (0.83)	105.1	0.03 (1.09)	2.3
0.39	0.75	0.00	0.00 (0.31)	−0.4	−0.07 (0.62)	−7.1	0.02 (0.57)	1.3
0.39	0.75	0.75	1.08 (0.55)	105.6	0.41 (1.20)	40.9	0.01 (0.69)	0.6
0.39	0.75	1.50	2.12 (0.95)	205.8	2.10 (3.10)	208.0	0.03 (0.88)	2.1
0.39	0.75	2.25	3.15 (1.41)	306.5	4.94 (7.01)	489.6	0.00 (1.12)	0.0
0.39	1.50	0.00	0.00 (0.15)	0.0	−0.03 (0.14)	−2.7	0.00 (0.29)	−0.2
0.39	1.50	0.75	0.53 (0.25)	51.9	0.16 (0.28)	16.2	−0.02 (0.36)	−1.6
0.39	1.50	1.50	1.06 (0.41)	103.1	0.85 (0.83)	84.4	−0.04 (0.46)	−2.8
0.39	1.50	2.25	1.57 (0.61)	152.7	1.93 (1.72)	191.5	−0.04 (0.56)	−3.4
0.39	2.25	0.00	0.00 (0.11)	0.0	−0.01 (0.07)	−1.4	0.00 (0.20)	−0.4
0.39	2.25	0.75	0.36 (0.16)	34.5	0.07 (0.13)	7.4	−0.02 (0.25)	−1.7
0.39	2.25	1.50	0.72 (0.27)	69.7	0.38 (0.37)	38.2	−0.04 (0.31)	−3.0
0.39	2.25	2.25	1.05 (0.39)	102.0	0.88 (0.73)	86.8	−0.04 (0.37)	−2.7
0.59	0.75	0.00	−0.01 (0.31)	−0.6	−0.07 (0.61)	−6.8	0.01 (0.38)	0.9
0.59	0.75	0.75	0.91 (0.50)	84.6	0.28 (1.10)	26.9	0.01 (0.46)	0.4
0.59	0.75	1.50	1.84 (0.87)	169.8	1.54 (2.76)	149.1	−0.02 (0.60)	−1.1
0.59	0.75	2.25	2.64 (1.27)	244.5	3.82 (5.61)	368.7	−0.03 (0.73)	−2.3

Table 4.1 (Continued)

β_{yx}	γ_x	$\gamma_{\varepsilon_{y(xz)}}$	$\dfrac{\gamma_{\varepsilon_{y(z)}}}{\gamma_{\varepsilon_{x(z)}}}$		$\dfrac{\kappa_{\varepsilon_{y(z)}}}{\kappa_{\varepsilon_{x(z)}}}$		$\dfrac{cor(\varepsilon_{x(z)},\varepsilon_{y(z)})_{1,2}}{cor(\varepsilon_{x(z)},\varepsilon_{y(z)})_{2,1}}$	
			MD (MAD)	%bias	MD (MAD)	%bias	MD (MAD)	%bias
0.59	1.50	0.00	0.00 (0.16)	0.1	−0.04 (0.16)	−3.5	0.00 (0.20)	−0.3
0.59	1.50	0.75	0.46 (0.22)	42.1	0.12 (0.28)	11.8	−0.02 (0.25)	−1.2
0.59	1.50	1.50	0.89 (0.36)	82.5	0.65 (0.69)	62.4	−0.02 (0.29)	−1.7
0.59	1.50	2.25	1.31 (0.52)	121.2	1.49 (1.39)	143.4	−0.03 (0.37)	−2.2
0.59	2.25	0.00	0.00 (0.11)	−0.3	−0.02 (0.08)	−1.6	−0.01 (0.14)	−0.6
0.59	2.25	0.75	0.31 (0.15)	28.1	0.06 (0.13)	5.5	−0.01 (0.17)	−0.5
0.59	2.25	1.50	0.59 (0.24)	54.8	0.29 (0.30)	28.2	−0.02 (0.21)	−1.4
0.59	2.25	2.25	0.87 (0.34)	80.4	0.68 (0.60)	65.2	−0.03 (0.25)	−2.2

Values in parentheses give the median absolute deviations (MAD) of observed biases.

the standard normal distribution. The assumption of normality of the error term is crucial for direction dependence measures that focus on higher moment differences. In contrast, no distributional assumptions are made when testing differences in HOCs. To quantify Type I error robustness to violations of the normality assumption, the error term was sampled from the same gamma-distributions as z. The intercept β_0 was fixed at zero throughout the study and the regression coefficients were, again, selected to account for small, medium, and large effects. Sample size was, again, fixed at $n = 200$. Simulation Study II consisted of 4 (magnitude of ρ_{xz}) × 3 (magnitude of β_{yx}) × 3 (magnitude of β_{yz}) × 4 (skewness of z) × 4 (skewness of $\varepsilon_{y(xz)}$) = 576 simulation conditions. Five-hundred samples were generated under each simulation condition. For each sample, the four difference measures given in (4.33)–(4.36) and the corresponding 95% bootstrap CIs (using 500 re-samples) were calculated using the percentile and the BCa method. The liberal robustness criterion proposed by Bradley (1978) was used to assess the robustness of nominal coverage rates, i.e. given an expected nominal coverage probability of 95%, a CI procedure was considered robust if the empirical coverage rates fell within the interval 92.5–97.5%.

To quantify the statistical power for the direction dependence CIs to detect the true model the same specifications were used as in Simulation Study I. That is, the power simulation consisted of 3 (skewness of x) × 3 (skewness of z) × 4 (skewness of $\varepsilon_{y(xz)}$) × 3 (magnitude of β_{yx}) × 3 (magnitude of β_{yz}) × 4 (magnitude of ρ_{xz}) = 1296 conditions with 500 samples per condition. Empirical power rates were determined in terms of model selection using the decision guidelines described above.

That is, the correctly specified model $\{x, z\} \to y$ was selected when direction dependence difference measures in (4.33)–(4.36) were significantly larger than zero.

4.4.2.1 Type I Error Coverage

Table 4.2 shows the observed coverage rates for the four considered direction dependence CIs including the observed proportions how often the CIs fell below or above the true value of zero. In general, empirical coverage rates were neither affected by the magnitude of β_{yz}, ρ_{xz}, nor γ_z. In other words, after partialing out the effect of z, direction dependence properties unique to x and y can be analysed independently of the causal, correlational, and distributional impact of z. Thus, simulation results are presented as a function of the causal effect β_{yx} and the distribution of $\varepsilon_{y(xz)}$. For normally distributed errors ($\gamma_{\varepsilon_{y(xz)}} = \kappa_{\varepsilon_{y(xz)}} = 0$), percentile CIs are overly conservative in statistical decisions, i.e. coverage rates are above the expected 95% which is in line with Dodge and Rousson's (2016) results observed for the simple bivariate case. In contrast, coverage rates for BCa CIs tend to be closer to the expected 95%. For skewness- and excess-kurtosis differences measures, coverage rates systematically decrease with non-normality of $\varepsilon_{y(xz)}$ no matter of the resampling approach. This effect is more pronounced for smaller effect sizes of β_{yx}. Most important, the coverage rates for skewness- and excess-kurtosis difference measures constantly fall below the true value. In other words, these are data situations in which researchers erroneously select the model $\{y, z\} \to x$ because the true predictor (x) appears to be closer to normality than the true outcome y. In contrast, resampling-based CIs for HOC-based measures are better able to protect the nominal coverage rate. For percentile CIs, coverage rates range from 0.989 to 0.998, coverage rates of BCa CIs vary from 0.887 to 0.972. In the majority of cases, the sampling distributions of $\Delta(\gamma)^*$ and $\Delta(\kappa)^*$ tend to be slightly asymmetric.

4.4.2.2 Statistical Power

Because overly conservative significance decisions under the null hypothesis observed for percentile CIs reduce the power to detect true differences, model selection based on percentile CIs is generally less powerful than model selection based on BCa CIs. For this reason, we only present results for BCa CIs. Simulation results for the observed power, i.e. scenarios in which variables deviate from normality and direction dependence procedures correctly identify the model $\{x, z\} \to y$ are summarized in Figures 4.1 and 4.2. Figure 4.1 shows the power for the main effects of the simulation study, i.e. aggregating across all remaining factors of the simulation experiment. Specifically, empirical power rates are given as functions of the regression slopes β_{yx} and β_{yz}, the skewnesses of x, z, and $\varepsilon_{y(xz)}$, and the correlation ρ_{xz}. In general, the probability of selecting the true model increases with the skewness of x. As expected, the power of skewness and excess-kurtosis-based measures decreases with the skewness of the error

Table 4.2 95% CI coverage rates of the four direction dependence measures as a function β_{yx} and $\gamma_{\varepsilon_{y(xz)}}$.

β_{yx}	$\gamma_{\varepsilon_{y(xz)}}$	$\Delta(\gamma)$	$\Delta(\kappa)$	$\Delta(\gamma)^*$	$\Delta(\kappa)^*$
Percentile Method					
0.14	0	0.997 (0.001; 0.001)	0.997 (0.002; 0.001)	0.997 (0.002; 0.001)	0.998 (0.000; 0.002)
0.14	0.75	0.406 (0.594; 0.000)	0.983 (0.017; 0.000)	0.997 (0.002; 0.001)	0.998 (0.000; 0.002)
0.14	1.50	0.006 (0.994; 0.000)	0.629 (0.371; 0.000)	0.994 (0.005; 0.001)	0.997 (0.000; 0.003)
0.14	2.25	0.000 (1.000; 0.000)	0.125 (0.875; 0.000)	0.989 (0.011; 0.000)	0.996 (0.000; 0.004)
0.39	0	0.998 (0.001; 0.001)	0.997 (0.002; 0.001)	0.998 (0.001; 0.001)	0.991 (0.000; 0.009)
0.39	0.75	0.601 (0.399; 0.000)	0.989 (0.011; 0.000)	0.997 (0.002; 0.001)	0.993 (0.000; 0.007)
0.39	1.50	0.041 (0.959; 0.000)	0.753 (0.247; 0.000)	0.995 (0.004; 0.001)	0.995 (0.000; 0.005)
0.39	2.25	0.002 (0.998; 0.000)	0.267 (0.733; 0.000)	0.991 (0.009; 0.000)	0.996 (0.001; 0.004)
0.59	0	0.998 (0.001; 0.001)	0.996 (0.002; 0.002)	0.998 (0.001; 0.001)	0.989 (0.000; 0.011)
0.59	0.75	0.786 (0.214; 0.000)	0.993 (0.007; 0.001)	0.997 (0.002; 0.001)	0.990 (0.000; 0.010)
0.59	1.50	0.190 (0.810; 0.000)	0.866 (0.134; 0.000)	0.994 (0.005; 0.001)	0.992 (0.001; 0.007)
0.59	2.25	0.036 (0.964; 0.000)	0.483 (0.517; 0.000)	0.992 (0.007; 0.000)	0.994 (0.001; 0.005)
BCa Method					
0.14	0	0.926 (0.036; 0.039)	0.907 (0.047; 0.046)	0.928 (0.036; 0.036)	0.942 (0.034; 0.024)
0.14	0.75	0.218 (0.782; 0.000)	0.823 (0.153; 0.023)	0.927 (0.042; 0.031)	0.926 (0.043; 0.030)
0.14	1.50	0.005 (0.995; 0.000)	0.324 (0.674; 0.002)	0.913 (0.058; 0.029)	0.920 (0.043; 0.037)
0.14	2.25	0.000 (1.000; 0.000)	0.041 (0.959; 0.000)	0.896 (0.077; 0.027)	0.922 (0.038; 0.040)
0.39	0	0.926 (0.036; 0.037)	0.911 (0.043; 0.046)	0.945 (0.028; 0.028)	0.972 (0.009; 0.020)
0.39	0.75	0.347 (0.653; 0.000)	0.851 (0.122; 0.028)	0.934 (0.036; 0.030)	0.961 (0.016; 0.023)
0.39	1.50	0.026 (0.974; 0.000)	0.434 (0.562; 0.003)	0.919 (0.050; 0.030)	0.921 (0.043; 0.036)
0.39	2.25	0.002 (0.998; 0.000)	0.097 (0.903; 0.000)	0.905 (0.068; 0.027)	0.887 (0.064; 0.049)
0.59	0	0.927 (0.036; 0.037)	0.907 (0.048; 0.045)	0.954 (0.023; 0.023)	0.963 (0.010; 0.027)
0.59	0.75	0.508 (0.490; 0.002)	0.871 (0.094; 0.034)	0.943 (0.031; 0.026)	0.962 (0.012; 0.026)
0.59	1.50	0.102 (0.898; 0.000)	0.570 (0.423; 0.007)	0.928 (0.045; 0.027)	0.951 (0.023; 0.026)
0.59	2.25	0.020 (0.980; 0.000)	0.214 (0.785; 0.001)	0.915 (0.058; 0.027)	0.925 (0.041; 0.035)

Values in parentheses give the proportions of how often the CI fell below or above the true value of zero.

term ($\gamma_{\varepsilon_{y(xz)}}$). In contrast, the power of HOC-based CIs is rather unaffected by the magnitude of $\gamma_{\varepsilon_{y(xz)}}$. These results are in good accordance with the theoretical findings described above. Non-normality of $\varepsilon_{y(xz)}$ increases the magnitude of non-normality of the outcome y which, in turn, decreases higher moment

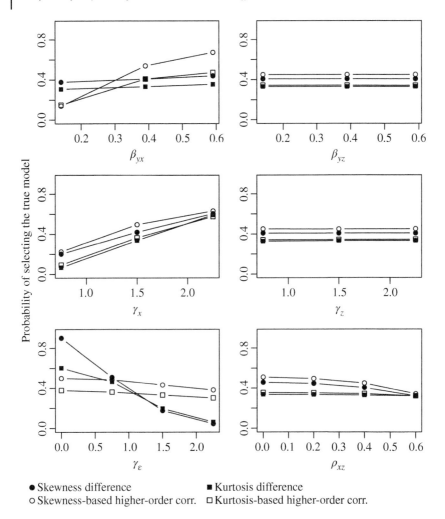

● Skewness difference   ■ Kurtosis difference
○ Skewness-based higher-order corr.   □ Kurtosis-based higher-order corr.

Figure 4.1 Empirical power of detecting the true model as a function of β_{yx}, γ_x, and γ_ε based on 95% BCa CIs of four direction dependence measures.

differences $\Delta(\gamma)$ and $\Delta(\kappa)$. In contrast, no distributional assumptions are made when comparing HOCs. Further, the power of HOC-based CIs systematically increases with the magnitude of β_{yx}, while the power of moment-based CIs is rather unaffected (at least for the considered effect size range). Note that quite similar results were observed by Dodge and Rousson (2000, 2016). That is, empirical power rates of moment-based procedures are rather constant across the range of effect sizes commonly encountered in practice and start to decline for very strong effects (such as $R^2 = 0.75$). Finally, due to the fact that covariate-adjusted

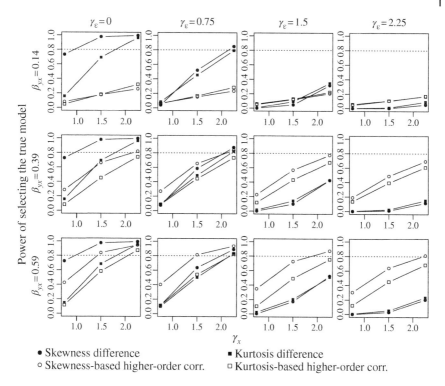

Figure 4.2 Empirical power of 95% BCa CIs of four direction dependence measures (dashed lines correspond to the 80% power cut-off).

predictor and outcome variables where used to determine the causal direction of effect, power curves of all BCa CIs are not affected by either the magnitude of β_{yz}, the skewness γ_z, nor the correlation ρ_{xz}.

In terms of the overall performance of the four BCa CIs, i.e. focusing on the power across all conditions for which $\gamma_x > 0$ and $\beta_{yx} > 0$ holds, testing skewness differences ($\Delta(\gamma)$) constitutes the most powerful procedure when the true error is normal (90.1% power), followed by testing differences in excess-kurtosis values ($\Delta(\kappa)$; 60.1% power), skewness-based HOCs ($\Delta(\gamma)^*$; 50.0% power), and kurtosis-based HOCs ($\Delta(\kappa)^*$; 37.9% power). The same rank order was observed for percentile CIs, i.e. testing the skewness difference (82.4% power) was more powerful than testing excess-kurtosis (43.3% power) and HOC differences ($\Delta(\gamma)^*$: 36.7%; $\Delta(\kappa)^*$: 27.0%). The rank order changes when true errors are non-normal, $\gamma_{\varepsilon_{y(xz)}} > 0$. Here, testing $\Delta(\gamma)^*$ (percentile: 28.0%; BCa: 43.3%) is more powerful than basing direction dependence decisions on $\Delta(\kappa)^*$ (percentile: 21.3%; BCa: 33.2%) or skewness differences $\Delta(\gamma)$ (percentile: 21.6%; BCa: 24.4%). Model selection using differences in excess-kurtosis values $\Delta(\kappa)$ has the lowest power (percentile: 15.5%; BCa: 24.1%).

Finally, Figure 4.2 gives the empirical power curves of moment- and HOC-based direction dependence measures (using 95% BCa CIs) as a function of those simulation factors that showed the largest impact on selecting the true model $\{x, z\} \to y$, i.e. the skewness of x, the skewness of $\varepsilon_{y(xz)}$, and the strength of the causal effect β_{yx}, with a focus on the well-established 80% power criterion (cf. Cohen, 1988). For $n = 200$, acceptable power can be achieved for 95% BCa CIs of $\Delta(\gamma)$ when the true predictor is moderately skewed ($\gamma_x = 1.5$) and the assumption of normality of the error term is fulfilled. In contrast, highly skewed predictors are required ($\gamma_x \geq 2.25$) when true errors are slightly skewed ($\gamma_{\varepsilon_{y(xz)}} = 0.75$). For highly skewed errors, observed power values fail to meet the 80% power criterion. Here, testing the difference in HOCs should be preferred. However, highly skewed predictors and large effect sizes ($\beta_{yx} = 0.59$) are required to achieve acceptable power with $n = 200$.

4.5 Data Example

To illustrate the application of the proposed direction dependence approach, we use data from a clinical trial on the effectiveness of acupuncture on chronic headache of primary care patients (Vickers, 2006; Vickers et al., 2004). Overall, 205 patients were randomly allocated to receive acupuncture treatment while 196 patients served as the control group. Patients completed a daily diary of headache and health-related quality of life for four weeks at baseline, three months, and one year after randomization. Headache severity was measured four times a day using a six point Likert scale (0 = no headache, ..., 5 = intense, incapacitating headache). Composite scores were calculated to represent headache severity at baseline, three months, and one year after randomization (higher scores reflect higher severity). Health-related quality of life was assessed using the SF-36 health status questionnaire (Ware & Sherbourne, 1992). In the present re-analysis, we focus on the relation between headache severity and the SF-36 sub-scale "energy/fatigue" based on 297 patients who provided valid data one year after randomization (i.e. 158 patients of the treatment and 139 patients of the control condition; 47 males and 250 females aged between 18 and 65, $M_{age} = 46.5$, $SD_{age} = 10.4$).

While Vickers et al. (2004) reported significant improvements of headache and energy in the treatment group, we further analyse the relation between headache severity and energy from a direction dependence perspective. Specifically, we posit a directional effect of headache severity on energy while adjusting for the covariates gender (0 = male, 1 = female), treatment status (0 = control, 1 = treatment), and headache severity and energy baseline scores. The reason for testing a causal hypothesis of the form headache $\to$ energy is that perceived

Table 4.3 Bivariate Pearson correlations and univariate descriptive measures.

Variable	(1)	(2)	(3)	(4)	(5)	(6)	M	SD	γ	κ
(1) Headache (baseline)	—	0.708	−0.209	−0.165	−0.069	−0.024	25.49	15.46	1.39	1.76
(2) Headache (12 months)		—	−0.074	−0.193	−0.203	−0.065	18.93	15.43	1.75	3.51
(3) Energy (baseline)			—	0.660	−0.104	−0.054	49.92	20.13	−0.26	−0.61
(4) Energy (12 months)				—	0.030	−0.130	54.86	20.69	−0.38	−0.53
(5) Treatment					—	−0.037	0.53	0.50	−0.13	−1.98
(6) Gender						—	0.84	0.37	−1.87	1.51

M = mean, SD = standard deviation, γ = skewness, κ = excess-kurtosis.

pain is often seen as an important factor affecting quality of life (Farahani & Assari, 2010). Univariate descriptive statistics and bivariate correlations for all variables are given in Table 4.3. All continuous variables were standardized prior to regression analyses. Regressing the full set of predictors on energy after 12 months suggested non-significant effects for treatment status and headache severity baseline scores. For reasons of parsimony, these two variables were excluded from subsequent regression analyses. To evaluate the validity of the initial regression model, a series of regression diagnostics were performed. Cook's distances and Bonferroni-adjusted largest studentized residuals identified one conspicuous observation with a Cook's distance of 0.091 (90% of the observations had a Cook's distance smaller than 0.008) and a studentized residual of 3.99 (Bonferroni corrected $p = 0.024$). Because outliers/highly influential observations may bias direction dependence decisions (Dodge & Rousson, 2016), we excluded this observation from the analysis. Parameter estimates of the final target model based on $n = 296$ observations are given in Table 4.4. In general, the model explained approximately 49% of the total variation in 12 months energy scores. Magnitude of 12 months energy significantly increases with energy baseline measures and significantly decreases with 12 months headache severity. In addition, female patients showed significantly lower levels of energy 12 months after randomization.

To evaluate the direction of effect of 12 months headache severity and 12 months energy, percentile and BCa CIs for difference in skewness ($\Delta(\gamma)$), excess-kurtosis ($\Delta(\kappa)$), and HOCs ($\Delta(\gamma)^*$ and $\Delta(\kappa)^*$) were performed while adjusting for baseline energy scores and gender. Two auxiliary models were calculated in which 12 months headache severity and energy scores were regressed on baseline energy and gender. Residuals of the two auxiliary regressions were used as "purified" measures of 12 months headache severity and 12 months energy.

Table 4.4 Results of the linear regression model predicting perceived energy 12 months after randomization.

Variable	β	SE	t-Value	p-Value
Constant	0.238	0.106	2.270	0.024
Headache (12 months)	−0.143	0.042	3.408	<0.001
Energy (baseline)	0.666	0.042	15.800	<0.001
Gender (female)	−0.294	0.114	−2.574	0.011
Multiple R^2	0.492			

SE = standard error.

In the first step, we evaluated distributional requirements for DDA. Figure 4.3 shows univariate histograms and the bivariate scatterplot for both variables. Adjusted 12 months headache severity scores significantly deviated from normality ($\gamma = 1.67$, $\kappa = 3.32$, Shapiro–Wilk $W = 0.86$, $p < 0.001$) while adjusted 12 months energy was close to the normal distribution ($\gamma = 0.11$, $\kappa = 0.038$, Shapiro–Wilk $W = 0.99$, $p = 0.596$). Regressing adjusted 12 months headache severity on adjusted 12 months energy, by necessity, leads to the same regression coefficient as given in Table 4.2, i.e. energy $= -0.143 \times$ headache $+ \varepsilon$ (with $\beta = -0.143$, SE $= 0.042$, $p < 0.001$, $R^2 = 0.038$). Residuals of this regression model were close to the normal distribution ($\gamma_{\varepsilon} = 0.075$, $\kappa_{\varepsilon} = -0.069$, Shapiro–Wilk $W = 0.99$, $p = 0.568$). The Breusch–Pagan test confirmed the absence of heteroscedasticity ($\chi^2(\mathrm{df} = 1) = 0.51$; $p = 0.476$) and the Harvey–Collier test indicated linearity of the energy–headache relation ($t(\mathrm{df} = 293) = 1.16$, $p = 0.246$). Further, the Hilbert–Schmidt Independence Criterion (HSIC) (Gretton et al., 2008) was used to evaluate the assumption of independence of predictor and error term. The HSIC is a kernel-based measure of independence which is provably omnibus in detecting any form of independence in the large sample limit (values close to zero indicate independence). Sen and Sen (2014) and Hidalgo, Wu, Engel, and Kosorok (2018) suggested a bootstrapping scheme to approximate the null distribution of the HSIC. Using 2000 resamples, we obtained a HSIC value of 0.118 together with a bootstrap p-value of 0.147 which suggests that independence holds for the predictor and the error term. Overall, we conclude that requirements are fulfilled for application of direction dependence CIs.

Table 4.5 gives the observed differences and the 95% percentile and BCa CIs (using 2000 resamples) for the four direction dependence measures. Since the residual distribution is close to normality (see above), we can base direction dependence decisions on the moment-based measures $\Delta(\gamma)$ and $\Delta(\kappa)$. Both measures are significantly greater than zero providing empirical support that

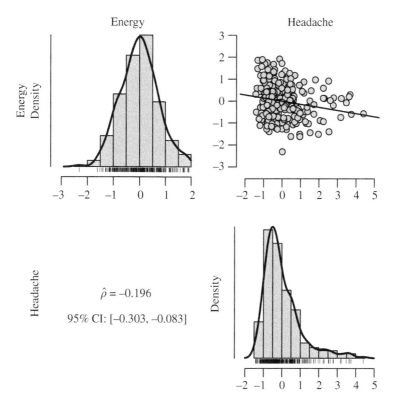

Figure 4.3 Univariate distributions (main diagonal) and bivariate scatterplot (upper right panel) of "purified" 12 months headache severity and energy scores including the bivariate linear regression line for 296 primary care patients.

12 months headache severity is more likely to be the predictor and 12 months energy is more likely to be the outcome variable. Both, percentile and BCa CIs for HOC-based measures, do not allow a clear cut decision which is in line with the observation that these measures require larger samples sizes to achieve acceptable power. Overall, we can conclude that the model headache → energy is better suited to approximate the underlying data-generating mechanism than the causally reversed model.

4.6 Discussion

The present study introduced asymmetric definitions of the partial correlation coefficient and extended direction dependence methods based on distributional characteristics of observed variables to multiple variable settings. Provided that (i)

Table 4.5 Results of resampling-based direction dependence tests.

	Obs. difference	95% Percentile CI	95% BCa CI	Decision
Skewness	1.559	[1.131; 1.871]	[1.233; 2.000]	$H \rightarrow E$
Excess-kurtosis	3.278	[1.585; 4.815]	[2.066; 5.881]	$H \rightarrow E$
Skewness-based HOC	0.085	[−0.145; 0.290]	[−0.117; 0.313]	Undecided
Kurtosis based HOC	−0.008	[−0.566; 0.577]	[−0.511; 0.646]	Undecided

CI = bootstrap confidence interval, E = adjusted energy, H = adjusted headache,
HOC = higher-order correlation.

the variables x and y are continuous, (ii) the data-generating mechanism is linear with independent additive error, and (iii) the causal effect is constant for all observations, asymmetry properties of the partial correlation enable researchers to probe the tentative causal relation between x and y while adjusting for additional exogenous variables. Simulation results suggest that non-normality of true errors reduces the power of moment-based direction dependence measures to detect the correct model and, at the same time, increases the risk of erroneously selecting the directionally mis-specified model. In contrast, HOC-based measures were robust to non-normal errors and should therefore be preferred whenever residuals of a tentative target model indicate deviations from normality. Further, while percentile bootstrap CIs of higher moment- and HOC-based direction dependence measures were generally found to be overly conservative, BCa CIs were better able to protect the nominal significance level and proved to be more powerful.

While the present study focused on the partial correlation coefficient, for completeness, it should be noted that asymmetry also exists for the semi-partial correlation. In the present context, two semipartial correlations can be computed: $\rho_{x(y|z)} = (\rho_{xy} - \rho_{xz}\rho_{yz})/\sqrt{1 - \rho_{yz}^2}$ measuring the linear association between x and y while holding z constant for y but not for x and $\rho_{y(x|z)} = (\rho_{xy} - \rho_{xz}\rho_{yz})/\sqrt{1 - \rho_{xz}^2}$ where z is held constant for x but not for y. Here, asymmetric facets for both coefficients are more of theoretical interest than of practical value. The reason for this is that additional assumptions have to be imposed on the pairwise relations of the triplet $\{x, y, z\}$ to be able to express higher powers of $\rho_{x(y|z)}$ and $\rho_{y(x|z)}$ as simple higher moment ratios. For third higher moments, for example, one obtains $\rho_{x(y|z)}^3 = \gamma_{\varepsilon_{y(z)}}/\gamma_x$ only if one assumes that $\rho_{xz} = 0$. Similarly, $\rho_{y(x|z)}^3 = \gamma_y/\gamma_{\varepsilon_{x(z)}}$ holds only if $\rho_{yz} = 0$. Both assumptions are very unlikely to hold in practice and direction dependence tests based on asymmetric facets of semi-partial correlations may be of limited use. Similar statements can be obtained for the fourth power of semi-partial correlations and the corresponding fourth higher moment ratios.

While Dodge and Rousson's (2000, 2001) original work mainly focused on studying the marginal behavior of variable distributions, Sungur (2005) suggested to evaluate causal hypotheses based on the behavior of joint variable distributions using copula-regression (note that the presented HOC-based measures also focus on the joint variable behavior; for details see the related Chapter 3 by Sungur, in this volume). Copula-based direction dependence approaches have also experienced rapid development. For example, Kim and Kim (2014) developed inferential procedures for asymmetric bivariate copula regressions (see also Kim & Kim, 2016). Extensions to study asymmetric dependencies in multivariate continuous data using multivariate skew normal copula-based regression have been proposed by Wei, Wang, and Kim (2016) and Wei and Kim (2018). Less, however, is known about the performance of copula-based direction dependence methods to detect the causal direction of effects. Simulation studies that compare the performance of these models and the direction dependence approach discussed here constitute important future endeavors.

4.6.1 Relation to Causal Inference Methods

Modern causal inference approaches combine the counterfactual framework of causation (Lewis, 1973) with structural equation models (SEMs; Holland, 1988; Pearl, 2009; Robins, 1986; Rubin, 1974). In essence, directed acyclic graphs (DAGs) are used to represent substantive causal theories in terms of causal relations among variables and no-confounding assumptions (ignorability conditions) are imposed on structural equations to endow parameters with valid causal interpretation. The conceptualization of this class of SEMs has been subject to controversy. While, in the psychometric literature, the term "structural equation modeling" is often used for modeling covariance structures for observed variables (cf. Jöreskog, 1977), SEMs for causal inference include no-confounding conditions as part of their definition (Pearl, 2009). Thus, Wang and Sobel (2013) suggested to distinguish between "regular" SEMs (rSEMs – models where no-confounding conditions are not part of the model definition) and "causal" SEMs (cSEMs – describing DAGs with ignorability conditions). Others suggested the notion of "Pearlian" DAGs to explicitly describe the attempt of elucidating quantitative conclusions and qualitative causal structure of constructs (cf. Dawid, 2008).

Adopting this SEM dichotomy, direction dependence methods may take a position in between because (i) in line with rSEMs, alternative explanations are explicitly considered and are part of the modeling process, and (ii) statistical connections between cSEMs and DDA exist. The first connection emerges when considering the instrumental variable (IV) approach as a common tool for confounder control (e.g. Angrist, Imbens, & Rubin, 1996; for a discussion of IV methods in observational data see Pokropek (2016)). IV models may serve as a remedy when measurement error in the predictor, omitted variable biases, or reverse causation are present (cf. Dunn, 2004; Smith, 1982; Wooldridge, 2002). Here, causal effects are obtained by using a third variable w (the instrument) that is known to be correlated with the tentative predictor x, i.e. $\mathrm{cov}(x, w) \neq 0$, and uncorrelated with the error, $\mathrm{cov}(w, \varepsilon_{y(x)}) = 0$. The IV estimator for the causal effect of x on y can be written as $\beta_{yx} = \mathrm{cov}(y, w)/\mathrm{cov}(x, w)$ (Durbin, 1954). A link between IV estimation and HOC-based direction dependence exists when one defines the instrument w as higher powers of the tentative predictor. That is, one uses $w = x^k$ for the model $x \to y$ and $w = y^k$ for $y \to x$ ($k \geq 2$) where the IV assumptions $\mathrm{cov}(x, x^k) \neq 0$ and $\mathrm{cov}(x^k, \varepsilon_{y(x)}) = 0$ will hold for the true model by definition.

For example, assuming $k = 2$ (together with $E[x] = E[y] = 0$ and $E[x^2] = E[y^2] = 1$) and making use of (4.21), one obtains $\beta_{yx} = \mathrm{cov}(y, x^2)/\mathrm{cov}(x, x^2) = \rho_{xy}$ as the IV estimator for $x \to y$ with $\mathrm{cov}(x, x^2) = \gamma_x$ and $\mathrm{cov}(y, x^2) = \rho_{xy}\gamma_x$. Using the IV approach in the causally reversed model $y \to x$ leads to $\beta_{xy} = \mathrm{cov}(x, y^2)/\mathrm{cov}(y, y^2) = \rho_{xy}^{-1}$ with $\mathrm{cov}(x, y^2) = \rho_{xy}^2\gamma_x$ and $\mathrm{cov}(y, y^2) = \gamma_y = \rho_{xy}^3\gamma_x$. Thus, for the simple bivariate case, the first and third ratio for the Pearson correlation in Eq. (4.27) can be re-cast as the IV estimates β_{yx} and β_{xy}^{-1} and the corresponding asymmetric sequence in (4.28) can be re-written as $(\rho_{xy}^3\gamma_x)^2 \leq (\rho_{xy}^2\gamma_x)^2 \leq (\rho_{xy}\gamma_x)^2 \leq \gamma_x^2$. In other words, direction dependence properties of observed variables make use of elements that arise from estimating causal effects with squared tentative predictors as instrumental variables.

The second connection concerns the independence assumption underlying direction dependence CIs and no-confounding assumptions in cSEMs. As defined in (4.1), the true model assumes stochastic independence of the predictor and the error term, i.e. $\mathrm{cov}[g_1(x), g_2(\varepsilon_{y(x)})] = 0$ for any absolutely integrable functions g_1 and g_2. Because one obtains $\mathrm{cov}(x, \varepsilon_{y(x)}) = \beta_{yu}\beta_{xu}\sigma_u^2$ in case of an unmeasured confounder u (assuming zero means for all variables) with β_{yu} and β_{xu} quantifying the magnitude of the confounder effects and σ_u^2 being the variance of u, stochastic independence implies the absence of confounders. Given that independence holds, the OLS regression coefficient is an unbiased estimator for the causal effect (e.g. Clogg & Haritou, 1997). In other words, if the independence assumption inherent to the presented direction dependence measures is fulfilled, conditions for proper interpretation of estimated effects as causal are satisfied. In the present context, testing the independence assumption implies identifying dependence structures in linearly uncorrelated variables (note that predictors and OLS residuals are always uncorrelated by construction). For example, in the presented empirical example, we made use of the HSIC to detect dependencies beyond first-order correlations. Taken together, by integrating independence and direction dependence properties it is possible to: (i) identify the directionality of the data mechanism, and (ii) empirically evaluate conditions that are necessary to endow parameter estimates with causal meaning.

References

Ames, A. J. (2013). Accuracy and precision of an effect size and its variance from a multilevel model for cluster randomized trials: A simulation study. *Multivariate Behavioral Research*, *48*(4), 592–618. doi:10.1080/00273171.2013.802978

Angrist, J. D., Imbens, G. W., & Rubin, D. B. (1996). Identification of causal effects using instrumental variables. *Journal of the American Statistical Association*, *91*, 444–455. doi:10.2307/2291629

Bradley, J. V. (1978). Robustness? *British Journal of Mathematical and Statistical Psychology*, *31*(2), 144–152. doi:10.1111/j.2044-8317.1978.tb00581.x

Bravais, A. (1846). Analyse mathématique sur les probabilités des erreurs de situation d'un point. *Mémoires présentés par divers savants á l'Académie royale des sciences de l'Institut de France*, *9*, 225–332.

Clogg, C. C., & Haritou, A. (1997). The regression method of causal inference and a dilemma confronting this method. In V. R. McKim & S. P. Turner (Eds.), *Causality in crisis: Statistical methods and the search for causal knowledge in the social science* (pp. 83–112). Notre Dame, IN: University of Notre Dame Press.

Cohen, J. (1988). *Statistical power analysis for the behavioral sciences* (2nd ed.). New Jersey, NJ: Lawrence Erlbaum Associates.

Collins, L. M., Schafer, J. L., & Kam, C.-M. (2001). A comparison of inclusive and restrictive strategies in modern missing data procedures. *Psychological Methods*, *6*(4), 330. doi:10.1037//1082-989x.6.4.330

Davison, A. C., & Hinkley, D. V. (1997). *Bootstrap methods and their application*. Cambridge, UK: Cambridge University Press.

Dawid, A. (2008). Beware of the DAG. *Journal of Machine Learning Research: Workshop and Conference Proceedings*, *6*, 59–86.

Dodge, Y., & Rousson, V. (2000). Direction dependence in a regression line. *Communications in Statistics—Theory and Methods*, *29*(9–10), 1957–1972. doi:10.1080/03610920008832589

Dodge, Y., & Rousson, V. (2001). On asymmetric properties of the correlation coeffcient in the regression setting. *The American Statistician*, *55*(1), 51–54. doi:10.1198/000313001300339932

Dodge, Y., & Rousson, V. (2016). Statistical inference for direction of dependence in linear models. In W. Wiedermann & A. von Eye (Eds.), *Statistics and causality: Methods for applied empirical research* (pp. 45–62). Hoboken, NJ: Wiley.

Dodge, Y., & Yadegari, I. (2010). On direction of dependence. *Metrika*, *72*(1), 139–150. doi:10.1007/s00184-009-0273-0

Dunn, G. (2004). *Statistical evaluation of measurement errors: Design and analysis of reliability studies* (2nd ed.). New York, NY: Oxford University Press.

Durbin, J. (1954). Errors in variables. *Review of the International Statistical Institute, 22*, 23–32. doi:10.2307/1401917

Efron, B. (1987). Better bootstrap confidence intervals. *Journal of the American Statistical Association, 82*(397), 171–185. doi:10.21236/ada150798

Falk, R., & Well, A. D. (1997). Many faces of the correlation coefficient. *Journal of Statistics Education, 5*(3), 1–18.

Farahani, M. A., & Assari, S. (2010). Relationship between pain and quality of life. In V. R. Preedy & R. R. Watson (Eds.), *Handbook of disease burdens and quality of life measures* (pp. 3933–3953). New York, NY: Springer.

Frisch, R., & Waugh, F. V. (1933). Partial time regressions as compared with individual trends. *Econometrica, 1*, 387–401. doi:10.2307/1907330

Gretton, A., Fukumizu, K., Teo, C. H., Song, L., Schölkopf, B., & Smola, A. J. (2008). A kernel statistical test of independence. *Advances in Neural Information Processing Systems, 20*, 585–592.

Hidalgo, S. J. T., Wu, M. C., Engel, S. M., & Kosorok, M. R. (2018). Goodness-of-fit test for nonparametric regression models: Smoothing spline anova models as example. *Computational Statistics & Data Analysis, 122*, 135–155. doi:10.1016/j.csda.2018.01.004

Holland, P. W. (1988). Causal inference, path analysis, and recursive structural equation models (with discussion). *Sociological Methodology, 18*, 449–493. doi:10.2307/271055

Jöreskog, K. G. (1977). Structural equation models in the social sciences: Specification, estimation and testing. In P. R. Krishnaiah (Ed.), *Application of statistics* (pp. 57–88). Amsterdam, NL: North-Holland.

Kendall, M. G., & Stuart, A. (1979). *The advanced theory of statistics: Inference and relationship* (2nd ed.). London, UK: Chares Griffin & Company.

Kim, D., & Kim, J.-M. (2014). Analysis of directional dependence using asymmetric copula-based regression models. *Journal of Statistical Computation and Simulation, 84*(9), 1990–2010. doi:10.1080/00949655.2013.779696

Kim, S., & Kim, D. (2016). Directional dependence analysis using skew-normal copula-based regression. In W. Wiedermann & A. von Eye (Eds.), *Statistics and causality: Methods for applied empirical research* (pp. 131–152). Hoboken, NJ: Wiley and Sons.

Lewis, D. (1973). Causation. *Journal of Philosophy, 70*, 556–567. doi:10.2307/2025310

Lovell, M. C. (1963). Seasonal adjustment of economic time series and multiple regression analysis. *Journal of the American Statistical Association, 58*(304), 993–1010. doi:10.1080/01621459.1963.10480682

Marszalek, J. M., Barber, C., Kohlhart, J., & Holmes, C. B. (2011). Sample size in psychological research over the past 30 years. *Perceptual and Motor Skills, 112*(2), 331–348. doi:10.2466/03.11.PMS.112.2.331-348

Muddapur, M. (2003). On directional dependence in a regression line. *Communications in Statistics—Theory and Methods, 32*(10), 2053–2057. doi:10.1081/STA-120023266

Nelsen, R. B. (1996). *An introduction to copulas* (2nd ed.). New York, NY: Springer.

Nelsen, R. B. (1998). Correlation, regression lines, and moments of inertia. *The American Statistician, 52*(4), 343–345. doi:10.2307/2685438

Pearl, J. (2009). *Causality: Models, reasoning, and inference* (2nd ed.). Cambridge, UK: Cambridge University Press.

Pearson, K. (1920). Notes on the history of correlation. *Biometrika, 13*(1), 25–45. doi:10.2307/2331722

Pokropek, A. (2016). Introduction to instrumental variables and their application to large-scale assessment data. *Large-Scale Assessments in Education, 4*, 1–20. doi:10.1186/s40536-016-0018-2

Pornprasertmanit, S., & Little, T. D. (2012). Determining directional dependency in causal associations. *International Journal of Behavioral Development, 36*(4), 313–322. doi:10.1177/0165025412448944

R Core Team. (2019). *R: A language and environment for statistical computing [Computer software manual]*. Vienna, Austria. Retrieved from http://www.R-project.org/

Robins, J. M. (1986). A new approach to causal inference in mortality studies with sustained exposure periods: Application to control of the healthy worker survivor effect. *Mathematical Modeling, 7*, 1393–1512. doi:10.1016/0270-0255(86)90088-6

Rodgers, J. L., & Nicewander, W. A. (1988). Thirteen ways to look at the correlation coefficient. *The American Statistician, 42*(1), 59–66. doi:10.2307/2685263

Rovine, M. J., & von Eye, A. (1997). A 14th way to look at a correlation coefficient: Correlation as the proportion of matches. *The American Statistician, 51*(1), 42–46. doi:10.2307/2684692

Rubin, D. B. (1974). Estimating causal effects of treatments in randomized and nonrandomized studies. *Journal of Educational Psychology, 66*, 688–701. doi:10.1037/h0037350

Sen, A., & Sen, B. (2014). Testing independence and goodness-of-fit in linear models. *Biometrika, 101*(4), 927–942. doi:10.1093/biomet/asu026

Smith, E. R. (1982). Beliefs, attributions, and evaluations: Nonhierarchical models of mediation in social cognition. *Journal of Personality and Social Psychology, 43*(2), 248–259. doi:10.1037/0022-3514.43.2.248

Sungur, E. A. (2005). A note on directional dependence in regression setting. *Communications in Statistics—Theory and Methods, 34*(9–10), 1957–1965. doi:10.1080/03610920500201228

Teuscher, F., & Guiard, V. (1995). Sharp inequalities between skewness and kurtosis for unimodal distributions. *Statistics & Probability Letters*, 22(3), 257–260. doi:10.1016/0167715294000741

Thoemmes, F. (2015). Empirical evaluation of directional-dependence tests. *International Journal of Behavioral Development*, 39(6), 560–569. doi:10.3403/02181143

Vickers, A. J. (2006). Whose data set is it anyway? Sharing raw data from randomized trials. *Trials*, 7(1), 1. doi:10.1186/1745-6215-7-15

Vickers, A. J., Rees, R. W., Zollman, C. E., McCarney, R., Smith, C. M., Ellis, N., … Van Haselen, R. (2004). Acupuncture for chronic headache in primary care: Large, pragmatic, randomised trial. *British Medical Journal*, 328(7442), 744. doi:10.1136/bmj.38029.421863.eb

von Eye, A., & DeShon, R. P. (2012). Directional dependence in developmental research. *International Journal of Behavioral Development*, 36(4), 303–312. doi:10.1177/0165025412439968

Wang, X., & Sobel, M. E. (2013). New perspectives on causal mediation analysis. In S. L. Morgan (Ed.), *Handbook of causal analysis for social research* (pp. 215–242). Dordrecht, NY: Springer.

Ware, J. E., & Sherbourne, C. D. (1992). The MOS 36-item short-form health survey (SF-36): I. Conceptual framework and item selection. *Medical Care*, 30, 473–483. doi:10.1097/00005650-199206000-00002

Wei, Z., & Kim, D. (2018). On multivariate asymmetric dependence using multivariate skew-normal copula-based regression. *International Journal of Approximate Reasoning*, 92, 376–391. doi:10.1016/j.ijar.2017.10.016

Wei, Z., Wang, T., & Kim, D. (2016). Multiple copula regression function and directional dependence under multivariate non-exchangeable copulas. In V.-N. Huynh, V. Kreinovich, & S. Sriboonchitta (Eds.), *Causal inference in econometrics* (pp. 171–184). Switzerland, Europe: Springer International Publishing.

Wiedermann, W. (2018). A note on fourth moment-based direction dependence measures when regression errors are non-normal. *Communications in Statistics: Theory and Methods*, 47, 5255–5264. doi:10.1080/03610926.2017.1388403

Wiedermann, W., & Hagmann, M. (2016). Asymmetric properties of the pearson correlation coefficient: Correlation as the negative association between linear regression residuals. *Communications in Statistics: Theory and Methods*, 45(21), 6263–6283. doi:10.1080/03610926.2014.960582

Wiedermann, W., & Li, X. (2018). Direction dependence analysis: A framework to test the direction of effects in linear models with an implementation in spss. *Behavior Research Methods*, 50(4), 1581–1601.

Wiedermann, W., Merkle, E. C., & Eye, A. (2018). Direction of dependence in measurement error models. *British Journal of Mathematical and Statistical Psychology*, 71, 117–145. doi:10.1111/bmsp.12111

Wiedermann, W., & Sebastian, J. (2019). Direction dependence analysis in the presence of confounders: Applications to linear mediation models. *Multivariate Behavioral Research, 12*, 1–21.

Wiedermann, W., & von Eye, A. (2015a). Direction dependence analysis: A confirmatory approach for testing directional theories. *International Journal of Behavioral Development, 39*(6), 570–580.

Wiedermann, W., & von Eye, A. (2015b). Direction of effects in multiple linear regression models. *Multivariate Behavioral Research, 50*(1), 23–40. doi:10.1080/00273171.2014.958429

Wooldridge, J. M. (2002). *Econometric analysis of cross section and panel data.* Cambridge, MA: MIT Press.

5

Recent Advances in Semi-Parametric Methods for Causal Discovery

Shohei Shimizu[1,2] and Patrick Blöbaum[3]

[1]*Faculty of Data Science, Shiga University, Hikone, Japan*
[2]*Center for Advanced Intelligence Project, Riken, Japan*
[3]*Tuebingen Causality Team, Amazon Research and Development Center, Tübingen, Germany*

5.1 Introduction

Causal discovery is a statistical and computational technique for recovering the hidden causal structure of variables in an unsupervised manner (Shimizu, 2016; Spirtes, Glymour, & Scheines, 1993; Zhang & Hyvärinen, 2016). In many empirical sciences including epidemiology, economics, and neuroscience, the causal mechanisms underlying various phenomena need to be studied. Thus, causal discovery has attracted much attention and is one of the most exciting topics in the fields of statistics and machine learning. Many new methods have recently been proposed for inferring the causal structure of variables, typically their causal directions. This chapter overviews such causal structure learning methods, particularly those based on semi-parametric assumptions on the functional forms and error distributions.

First, we briefly overview the causal discovery framework based on the structural causal model (SCM) (Pearl, 2000). This framework uses special types of equations known as structural equations (Bollen, 1989) to represent how the values of variables are determined.

An example of such structural equations is given by

$$x_1 = f_1(z_1, e_1) \tag{5.1}$$

$$x_2 = f_2(x_1, z_1, e_2), \tag{5.2}$$

where the variables x_1 and x_2 are observed variables, z_1 is a hidden variable, e_1 and e_2 are error variables. The error variable e_1 denotes all the direct causes of x_1 other than z_1. Similarly, the error variable e_2 denotes all the direct causes of x_2

Direction Dependence in Statistical Modeling: Methods of Analysis, First Edition.
Edited by Wolfgang Wiedermann, Daeyoung Kim, Engin A. Sungur, and Alexander von Eye.
© 2021 John Wiley & Sons, Inc. Published 2021 by John Wiley & Sons, Inc.

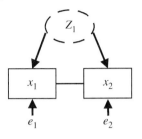

Figure 5.1 An example causal graph.

other than x_1 and z_1. These structural equations mean that the left-hand sides of the equations are defined by their right-hand sides, not vice versa.

The causal structure of variables x_1 and x_2 for Eqs. (5.1) and (5.2) is graphically represented as given in Figure 5.1, where observed variables x_1 and x_2 are represented by boxes, a hidden common cause z_1 is represented by a dotted circle. The error variables e_1 and e_2 are unobserved, although they are not represented by dotted circles due to a convention. Such a graph representing the causal structure of variables is called a causal graph. Causal discovery aims to infer or recover the causal graph when the graph structure is unknown.

Next, we discuss the basic idea of causal discovery. Causal discovery methods combine assumptions on the causal structure of variables with their data to infer the structure. Those assumptions restrict the search space of causal structures. Then, the set of best-fitting models to the data among the models satisfying the assumptions is selected after inconsistent models to the data are discarded. If no assumptions are made, any meaningful inference could not be made.

A typical assumption is that the causal structure of variables is represented by a directed acyclic graph, abbreviated as DAG. Another typical assumption is that there is no hidden or unobserved common causes, which means that all the common causes of observed variables are observed. Then, the causal relations of observed variables are represented in the framework of SCMs as follows

$$x_i = f_i(\text{pa}(x_i), e_i)(i = 1, \dots, p), \tag{5.3}$$

where x_i $(i = 1, \dots, p)$ are observed variables, $\text{pa}(x_i)$ denotes the parents of x_i $(i = 1, \dots, p)$, and e_i $(i = 1, \dots, p)$ are error variables that are independent. The independence assumption implies that there are no hidden common causes.

Under these two typical assumptions, for two variable cases, we have the following three candidate causal models of observed variables x_1 and x_2, for example. In the first model (a), x_1 causes x_2 as follows.

$$x_1 = e_1 \tag{5.4}$$

$$x_2 = f_2(x_1, e_2), \tag{5.5}$$

where error variables e_1 and e_2 are independent, which implies that there is no hidden common causes between x_1 and x_2. On the other hand, in the second model

(b), x_2 causes x_1 as follows.

$$x_1 = f_1(x_2, e_1) \tag{5.6}$$
$$x_2 = e_2, \tag{5.7}$$

where error variables e_1 and e_2 are independent as well. Finally, in the third model (c), x_1 and x_2 have no causal relation as follows.

$$x_1 = e_1 \tag{5.8}$$
$$x_2 = e_2, \tag{5.9}$$

where error variables e_1 and e_2 are independent, and thus x_1 and x_2 are independent.

If two observed variables x_1 and x_2 are implied to be independent from the data, the third model is selected. This is because in the first and second models x_1 and x_2 are dependent since either of x_1 and x_2 causes the other, and only in the third model they are independent. On the other hand, if the two variables x_1 and x_2 are implied to be dependent from the data, the third model is inconsistent to the data, and the first and second models are considered to be not inconsistent to the data.

Thus, if only independence of observed variables is used to infer the underlying causal graph, the first and second models cannot be distinguished between. The reason why only independence of x_1 and x_2 is used is that no assumption on the functions f_1 and f_2 is made. If some assumptions on the functional forms are made, more information can be obtained from the data.

In the subsequent sections, we review such methods with parametric assumptions on the functional forms but allowing an infinite number of error distributions, i.e. causal discovery methods with semi-parametric assumptions.

Major application areas of causal discovery methods include economics (Coad & Binder, 2014; Moneta, Entner, Hoyer, & Coad, 2013), epidemiology (Rosenström et al., 2012), and neuroscience (Mills-Finnerty, Hanson, & Hanson, 2014; Smith et al., 2011). In those areas, it is ethically difficult or costly to actually intervene on variables. Therefore, it is effective to find probable causal hypotheses using causal discovery methods. Other application areas include psychology (von Eye & DeShon, 2012), chemistry (Campomanes et al., 2014), meteorology (Niyogi, Kishtawal, Tripathi, & Govindaraju, 2010), and genetics (Ozaki, Toyoda, Iwama, Kubo, & Ando, 2011).

5.2 Linear Non-Gaussian Methods

5.2.1 LiNGAM

Although the independence information of two observed variables was not capable of distinguishing between the first and second models in Figure 5.2 above, it

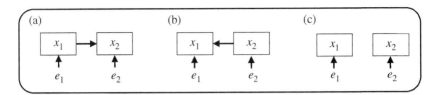

Figure 5.2 Three candidate causal models.

can be shown that it is possible to distinguish between the two models if the functional relation is linear and error variables e_1 and e_2 follow some non-Gaussian continuous distributions (Dodge & Rousson, 2001; Shimizu, Hoyer, Hyvärinen, & Kerminen, 2006).

Shimizu et al. (2006) proposed a non-Gaussian SCM with no hidden common causes, known as a linear non-Gaussian acyclic model, abbreviated as LiNGAM:

$$x_i = \sum_{j \in \text{pa}(x_i)} b_{ij} x_j + e_i, \tag{5.10}$$

where e_i $(i = 1, ..., p)$ are continuous unobserved error variables, b_{ij} are the connection strengths from x_j to x_i $(i, j = 1, ..., p)$, pa(x_i) denotes the parents of x_i, i.e. the direct causes of x_i in the causal graph $(i = 1, ..., p)$. The causal graph, the graphical representation of causal relations of the variables x_i $(i = 1, ..., p)$ is assumed to be a DAG. Moreover, the error variables e_i follow non-Gaussian distributions, with zero mean and non-zero variance, and are independent of each other. Note that the independence assumption between error variables e_i $(i = 1, ..., p)$ implies that there are no hidden common causes between x_i $(i = 1, ..., p)$.

In matrix form, the LiNGAM model in Eq. (5.10) is written as

$$x = Bx + e, \tag{5.11}$$

where the connection strength matrix **B** collects the connection strengths b_{ij} $(i, j = 1, ..., p)$, and the vectors x and e collect the observed variables x_i $(i = 1, ..., p)$ and the error variables e_i $(i = 1, ..., p)$, respectively.

LiNGAM has been proven to be identifiable (Shimizu et al., 2006), i.e. the connection strength matrix **B** can be uniquely identified based on the data x only. Once the connection strength matrix **B** has been identified, the causal graph can be drawn based on the zero/non-zero pattern of the elements of **B**, i.e. if b_{ij} is non-zero, a directed edge is drawn from x_j to x_i, and otherwise no edge is drawn from x_j to x_i.

To illustrate this, the three models in Figure 5.2 are written in the following forms. The first model with x_1 causing x_2 is written as follows:

$$\underbrace{\begin{bmatrix} x_1 \\ x_2 \end{bmatrix}}_{x} = \underbrace{\begin{bmatrix} 1 & 0 \\ b_{21} & 1 \end{bmatrix}}_{A} \underbrace{\begin{bmatrix} e_1 \\ e_2 \end{bmatrix}}_{s}. \tag{5.12}$$

The second model with x_2 causing x_1 is written as follows:

$$\underbrace{\begin{bmatrix} x_1 \\ x_2 \end{bmatrix}}_{x} = \underbrace{\begin{bmatrix} 1 & b_{12} \\ 0 & 1 \end{bmatrix}}_{A} \underbrace{\begin{bmatrix} e_1 \\ e_2 \end{bmatrix}}_{s}. \tag{5.13}$$

The third model with x_1 and x_2 being independent is written as follows:

$$\underbrace{\begin{bmatrix} x_1 \\ x_2 \end{bmatrix}}_{x} = \underbrace{\begin{bmatrix} 1 & 0 \\ 0 & 1 \end{bmatrix}}_{A} \underbrace{\begin{bmatrix} e_1 \\ e_2 \end{bmatrix}}_{s}. \tag{5.14}$$

All the three models are viewed as independent component analysis (ICA) models with the matrix A being the mixing matrix and with the vector s being the non-Gaussian independent component vector. The mixing matrix A of ICA is known to be identifiable up to the permutation and signs of its columns (Comon, 1994; Hyvärinen, Karhunen, & Oja, 2001).

The zero/non-zero patterns of the three mixing matrices above are different from each other. The three patterns cannot be equal to each other for every combination of the column permutation and sign changes. Thus, it is possible to find which model of the three is the most consistent with the data based on the estimated mixing matrix. This is the basic idea of the model identification. Estimation methods based on independence and non-Gaussianity of error variables have been proposed (Shimizu et al., 2006, 2011; Sogawa, Shimizu, Kawahara, & Washio, 2010). In the following, we provide a brief overview of some of the extensions of LiNGAM.

5.2.2 Hidden Common Causes

We first discuss an extension of LiNGAM to cases with hidden common causes. The independence assumption between error variables e_i ($i = 1, \ldots, p$) in LiNGAM given in Eq. (5.11) implies that there are no hidden common causes (Shimizu et al., 2006). A hidden common cause is an unobserved variable that directly causes more than one observed variable. However, hidden common causes exist in many applications. If such hidden common causes are ignored, the estimation results obtained may be seriously biased (Bollen, 1989; Pearl, 2000; Spirtes et al., 1993).

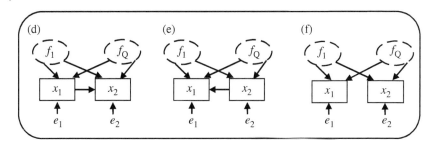

Figure 5.3 Three candidate causal models in the presence of hidden common causes.

Hoyer, Shimizu, Kerminen, and Palviainen (2008) proposed LiNGAM with hidden common causes, and the model can be formulated as follows:

$$x = \mathbf{B}x + \Lambda f + e, \tag{5.15}$$

where the difference obtained from LiNGAM in Eq. (5.11) represents the existence of the hidden common cause vector f. The vector f collects the non-Gaussian hidden common causes f_q with zero mean and unit variance ($q = 1, ..., Q$). Without loss of generality, the hidden common causes $f_q (q = 1, ..., Q)$ are assumed to be independent of each other, since any dependent hidden common causes can be remodeled by means of linear combinations of independent unobserved variables, provided that the underlying model is linear acyclic and the error variables corresponding to the observed variables and hidden common causes are independent (Hoyer, Shimizu, et al., 2008). The matrix Λ collects λ_{iq}, which denote the connection strengths from f_q to $x_i (i = 1, ..., p; q = 1, ..., Q)$.

It has been shown (Hoyer, Shimizu, et al., 2008) that one can distinguish between the following three models when assuming faithfulness (Spirtes et al., 1993) of x_i and f_q. In the first model (d) in Figure 5.3, x_1 causes x_2 in the presence of hidden common causes f_q ($q = 1, ..., Q$) as follows:

$$x_1 = \sum_{q=1}^{Q} \lambda_{1q} f_q + e_1 \tag{5.16}$$

$$x_2 = b_{21} x_1 + \sum_{q=1}^{Q} \lambda_{2q} f_q + e_2. \tag{5.17}$$

In the second model (e), x_2 causes x_1 in the presence of hidden common causes f_q ($q = 1, ..., Q$) as follows:

$$x_1 = b_{12} x_2 + \sum_{q=1}^{Q} \lambda_{1q} f_q + e_1 \tag{5.18}$$

$$x_2 = \sum_{q=1}^{Q} \lambda_{2q} f_q + e_2. \tag{5.19}$$

The coefficients b_{21} and b_{12} represent the causal effect of x_1 on x_2 and that of x_2 on x_1, respectively.

In the third model (f), x_1 and x_2 do not cause each other, whereas hidden common causes f_q ($q = 1, \ldots, Q$) exist as follows:

$$x_1 = \sum_{q=1}^{Q} \lambda_{1q} f_q + e_1 \tag{5.20}$$

$$x_2 = \sum_{q=1}^{Q} \lambda_{2q} f_q + e_2. \tag{5.21}$$

The corresponding causal graphs are provided in Figure 5.3.

To illustrate the idea of identifying the causal graph, the three models are reformulated with Q being one as follows. The first model with x_1 causing x_2 is written as follows:

$$\underbrace{\begin{bmatrix} x_1 \\ x_2 \end{bmatrix}}_{x} = \underbrace{\begin{bmatrix} 1 & 0 & \lambda_{11} \\ b_{21} & 1 & b_{21}\lambda_{11} + \lambda_{21} \end{bmatrix}}_{A} \underbrace{\begin{bmatrix} e_1 \\ e_2 \\ f_1 \end{bmatrix}}_{s}. \tag{5.22}$$

The second model with x_2 causing x_1 is written as follows:

$$\underbrace{\begin{bmatrix} x_1 \\ x_2 \end{bmatrix}}_{x} = \underbrace{\begin{bmatrix} 1 & b_{12} & b_{12}\lambda_{21} + \lambda_{11} \\ 0 & 1 & \lambda_{21} \end{bmatrix}}_{A} \underbrace{\begin{bmatrix} e_1 \\ e_2 \\ f_1 \end{bmatrix}}_{s}. \tag{5.23}$$

The third model with x_1 and x_2 not causing each other is written as follows:

$$\underbrace{\begin{bmatrix} x_1 \\ x_2 \end{bmatrix}}_{x} = \underbrace{\begin{bmatrix} 1 & 0 & \lambda_{11} \\ 0 & 1 & \lambda_{21} \end{bmatrix}}_{A} \underbrace{\begin{bmatrix} e_1 \\ e_2 \\ f_1 \end{bmatrix}}_{s}. \tag{5.24}$$

All the three models also are viewed as ICA models (Hyvärinen et al., 2001) with the matrix A being the mixing matrix and with the vector s being the non-Gaussian independent component vector. Note that this ICA model has more independent components than observed variables and hence is called overcomplete ICA model (Lewicki & Sejnowski, 2000). The mixing matrix A of overcomplete ICA is known to be identifiable up to the permutation and signs of its columns (Eriksson & Koivunen, 2004).

Again, the zero/non-zero patterns of the three mixing matrices above are different from each other. The patterns cannot be equal to each other for every combination of the column permutation and sign changes. Thus, it is possible to find which model of the three is the most consistent with the data based on the estimated mixing matrix similarly to the case with no hidden common causes above. Hoyer, Shimizu, et al. (2008) and Henao and Winther (2011) proposed estimation methods with modelling hidden common causes as unobserved variables. Shimizu and Bollen (2014) proposed an alternative Bayesian estimation approach avoiding explicitly modelling hidden common causes in the model instead by modeling their sums. Tashiro, Shimizu, Hyvärinen, and Washio (2014) considered an estimation method robust against hidden common causes, i.e. an approach for learning causal relations of observed variables between which hidden common causes do not exist.

5.2.3 Time Series

Hyvärinen, Zhang, Shimizu, and Hoyer (2010) considered analyzing both lagged and instantaneous causal effects in time series data. Such an approach is useful if the measurements may have a lower time resolution than the causal influences known as a structural vector autoregressive model in econometrics (Swanson & Granger, 1997). LiNGAM is used for modeling instantaneous causal effects, while a classic auto-regressive model is used for modeling lagged causal effects. The combination of the two models leads to the following model:

$$x(t) = \sum_{\tau=0}^{h} \mathbf{B}_{\tau} x(t - \tau) + e(t), \qquad (5.25)$$

where $x(t)$ and $e(t)$ are the observed variable vectors and the error variable vectors at time point t, respectively. $\mathbf{B}_{\tau}$ denotes the connection strength matrices having a time lag τ. Note that the time lag τ starts from zero.

Hyvärinen et al. (2010) showed that the model in Eq. (5.25) is identifiable if $e_i(t)$ are non-Gaussian as well as mutually and temporally independent. A simple estimation method for this model is to fit a classic auto-regressive model on $x(t)$ and apply LiNGAM on the residuals (Hyvärinen et al., 2010). The LiNGAM model can be replaced by its cyclic version (Lacerda, Spirtes, Ramsey, & Hoyer, 2008) or its extension to cases with hidden common causes (Hoyer, Shimizu, et al., 2008) to allow cyclic instantaneous causal relations or instantaneous hidden common causes, respectively. Furthermore, it would be possible to replace the autoregressive model with a time-series model allowing hidden common causes (Malinsky & Spirtes, 2018).

In practice, data are often obtained by subsampling or temporally aggregating the original data generating processes. Gong et al. (Gong, Zhang, Schölkopf, Glymour, & Tao, 2017; Gong, Zhang, Schoelkopf, Tao, & Geiger, 2015) considered several sufficient identifiability conditions to infer the original causal relations from the subsampled data and from temporally aggregated data, respectively.

5.2.4 Multiple Data Sets

In some application domains, data are obtained under differing conditions: under different experimental conditions, for different subjects or at different time points. In other words, multiple data sets are obtained, as opposed to a single data set. Ramsey, Hanson, and Glymour (2011), Shimizu (2012), Kadowaki, Shimizu, and Washio (2013), and Schaechtle, Stathis, Holloway, and Bromuri (2013) proposed methods for estimating a common causal structure for multiple data sets.

For example, the key idea of Shimizu (2012) is that we first estimate a variable that can be at the top of a causal ordering of variables in every dataset by finding such a variable that jointly minimizes its independence of its regression residuals over all the datasets. Then we remove the effect of the variable at the top of the causal ordering from the other variables by regressing it out. We iterate this procedure until all the variables are ordered.

Ramsey et al. (2011) demonstrated the excellent performance of such an estimation approach on simulated functional magnetic resonance imaging (fMRI) data created by Smith et al. (2011).

5.2.5 Other Methodological Issues

Shimizu, Hoyer, and Hyvärinen (2009) and Zhang et al. (2018) investigated the causal analysis of latent factors, as opposed to observed variables. Kawahara, Bollen, Shimizu, and Washio (2010) and Entner and Hoyer (2002) proposed a LiNGAM analysis of groups of variables, instead of simply single variables, where grouping of variables is learned from data or certain background knowledge is used to divide variables into groups a priori. Hoyer, Hyvärinen et al. (2008) proposed a non-Gaussian method that is robust against the Gaussianity of error variables. Blöbaum and Shimizu (2017) proposed a framework to understand the prediction mechanisms of predictive models using causal discovery methods. Wenjuan, Lu, and Chunchen (2018) proposed to combine LiNGAM model and logistic regression model to analyze mixed variable cases in which both continuous and discrete variables exist in data. Entner, Hoyer, and Spirtes (2012) and Silva and Shimizu (2017) considered methods for learning instrumental variables based on non-Gaussianity.

5.3 Nonlinear Bivariate Methods

LiNGAM provides an efficient way to distinguish between cause and effect in linear settings, but struggles in nonlinear settings. In this section, we discuss different approaches to address the problem of inferring the causal direction between two variables if the structural equation has a nonlinear form. In order to simplify the notation, we assume that Figure 5.2a represents the true causal relationship and rewrite Eq. (5.5) as

$$y = \phi(x, e), \tag{5.26}$$

where x is the cause, y is the effect and $x \perp e$. Here, recall an independence between x and e implies that we assume there are no hidden common causes of x and y.

5.3.1 Additive Noise Models

If the function ϕ in Eq. (5.26) is nonlinear, the way how the noise influences the effect can become quite complicated and has typically a non-additive form. Therefore, a common assumption, particularly in standard regression problems, is to assume that Eq. (5.26) has the form

$$y = \phi(x, e) = f(x) + e, \tag{5.27}$$

where x, e are continuous and $x \perp e$. If f is linear and e non-Gaussian, this is equivalent to the assumptions in LiNGAM. However, here we allow that f is nonlinear and e can follow an arbitrary distribution. In the literature, this is also called an *Additive Noise Model* (ANM), where the key factors are the additive influence of the noise and the assumed independence between input and noise.

The work by Hoyer, Janzing, Mooij, Peters, and Schölkopf (2009) has shown that the model defined by Eq. (5.27) can, with some exceptions, be uniquely identified, i.e. it is only possible to fit an ANM from x to y such that the residual is independent of the input. This particularly implies that, as opposed to LiNGAM, x and e can be Gaussian distributed if f is nonlinear. Accordingly, the problem becomes unidentifiable if f is linear and x and e are Gaussian.

In practice, the causal direction between two variables based on the ANM assumption can be inferred by a two-step method. First, fit a regression model in both possible directions and second, test for independence between residuals and inputs. The residual should only be independent of the input in the true direction. For the independence test, the authors proposed to use a kernel based method, the Hilbert–Schmidt independence criterion (Gretton et al., 2008). Although the performance of this approach depends on two different components, it is fairly robust against the choice of regression model and type of independence test. Due to its simplicity, the ANM method can be easily implemented and combined with any kind of standard regression methods and independence measures beyond simple correlations.

5.3.1.1   Post-Nonlinear Models

The model class represented by Eq. (5.27) can be further extended to the form of a *post-nonlinear* (PNL) model (Zhang & Hyvärinen, 2009)

$$y = \phi(x, e) = g(f(x) + e), \tag{5.28}$$

where x, e are continuous, g is a nonlinear function and $x \perp e$. Due to the additional nonlinearity of g, more complex relationships between $f(x)$ and e can be captured. Similar as for ANM, Zhang and Hyvärinen (2009) showed that a PNL model can only be fitted from x to y in the generic case, except g, f and the distribution $p(x)$ are adjusted in a specific way. In contrast to an ANM, a PNL model is more difficult to evaluate since standard regression methods are not applicable anymore.

5.3.1.2   Discrete Additive Noise Models

In Eqs. (5.27) and (5.28), we assume that x and e are continuous. However, it turns out that an identifiability of ANMs can likewise be shown if input and noise variables are discrete (Peters, Janzing, & Schölkopf, 2010), i.e.

$$y = \phi(x, e) = f(x) + e,$$

where x, $e \in \mathbb{Z}$, $f : \mathbb{Z} \to \mathbb{Z}$, and $x \perp e$. Also in the case of a discrete ANM, a model can, under certain conditions stated in Peters et al. (2010), only be fitted from x to y such that the residual is independent of the input. In practice, these conditions are more likely to be violated than the ones in case of an ANM and a PNL model. Accordingly, the assumptions of a discrete ANM are generally stronger.

5.3.2   Independence of Mechanism and Input

In this section, we aim to address the causal discovery problem by making use of a different type of assumption that implicitly put restrictions on the relation between the form of ϕ and the distributions of $p(x)$ and $p(e)$ in Eq. (5.26). Here, we postulate that mechanisms in nature represent independent processes and, thus, have no information about each other. While this formulation is rather abstract, a simple interpretation of this postulate states that, for instance, the process of how the noise e in an ANM is generated is independent of the process of how the cause x is generated. Therefore, knowing the distribution $p(e)$ would not give any information about the distribution $p(x)$, since they follow two independent generation processes.

While there are many ways to formalize the concept of independent mechanisms, we focus on a formulation in terms of an algorithmic independence between $p(x)$ and $p(y \mid x)$ in the sense that $p(x)$ contains no information about the conditional distribution $p(y \mid x)$. With respect to Eq. (5.26), this means that the functional relationship ϕ and the distribution of the input $p(x)$ follow independent mechanisms by nature, and thus, contain no information about each other. The general idea behind this independence postulate is similar to the "modularity and autonomy" in structural equation models, which is one of the fundamental assumption in causality. It states that structural equations remain unchanged when other parts of a system change (, 2000), i.e. changing or intervening on the distribution of $p(x)$ in Eq. (5.26) does not change ϕ.

An important implication of the stated independence postulate is that the distribution of the effect $p(y)$, on the other hand, contains information about $p(x \mid y)$. The succeeding sections discuss some ways of exploiting this asymmetry between cause and effect for causal discovery. For a more foundational insight, we refer the readers to the works by Janzing and Schölkopf (2010) and Lemeire and Janzing (2012), who formalize the postulate via algorithmic information and state that knowing $p(x)$ does not enable a shorter description of $p(y \mid x)$ and vice versa.

5.3.2.1 Information-Geometric Approach for Causal Inference

The information-geometric approach for causal inference (IGCI) (Daniušis et al., 2010; Janzing et al., 2012) formalize the independence postulate in the deterministic setting

$$y = f(x)$$

by assuming that f is nonlinear and that the slope f' changes independently of the density of $p(x)$. This can be formally stated as

$$\mathrm{Cov}[\log(f'), p(x)] = 0, \tag{5.29}$$

where $x \mapsto \log(f'(x))$ and $x \mapsto p(x)$ are defined as random variables on the probability space $[0, 1]$ with respect to a uniform measure. In this sense, the (logarithmic) slope of the function and the input are assumed to be uncorrelated. If f is independent of $p(x)$, then changes in f' independently of $p(x)$ result in vanishing correlations. Typically, this assumption would be violated if f is adjusted to $p(x)$ or vice versa, e.g. due to an intelligent design by first observing $p(x)$ and then adjusting f. More discussions about when this form of the independence postulate is violated can be found in Janzing et al. (2012).

Based on the assumption stated in Eq. (5.27), (Daniušis et al., 2010; Janzing et al., 2012) showed that

$$\mathrm{Cov}[\log(f^{-1'}), p(y)] > 0, \tag{5.30}$$

where f^{-1} denotes the inverse of f. If f would be linear, Eq. (5.30) becomes zero. Using the logarithm of the slope in Eqs. (5.29) and (5.30) allows an information-geometric interpretation. By comparing both equations, we see that there is a correlation between $p(y)$ and $f^{-1'}$, but not between $p(x)$ and f', which follows the general idea of the independence postulate. This asymmetry exploited by the IGCI approach for inferring the causal direction.

In practice, the Kullback–Leibler divergence between a reference measure and $p(x)$ is compared with the Kullback–Leibler divergence between the same reference measure and $p(y)$. Typical choices for reference measures are non-informative distributions, such as the uniform or Gaussian distribution.

While IGCI addresses nonlinear deterministic relationships, the approaches described in the next sections allow nonlinear and nondeterministic relationships.

5.3.2.2 Causal Inference with Unsupervised Inverse Regression

The underlying theory of IGCI is based on an asymmetry between the information contained in $p(x)$ and $p(y)$ about f and f^{-1}, respectively. In the same spirit, Schölkopf et al. (2013) hypothesized that the effect distribution $p(y)$ may contain information about the cause given the effect (the conditional $p(x\,|\,y)$) that can be employed for predicting the cause from the effect, but the cause distribution $p(x)$ alone does not contain any information about the effect given the cause (the conditional $p(y\,|\,x)$) that helps to predict the effect from the cause.

This idea is further exploited for causal discovery by a technique called CURE (Sgouritsa, Janzing, Hennig, & Schölkopf, 2015), which utilize unsupervised inverse regression. The idea of CURE is to estimate a prediction model for both possible causal directions in an unsupervised manner, i.e. only the input data is used for the training of the prediction models. Then, an unsupervised regression of y on x by only using information from $p(x)$ should perform worse than a regression of x on y by only using information from $p(y)$. CURE implements this idea using a modification of Gaussian process regression. While the general assumptions are less restrictive as in IGCI, the biggest drawback of CURE is a very high computational cost, which makes it impractical in larger data sets.

5.3.2.3 Approximation of Kolmogorov Complexities via the Minimum Description Length Principle

Using algorithmic information theory, Janzing and Schölkopf (2010) has shown that the algorithmic independence of $p(x)$ and $p(y\,|\,x)$ implies

$$K(p(x)) + K(p(y\mid x)) \le K(p(y)) + K(p(x\mid y)), \tag{5.31}$$

where K denotes the description length of a distribution in terms of its Kolmogorov complexity. Equation (5.31) can be interpreted in the sense that $p(x, y)$ has a simpler description in causal direction than in anticausal direction.

Although Kolmogorov complexities are uncomputable, they can be approximated by, for instance, using the minimum description length principle (Rissanen, 1978). This is utilized for causal discovery in the work by Marx and Vreeken (2017), who propose a method that is based on a global and multiple local regression models fit in both possible causal directions. Based on the fitted models, the complexities of the directions are approximated by the description length of the fitted regression models and the description length of the error. This method is called SLOPE.

5.3.2.4 Regression Error Based Causal Inference

Another way to exploit asymmetries based on the algorithmic independence is provided by an approach called Regression Error based Causal Inference (RECI) (Blöbaum, Janzing, Washio, Shimizu, & Schölkopf, 2018), which is based on a similar assumption as Eq. 5.29 in IGCI, but allows a nondeterministic relationship of the form stated in Eq. 5.26. While structural equation models generally assume that $x \perp e$, RECI decomposes Eq. (5.26) in the form of

$$y = \phi(x, e) = f(x) + \alpha \tilde{e}, \tag{5.32}$$

where f is a nonlinear function, α is a scaling factor of the residual $\tilde{e}$ and, in contrast to ANMs, x and $\tilde{e}$ can be dependent. Note that due to the allowed dependency between x and $\tilde{e}$, Eq. (5.26) can always be represented as Eq. ((5.32)).

RECI's core assumption is based on a modified version of Eq. (5.29), which incorporates the dependency between x and $\tilde{e}$ and is formalized by

$$\mathrm{Cov}[f', p(x)\mathrm{Var}[\tilde{e} \mid x]] = 0. \tag{5.33}$$

Compared to Eq. (5.29), here $x \mapsto f'(x)$ and $x \mapsto p(x)\mathrm{Var}[\tilde{e} \mid x]$ are considered as uncorrelated random variables. The general justification of this assumption is similar to that of Eq. (5.29), although the map $x \mapsto p(x)\mathrm{Var}[\tilde{e} \mid x]$ is now a property of $p(x)$ and $p(y|x)$ and not only of $p(x)$ as in Eq. (5.29). Note that if $\mathrm{Var}[\tilde{e} \mid x]$ is constant, Eq. (5.33) reduces to Eq. ((5.29)).

Under this assumption, Blöbaum et al. (2018) showed that

$$\lim_{\alpha \to 0} \frac{\mathbb{E}[\mathrm{Var}[x \mid y]]}{\mathbb{E}[\mathrm{Var}[y \mid x]]} > 1, \tag{5.34}$$

where x and y are normalized on $[0, 1]$. The inequality stated in Eq. (5.34) can be interpreted in the way that the least-squares regression error of regressing x on y is greater than the error of regressing y on x in a "near deterministic" setting. Whether Eq. (5.34) holds for only for small α or whether the asymmetries also occurs for large noise cannot be concluded from the theoretical insights. However, experiments in real-world data suggest that the asymmetries often appear even for realistic noise levels.

Compared to SLOPE, RECI and SLOPE only require the regression in both possible causal directions, where the inference of the causal direction in RECI is based on a comparison of regression errors and in SLOPE it is based on a comparison of the description lengths.

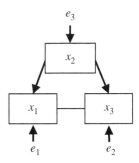

Figure 5.4 The skeleton of a causal graph with an undirected edge between x_1 and x_3. This could be resolved with techniques for the bivariate setting.

5.3.3 Applications to Multivariate Cases

The discussed methods in this section are focused on causal discovery in a bivariate setting and assume that there are no hidden common causes. Generally, most of the discussed methods are fairly robust to a violation of this assumption, but provide no identifiability guarantee in confounded settings. However, we might still be interested in using these methods for resolving undirected edges in a skeleton of a causal graph, which was, for instance, obtained by methods such as FCI or PC (Spirtes et al., 1993). An example is illustrated in Figure 5.4. In this example, we cannot directly apply the aforementioned bivariate methods on x_1 and x_3 due to the common cause x_2. However, instead we can treat x_2 simply as an additional cause. An extension of most methods to multivariate cases it therefore straightforward.

For example, in case of ANMs, instead of regressing x_1 on x_3 and vice versa, we can simply regress x_1 on x_2, x_3 and, accordingly, regress x_3 on x_1, x_2. The idea of treating x_2 as an additional cause can also be used in the methods that are based on the independence postulate, seeing that the formalization only needs to state a general independence between the distributions of all causes and the conditional of the effect given theses causes. Note that this is only a general idea of how observed common causes could be incorporated into bivariate methods and might not be a priori mathematically justified by the underlying theories. Further insights on this matter can be found in Peters, Mooij, Janzing, and Schlkopf (2012).

5.4 Conclusion

This chapter discussed that a wider variety of causal structures can be identified or uniquely estimated under semi-parametric assumptions on functional forms and error distributions than when using classical methods based on conditional independence. Non-Gaussian data is encountered in many applications, including the social sciences and the life sciences. Such non-Gaussianity rises when functional forms are nonlinear or error variables follow non-Gaussian distributions. The semi-parametric approaches discussed in this chapter may be a suitable approach in such applications.

References

Blöbaum, P., Janzing, D., Washio, T., Shimizu, S., & Schölkopf, B. (2018, April). Cause-effect inference by comparing regression errors. In *Proceedings of the 21st International Conference on Artificial Intelligence and Statistics (AISTATS 2018)*, Playa Blanca, Lanzarote, Canary Islands.

Blöbaum, P. & Shimizu, S. (2017, September). Estimation of interventional effects of features on prediction. In *Proceedings 2017 IEEE 27th International Workshop on Machine Learning for Signal Processing (MLSP2017)* (pp. 1–6), Tokyo, Japan.

Bollen, K. (1989). *Structural equations with latent variables*. New York, NY: Wiley & Sons.

Campomanes, P., Neri, M., Horta, B. A. C., Roehrig, U. F., Vanni, S., Tavernelli, I., & Rothlisberger, U. (2014). Origin of the spectral shifts among the early intermediates of the rhodopsin photocycle. *Journal of the American Chemical Society*, *1360*(10), 3842–3851.

Coad, A., & Binder, M. (2014). Causal linkages between work and life satisfaction and their determinants in a structural var approach. *Economics Letters*, *124*(2), 263–268.

Comon, P. (1994). Independent component analysis, a new concept? *Signal Processing*, *36*, 62–83.

Daniušis, P., Janzing, D., Mooij, J., Zscheischler, J., Steudel, B., Zhang, K., & Schölkopf, B. (2010, July). Inferring deterministic causal relations. In *Proceedings of the 26th Conference on Uncertainty in Artificial Intelligence* (pp. 143–150). Corvallis, OR, July 2010. Catalina Island, California: AUAI Press.

Dodge, Y., & Rousson, V. (2001). On asymmetric properties of the correlation coefficient in the regression setting. *The American Statistician*, *55*(1), 51–54.

Entner, D., Hoyer, P., & Spirtes, P. (2012, April). Statistical test for consistent estimation of causal effects in linear non-gaussian models. In *Proceedings of the 15th International Conference on Artificial Intelligence and Statistics (AISTATS2012)* (pp. 364–372), La Palma, Canary Islands.

D. Entner & P. O. Hoyer (2002). Estimating a causal order among groups of variables in linear models. In *Proceedings of the 22nd International Conference on Artificial Neural Networks (ICANN2012)* (pp. 83–90), La Palma, Canary Islands.

Eriksson, J., & Koivunen, V. (2004). Identifiability, separability, and uniqueness of linear ICA models. *IEEE Signal Processing Letters*, *11*, 601–604.

Gong, M., Zhang, K., Schoelkopf, B., Tao, D., & Geiger, P. (2015, July). Discovering temporal causal relations from subsampled data. In *Proceedings of 32nd*

International Conference on Machine Learning (ICML2015) (pp. 1898–1906), Lille, France.

Gong, M., Zhang, K., Schölkopf, B., Glymour,C., & Tao, D. (2017) Causal discovery from temporally aggregated time series. In *Proceedings of the 33rd Conference on Uncertainty in Artificial Intelligence (UAI2017)*, Sydney, Australia.

Gretton, A., Fukumizu, K., Teo, C. H., Song, L., Schölkopf, B., & Smola, A. J. (2008). A kernel statistical test of independence. *Advances in Neural Information Processing Systems*, *20*, 585–592.

Henao, R., & Winther, O. (2011). Sparse linear identifiable multivariate modeling. *Journal of Machine Learning Research*, *12*, 863–905.

Hoyer, P. O., Hyvärinen, A., Scheines, R., Spirtes, P., Ramsey, J., Lacerda, G., & Shimizu, S. (2008). Causal discovery of linear acyclic models with arbitrary distributions. In *Proceedings of the 24th Conference on Uncertainty in Artificial Intelligence* (pp. 282–289).

Hoyer, P. O., Janzing, D., Mooij, J., Peters, J., & Schölkopf, B. (2009). Nonlinear causal discovery with additive noise models. *Advances in Neural Information Processing Systems*, *21*, 689–696.

Hoyer, P. O., Shimizu, S., Kerminen, A., & Palviainen, M. (2008). Estimation of causal effects using linear non-Gaussian causal models with hidden variables. *International Journal of Approximate Reasoning*, *49*(2), 362–378.

Hyvärinen, A., Karhunen, J., & Oja, E. (2001). *Independent component analysis*. New York, NY: Wiley & Sons.

Hyvärinen, A., Zhang, K., Shimizu, S., & Hoyer, P. O. (2010). Estimation of a structural vector autoregressive model using non-Gaussianity. *Journal of Machine Learning Research*, *11*, 1709–1731.

Janzing, D., Mooij, J., Zhang, K., Lemeire, J., Zscheischler, J., Daniušis, P., … Schölkopf, B. (2012). Information-geometric approach to inferring causal directions. *Artificial Intelligence*, *182*, 1–31.

Janzing, D., & Schölkopf, B. (2010). Causal inference using the algorithmic Markov condition. *IEEE Transactions on Information Theory*, *56*(10), 5168–5194.

Kadowaki, K., Shimizu, S., & Washio, T. (2013). Estimation of causal structures in longitudinal data using non-gaussianity. In *Proceedings of the 23rd IEEE International Workshop on Machine Learning for Signal Processing (MLSP2013)* (pp. 1–6).

Kawahara, Y., Bollen, K., Shimizu, S., & Washio, T. (2010). GroupLiNGAM: Linear non-Gaussian acyclic models for sets of variables. *ArXiv:1006.5041*.

Lacerda, G., Spirtes, P., Ramsey, J., & Hoyer, P. O. (2008, August). Discovering cyclic causal models by independent components analysis. In *Proceedings of the 24th Conference on Uncertainty in Artificial Intelligence (UAI2008)* (pp. 366–374), Monterey, California, USA.

Lemeire, J., & Janzing, D. (2012). Replacing causal faithfulness with algorithmic independence of conditionals. *Minds and Machines, 23*(2), 227–249.

Lewicki, M., & Sejnowski, T. J. (2000). Learning overcomplete representations. *Neural Computation, 12*(2), 337–365.

Malinsky, D. & Spirtes, P. (2018, August). Causal structure learning from multivariate time series in settings with unmeasured confounding. In *Proceedings of 2018 ACM SIGKDD Workshop on Causal Discovery* (pp. 23–47), Anchorage, Alaska.

Marx, A. & Vreeken, J. (2017, November). Telling cause from effect using MDL-based local and global regression. In *2017 IEEE International Conference on Data Mining (ICDM)* (pp. 307–316), New Orleans. doi:10.1109/ICDM.2017.40.

Mills-Finnerty, C., Hanson, C., & Hanson, S. J. (2014). Brain network response underlying decisions about abstract reinforcers. *NeuroImage, 103*, 48–54.

Moneta, A., Entner, D., Hoyer, P. O., & Coad, A. (2013). Causal inference by independent component analysis: Theory and applications. *Oxford Bulletin of Economics and Statistics, 75*(5), 705–730.

Niyogi, D., Kishtawal, C., Tripathi, S., & Govindaraju, R. S. (2010). Observational evidence that agricultural intensification and land use change may be reducing the Indian summer monsoon rainfall. *Water Resources Research, 46*, W03533.

Ozaki, K., Toyoda, H., Iwama, N., Kubo, S., & Ando, J. (2011). Using non-normal SEM to resolve the ACDE model in the classical twin design. *Behavior Genetics, 41*(2), 329–339.

Pearl, J. (2000). *Causality: Models, reasoning, and inference.* Cambridge, UK: Cambridge University Press. (2nd ed. 2009)

J. Peters, D. Janzing, & B. Schölkopf (2010, May). Identifying cause and effect on discrete data using additive noise models. In *JMLR Workshop and Conference Proceedings, AISTATS 2010 (Proceedings of the 13th International Conference on Artificial Intelligence and Statistics)* (Vol. 9, pp. 597–604), Chia Laguna Resort, Sardinia, Italy.

Peters, J., Mooij, J., Janzing, D., & Schlkopf, B. (2012) Identifiability of causal graphs using functional models. *CoRR*, abs/1202.3757, 01.

Ramsey, J. D., Hanson, S. J., & Glymour, C. (2011). Multi-subject search correctly identifies causal connections and most causal directions in the DCM models of the Smith et al. simulation study. *NeuroImage, 58*(3), 838–848.

Rissanen, J. (1978). Modeling by shortest data description. *Automatica, 14*(5), 465–471. doi:10.1016/0005-1098(78)90005-5 Retrieved from http://www.sciencedirect.com/science/article/pii/0005109878900055

Rosenström, T., Jokela, M., Puttonen, S., Hintsanen, M., Pulkki-Råback, L., Viikari, J. S., … Keltikangas-Järvinen, L. (2012). Pairwise measures of causal direction in the epidemiology of sleep problems and depression. *PLoS ONE, 70*(11), e50841.

Schaechtle, U., Stathis, K., Holloway, R., & Bromuri, S. (2013, August). Multi-dimensional causal discovery. In *Proceedings of the 23rd International Joint Conference on Artificial Intelligence (IJCAI2013)*, Beijing, China.

Schölkopf, B., Janzing, D., Peters, J., Sgouritsa, E., Zhang, K., & Mooij, J. (2013). *Semi-supervised learning in causal and anticausal settings, chapter 13, Festschrift in Honor of Vladimir Vapnik* (pp. 129–141). New York, NY: Springer-Verlag. doi:10.1007/978-3-642-41136-6_13

Sgouritsa, E., Janzing, D., Hennig, P., & Schölkopf, B. (2015). Inference of cause and effect with unsupervised inverse regression. *Artificial Intelligence and Statistics, 38*, 847–855.

Shimizu, S. (2012). Joint estimation of linear non-Gaussian acyclic models. *Neurocomputing, 81*, 104–107.

Shimizu, S. (2016). Non-Gaussian structural equation models for causal discovery. In W. Wiedermann & A. von Eye (Eds.), *Statistics and causality: Methods for applied empirical research* (pp. 153–184). New York, NY: Wiley & Sons.

Shimizu, S., & Bollen, K. (2014). Bayesian estimation of causal direction in acyclic structural equation models with individual-specific confounder variables and non-Gaussian distributions. *Journal of Machine Learning Research, 15*, 2629–2652.

Shimizu, S., Hoyer, P. O., & Hyvärinen, A. (2009). Estimation of linear non-gaussian acyclic models for latent factors. *Neurocomputing, 72*, 2024–2027.

Shimizu, S., Hoyer, P. O., Hyvärinen, A., & Kerminen, A. (2006). A linear non-Gaussian acyclic model for causal discovery. *Journal of Machine Learning Research, 7*, 2003–2030.

Shimizu, S., Inazumi, T., Sogawa, Y., Hyvärinen, A., Kawahara, Y., Washio, T., … Bollen, K. (2011). DirectLiNGAM: A direct method for learning a linear non-Gaussian structural equation model. *Journal of Machine Learning Research, 12*, 1225–1248.

Silva, R., & Shimizu, S. (2017). Learning instrumental variables with structural and non-gaussianity assumptions. *Journal of Machine Learning Research, 18*(1), 4321–4369.

Smith, S. M., Miller, K. L., Salimi-Khorshidi, G., Webster, M., Beckmann, C. F., Nichols, T. E., … Woolrich, M. W. (2011). Network modelling methods for FMRI. *NeuroImage, 54*(2), 875–891.

Y. Sogawa, S. Shimizu, Y. Kawahara, & T. Washio (2010, July). An experimental comparison of linear non-Gaussian causal discovery methods and their variants. In *Proceedings of 2010 International Joint Conference on Neural Networks (IJCNN2010)* (pp. 768–775), Barcelona, Spain.

Spirtes, P., Glymour, C., & Scheines, R. (1993). *Causation, prediction, and search.* Cambridge, UK: Springer-Verlag, MIT Press. (2nd ed. 2000)

Swanson, N. R., & Granger, C. W. J. (1997). Impulse response functions based on a causal approach to residual orthogonalization in vector autoregressions. *Journal of the American Statistical Association*, 357–367.

Tashiro, T., Shimizu, S., Hyvärinen, A., & Washio, T. (2014). Parcelingam: A causal ordering method robust against latent confounders. *Neural Computation*, *260*(1), 57–83.

von Eye, A., & DeShon, R. P. (2012). Directional dependence in developmental research. *International Journal of Behavioral Development*, *360*(4), 303–312.

Wenjuan, W., Lu, F., & Chunchen, L. (2018, July) Mixed causal structure discovery with application to prescriptive pricing. In *Proceedings of 27rd International Joint Conference on Artificial Intelligence (IJCAI2018)* (pp. 5126–5134), Stockholm, Sweden.

Zhang, K., Gong, M., Ramsey, J., Batmanghelich, K., Spirtes, P., & Glymour, C. (2018, August). Causal discovery in the presence of measurement error: Identifiability conditions. In *Proceedings of 34th Conference on Uncertainty in Artificial Intelligence (UAI2018)* (pp. 1–10), Monterey, California.

K. Zhang & A. Hyvärinen (2009, June). On the identifiability of the post-nonlinear causal model. In *Proceedings of 25th Conference on Uncertainty in Artificial Intelligence (UAI2009)* (pp. 647–655), Montreal, Canada.

Zhang, K., & Hyvärinen, A. (2016). Nonlinear functional causal models for distinguishing causes form effect. In W. Wiedermann & A. von Eye (Eds.), *Statistics and causality: Methods for applied empirical research*. New York, NY: Wiley & Sons.

6

Assumption Checking for Directional Causality Analyses

Phillip K. Wood

Department of Psychological Sciences, University of Missouri, Columbia, MO, USA

In order for any statistical technique to be interpreted, it is important to check whether the data set under consideration appears to meet the assumptions of the chosen statistical model. The critical role of such model assumptions in directional dependence has been highlighted by several authors since its original presentation by Dodge and Rousson (2000, 2001). Within the directional dependence literature, some of these assumption checks have already been proposed (e.g. Pornprasertmanit & Little, 2012; Thoemmes, 2015; von Eye & Wiedermann, 2015), for example. This chapter reviews these assumption checks for directional dependence tests so far, but also includes other influence diagnostics which have not been considered but which are common in standard regression analyses. In addition, it is noted that, for some influence diagnostics, the observations identified as influential are not necessarily the same under both a target and alternative model of causal direction which will be illustrated below.

Although directional dependence models are linear regression models, not all variables used in regression models may be used in directional dependence tests. Directional dependence tests assume that the variables under consideration are continuous variables assessed on a single homogeneous population. Categorical variables such as dummy, effect, or contrast coding schemes are, by definition, non-normally distributed and an inappropriate use of direction dependence tests would erroneously identify variables such as sex as "causing" any other continuously distributed variable. Even continuously measured variables, however, which are heavily associated with such categorical variables, run the risk of identifying erroneous directions of effect given the resulting bimodal or skewed distributions which would result. For example, measures of alcohol consumption might appear non-normally distributed given that substantial sex differences exist in alcohol consumption. For this reason, directional dependence analyses often

Direction Dependence in Statistical Modeling: Methods of Analysis, First Edition.
Edited by Wolfgang Wiedermann, Daeyoung Kim, Engin A. Sungur, and Alexander von Eye.
© 2021 John Wiley & Sons, Inc. Published 2021 by John Wiley & Sons, Inc.

include covariates as a way for adjusting for such effects. Use of such covariates, however, obligates the researcher to assess whether it is reasonable to assume that the covariate is linearly related to the variables in the directional dependence analysis. For the case of both continuous and categorical covariates, for example, the question of whether regression effects are homogeneous across levels of a categorical variable is an important part of evaluating the appropriateness of the model. As a more general point, however, the question of whether all relevant covariates have been correctly identified represents an important limitation in the traditional use of manifest variable covariates as a way of adjusting scores to produce a conditionally homogeneous data set. Although identification of all possible covariates as a requisite to the correct application of direction dependence is a counsel of perfection, the use of factor scores from different types of longitudinal structural models described below may constitute a promising alternative.

Directional dependence models have also employed increasingly complex psychometric considerations. It may be the case, for example, that observed patterns of non-normality in a given continuous manifest variable may be an artifact of measurement error associated with the manifest variable (Wiedermann, Merkle, & von Eye, 2018). von Eye and Wiedermann (2014), for this reason, proposed that researchers consider the use of factor scores as a way of ascertaining estimates of individuals' level of the constructs of interest. Figure 6.1, for example, provides a path diagram for the items of two scales in a data set to be considered later in the chapter. This diagram shows that the individual items of each scale contain measurement error, that the covariance between the latent constructs Enhance and Social are adjusted for the effect of Sex, that one of the items ("It's fun" and labeled with the letter "R") cross-loads on both constructs and that two of the items in the Social factor have correlated errors of measurement ("Helps Me Enjoy and Party" with "To Be Sociable"). As such, the structural measurement model shown in the figure takes into account psychometric properties now considered in the directional dependence of two manifest composite variables. Although use of factor scores as one way of having the observed distribution of behavior be less susceptible to the idiosyncratic characteristics of a particular observed variable, there has been little assessment of the extent to which the use of factor scores results in a smaller or more consistent set of observations being identified as influential.

The rotational indeterminacy of factor scores, however, means that more than one set of factor scores may be estimated within longitudinal models. Although cross-lagged panel models have often been proposed for longitudinal data, such models may produce misleading conclusions if the constructs under consideration contain substantial trait-like time invariant components (Hamaker, Kuiper, & Grasman, 2015). As Hamaker et al. point out, if the focus of interest is on individual developmental trajectories, latent difference scores are preferable to cross-lagged

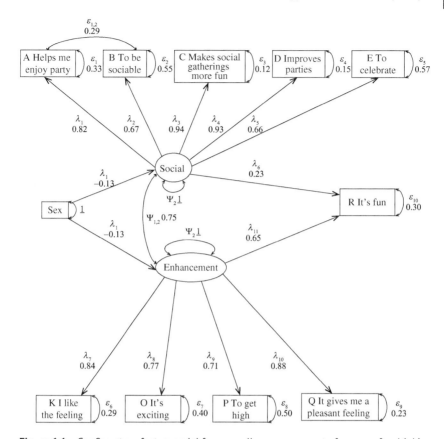

Figure 6.1 Confirmatory factor model for pre-college assessment of reasons for drinking.

associations when the influence of one construct on another is of central interest (p. 107). Analytically, latent difference scores may be thought of as a general covariate adjustment for all systematic stable inter-individual differences (i.e. general trait effects) associated with the construct (Figure 6.2).

Within the more general context of a multi-wave longitudinal study, however, use of a latent state-trait model (Schmitt & Steyer, 1993; Steyer, Mayer, Geiser, & Cole, 2015; Steyer, Schwenkmezger, & Auer, 1990) may be used to assess directional dependence associations between the factor scores of trait-level constructs. A latent state-trait model for the data set to be considered later in the chapter is shown in Figure 6.3.

The observed distributions of trait factor scores may not be as influenced by the momentary perturbations specific to a given prior measurement occasion and, as with the factor scores of item-level responses, the repeated assessments of the trait may thereby result in an improved resolution of the underlying distribution of the

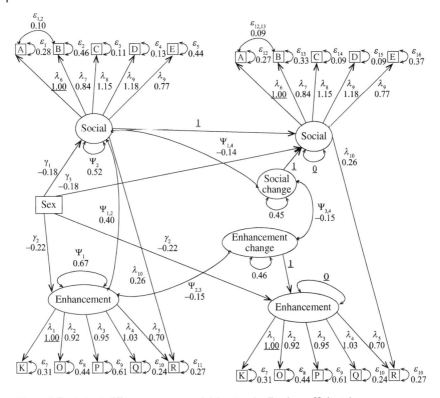

Figure 6.2 Latent difference score model (unstandardized co-efficients).

traits under consideration and, perhaps fewer influential observations as a result. Conversely, direction dependence associations between the constructs at each individual wave may be used to inform state-specific associations adjusted for the effects of the general traits of the constructs under consideration and, as such, represents a covariate adjustment for all systematic trait-level inter-individual differences.

Checking model assumptions does not represent an invalidation of the directional dependence approach nor is it a way to improve model fit for a favored directional hypothesis. Interpretation of an initial directional finding by itself does not provide much in the way of empirical support for a given research claim absent an evaluation of the counterclaims a reasonable skeptic may advance, however. Although in some situations addressing such counterclaims requires the collection of additional data, in others assumption checks provide a way in which a researcher can address the objections or possible limitations of a directional test.

Path diagrams may be used to portray various causal mechanisms and, for this reason, a brief discussion of epistemic causality is given to highlight the varied

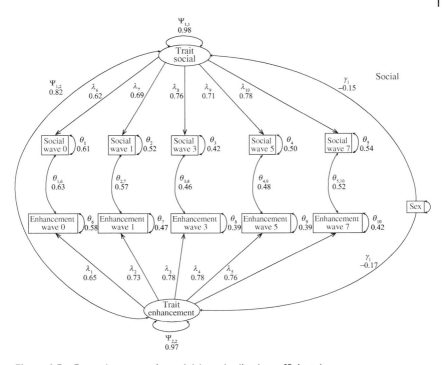

Figure 6.3 Example state trait model (standardized co-efficients).

nature of such relationships. After this, a large example data set is presented as an example for exploration of various types of hypotheses of directional dependence. An initial example analysis of directional dependence involving two observed variables is considered and the influence diagnostics under consideration are introduced. After this, directional dependence models based on a confirmatory factor measurement model, a latent difference score model and a state-trait longitudinal model are considered. Across all models, the degree to which different influential observations are identified across the target and alternate models and the implications of removing such observations from the analysis is evaluated.

6.1 Epistemic Causality

It is also helpful to recall that the type of causality expressed in a given structural equation model can characterize a variety of causal mechanisms including, as one type, the time-delayed causal effects often considered in cross-lagged and Granger models (as discussed, for example by von Eye and Wiedermann (2015)). Such time-bound causal effects are, however, a subset within a larger universe of causal relationships which may act contemporaneously or over the course of time and include transmission rates, conditional probability, measures of propensity or processes such as metabolic rates. The existence of such a variety of causal mechanisms has led some to characterize such structural modeling as an operationalization of "epistemic causality" (Williamson, 2006a,b, 2009) to refer generally to several types of relationships between constructs in addition to the time-bound associations usually considered including transmission rates, conditional probabilities, measures of propensity, or processes such as metabolic rates. Under this view, these types of causality act in conjunction with evidence and background knowledge to constrain one's beliefs about the degree and pattern of relationships between constructs and these systematic constraints enable the general expression of relationships as a causal graph.

The importance of epistemic causality as indicating a variety of possible causal mechanisms is particularly relevant when the various types of research designs are considered in directional dependence. Various types of causal mechanisms may be possible when considering a causal relationship between two manifest variables, for example. It may be that one variable may be a stable inter-individual difference which affects a change in the conditional probability of the other variable, for example. However, it may be that assessment at one point in time represents a transient relationship between two variables not observed at a subsequent measurement occasion. It is, of course, also possible that both types of causal mechanisms may be at play. For example, the relationship between the traits of optimism and resilience may demonstrate a causal mechanism but that levels of optimism which vary from measurement occasion to measurement occasion may also demonstrate a causal relationship specific to a given measurement occasion alone.

6.1.1 Example Data Set

The patterns of directional dependence will be first examined using two subscales of the Cooper's (1994) Motives for Drinking Scale. Subscales for this measure were originally based on an exploratory oblique factor analysis of several likert-format questions dealing with a variety of motivations for drinking. The Social subscale deals with drinking to enjoy or improve social situations and includes items such as "Drinking makes social gatherings more fun," and "It helps you enjoy a party"). The Enhancement subscale, by contrast, contains items which deal with enjoyable feelings associated with alcohol (e.g. "Because it gives you a pleasant feeling," "Because it's fun."). These two subscales have been shown to be particularly salient for young adults and are correlated moderately (0.68–0.78 across studies). There is some reason to believe that these two drinking motives are differentially related to alcohol consumption, with social motives for consumption related to more moderate drinking and enhancement motives more closely tied to heavy consumption (Kuntsche, Knibbe, Gmel, & Engels, 2005) and that social motives influence later reports of enhancement motives (Crutzen, Kuntsche, & Schelleman-Offermans, 2013). In addition, evidence exists that enhancement motives for drinking are more strongly tied to subsequent binge drinking while social motives are not (Lannoy, Dormal, Billieux, & Maurage, 2019). Although longitudinal research suggests that social motives cause enhancement motives, a competing model in which enhancement motives cause social motives may also be reasonable given the normative role of often heavy alcohol consumption for many college students during social events.

The IMPACTS data set for these analysis is taken from a larger longitudinal study of 3720 first time college students at a large Midwestern university who were administered the Motivations for Drinking measure in the summer before entrance into college and again in the Fall for four years. (Further details of the study are given in Sher and Rutledge (2007).) Although the initial sample of students constituted a near complete assessment of all eligible freshmen, over the course of the four years missing data occurred between 69 and 60% of the original sample.

6.2 Assessment of Functional Form: Loess Regression

The simplest directional dependence assessment which can be conducted is based on a direct analysis of the composite scores for the two sub-tests. As an illustration, data from the pre-college assessment sub-tests will be considered. Given that assessments of directional dependence involve the evaluation of two competing regression models, traditional checks of whether the assumptions of the proposed regression models offer a set of concrete tests of such assumptions. These assumptions include assessments of the functional form of the regression models. Directional data packages such as the Direction Dependence Analysis package in R (Wiedermann & Li, 2019) include the option of modeling polynomial regressions. Loess regressions represent another frequently used assessment of functional form in regression which permit exploration of a wide variety of possible functional relationships.

Loess regression, a method of generating a smooth curve through data points was first proposed by Cleveland (1979). In this type of regression, predictor values are first arranged in increasing order. A subset size for the data is chosen (known as a window-size) and a local linear regression is calculated based on the first through window-size observations. The process is then repeated, but based on the second through window-size + 1 observations. The process is then repeated until all observations have been included and the resulting regression line is produced by smoothing the results from the several regressions. Locally weighted regression (Cleveland & Devlin, 1988) is an extension of this model which uses local polynomial regressions rather than linear regressions. Loess regression and locally weighted regression are often referred to interchangeably and the smoothed curves from the two approaches differ little. Given that the computational burden of locally weighted regression is no longer great in modern computing it is often the default choice.

Figure 6.4 shows the Loess regression predicting pre-college Social Motives from pre-college Enhancement Motives. In this model scores were first residualized on sex to create a more homogeneous pattern of scores. As can be seen from the figure, the relationship between the two variables appears to be largely linear, with no marked plateaus or regions of nonlinear association. Given that two competing regression models are under consideration in directional dependence analyses, it is reasonable to estimate Loess regressions for both candidate models. Although

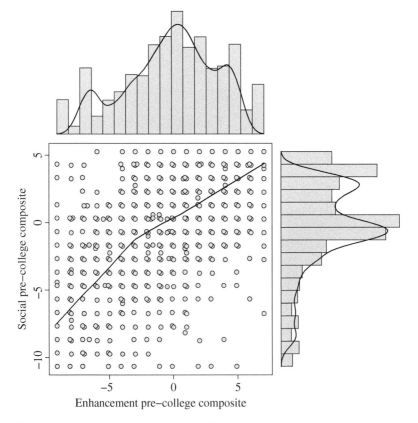

Figure 6.4 Loess regression of social motives on enhancement motives, pre-college assessment only.

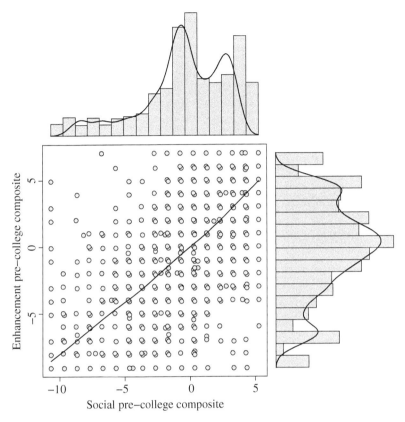

Figure 6.5 Loess regression of enhancement motives on social motives – pre-college assessment.

in small samples, the scatter-plot in the figure can be used to identify a piling up of scores at the top or bottom of the distributions, or regions of departures from linearity, in large data sets, such associations may be obscured. For this reason, it is helpful to also consider the Loess regression of a regression model with the roles of predictor and criterion reversed, as shown in Figure 6.5. As can be seen, in this particular case no significant departure from a general linear relationship between the variables is found. Although Loess regression is a useful tool for examining the functional relationship between a predictor and criterion, the general pattern of association recovered by a Loess regression can be an artifact of influential or outlying observations or some other phenomenon such as differential precision of measurement. Such alternative explanations can seem especially plausible in small samples.

6.3   Influential and Outlying Observations

The finding above, namely that both sub-tests appear to demonstrate significant departures from normality regardless of the candidate model under consideration (for enhancement skewness and kurtosis are −0.36 and 2.37, respectively and, for social skewness and kurtosis are −0.85 and 3.35, respectively), it not unexpected and is consonant with von Eye and DeShon's (2012a,b) observation that empirical data often deviate from normality but that directional dependence should work reasonably well in general. Although the general functional form of the relationship between these two variables appears linear, both of the regressions could be misspecified due to an out-sized effect of a small number of observations. These can take the form of influential observations (i.e. observations which have extreme values on the predictor variable(s) of the model). If not included in estimation, such observations could result in significantly different parameter estimates.

Alternatively, other observations may be outlying, meaning that the residual associated with that observation is significantly larger than would be expected on the basis of chance. Outlying observations, if included in the analysis, can result in an over-estimate of the model's mean square error or could result in significantly different regression parameters as well. Given that the distribution of errors is key in determining direction of effect in directional dependence models, it is also possible that inclusion of such outlying observations would also affect estimates of a model's skewness and kurtosis. Use of regression diagnostics in previous authors have recommended examination of the some of more widely used influence diagnostics to assess whether a proposed claim of directional dependence may be thought of as an artifact of such influential or outlying observations. Pornprasertmanit and Little (2012) noted that tests of directional dependence assume that errors of prediction in the correctly specified regression are linear, with normally distributed errors and no outliers. Although they also mentioned the assumption that no unobserved explanatory variables are thought to exist, Wiedermann and Li (2018) have subsequently extended the examination of directional hypotheses to consider confounding variables as well.

In Section 6.4, a directional dependence analysis of two composite variables will be presented. After this, a description of the specific influence diagnostics used in this present chapter will be presented which is then followed by a discussion of the effects of excluding influential and outlying observations from directional dependence tests.

6.4 Directional Dependence Based on All Available Data

When directional dependence tests are applied to simple composite scores for Social and Enhancement reasons for drinking, more support is found for the hypothesis that social reasons cause enhancement reasons than for the reverse. A summary of the direction dependence tests for the composite score is given in the row labeled "All Available Data" in Table 6.1. Specifically, skewness and co-skewness as well as kurtosis and co-kurtosis measures were used to evaluate distributional properties of variable distributions, Breusch–Pagan tests and the Hilbert–Schmidt Independence Criterion (HSIC) were used to evaluate the independence of predictors and errors (for an overview of decision guidelines to perform direction dependence analysis see the related Chapter 2 by Wiedermann, Li, and von Eye, in this volume). As shown in the table, both composites appear to be negatively skewed, but the Social composite is more negatively skewed than the Enhancement composite. The Social composite appears to be mildly leptokurtic, while the Enhancement subtest appears platykurtic. Four directional dependence tests are shown in the table. The difference in observed skewness appears to favor the target model that Social causes Enhancement (0.50). Although the same test based on kurtosis appears to favor the alternate model that Enhancement causes Social, this must be interpreted with caution as the kurtosis measures are opposite in sign. The higher-order correlation tests based on skewness and kurtosis (that is, tests of co-skewness and co-kurtosis differences) appear to favor the target model as well. Although the Breusch–Pagan and robust Breusch–Pagan tests are significant for both target and alternate models, the magnitude of the Breusch–Pagan test for the alternate model is substantially larger, indicating support for the target model. Similarly, although the Hilbert–Schmidt Independence Criteria are significant for both target and alternate models, the magnitude of the criterion for the target model appears substantially smaller than for the alternate model indicating support for the target model Social → Enhancement.

As stated above, a subsequent question that a reasonable skeptic might raise concerning this conclusion is that the observed effect is due to the presence of influential and/or outlying observations. Assumption checking diagnostics considered by Pornprasertmanit and Little (2012) (and described in more detail below) are often employed in evaluation of linear regression and include (i) studentized deleted residuals, (ii) the lever, and (iii) Cook's D. Because Cook's D may be thought of as a composite of both the lever and studentized deleted residuals, we will not

Table 6.1 Direction analysis of pre-college composite scores (wave 0).

	Predictor variable	Skewness	Kurtosis	BP-test	Robust BP-test	HSIC
All Available Data	Social	−0.85	0.35	**8.24 (p = 0.004)**	**6.05 (p = 0.01)**	**4.97 (p = 0.001)**
	Enhance	−0.36	−0.63	192.20 (**p < 0.0001**)	126.30 (**p < 0.0001**)	13.60
	Difference (95% CI)	**0.50 (0.44–0.56)**	−0.29 (−0.52 to −0.04)			
	Higher-order correlation test (95% CI)	**0.15 (0.11–0.19)**	**1.82 (1.36–2.29)**			
Influential observations excluded based on Social > Enhance	Social	0.07	−1.30	6.27 (p = 0.01)	8.07 (p = 0.004)	1.16 (**p < 0.001**)
	Enhance	0.68	0.51	**2.69 (p = 0.10)**	**5.44 (p = 0.02)**	**0.62 (p = 0.05)**
	Difference (95% CI)	−0.61 (−1.43 to −0.15)	0.79 (−0.06 to 1.56)			
	Higher-order correlation test (95% CI)	−0.28 (−0.85 to 0.01)	0.70 (−0.73 to 4.01)			
Influential observations excluded based on Enhance > Social	Social	0.69	−0.86	**7.92 (p = 0.005)**	5.60 (p = 0.02)	1.15 (**p < 0.001**)
	Enhance	1.03	−0.29	**0.03 (p = 0.86)**	**0.02 (p = 0.89)**	**0.58 (p = 0.05)**
	Difference (95% CI)	**−0.34 (−0.78 to −0.03)**	0.57 (0.11–1.62)			
	Higher-order correlation test (95% CI)	**−0.22 (−0.65 to −0.02)**	1.31 (−1.30 to 13.02)			

consider it jointly with the other influence diagnostics. In this chapter, however, two additional influence diagnostics are also considered which appear to identify different subsets of possibly influential observations: DFBETA and DFFITS. Each of these will be discussed in turn below.

6.4.1 Studentized Deleted Residuals

If it is the case that a single observation is unusually aberrant in terms of a deviation from their predicted criterion values, it makes the most conceptual sense to calculate the residual for observations based on all data except the observation under investigation. Doing so prevents a "slightly off" regression equation from being used on the conforming observations and also prevents those individuals who are members of the group for which the single regression line actually holds from being falsely identified as individuals with significantly deviant criterion values. The studentized deleted residual, usually denoted as t_i, is given as:

$$t_i = \frac{e_i}{\sqrt{\text{MSE}_i(1 - h_{ii})}}$$

where e_i is the of prediction associated with observation i based on a regression which excludes the ith observation, MSE_i is the mean square error based on the regression associated with all but the ith observation, and h_{ii} is the leverage statistic associated with the ith observation.

6.4.2 Lever

Another consideration of the data which could arise from examination of regression data is to consider whether some value(s) on predictor variables may unduly affect the observed regression. The leverage statistic represents one such measure of influence. Conceptually, the leverage statistic may be motivated by thinking of the observed regression weight as an aggregate of several individual "votes" as to what the regression weight should be. Each "vote," however, is not counted equally. Some votes may be very influential in determining the observed regression weight, while other votes may not "count" at all. Influence analysis via the lever, then, seeks to determine if a relatively small number of individuals contain unusual values on the predictor variables which could lead to a calculated regression weight which is untypical for the individuals sampled.

One may think conceptually of the lever as a measure of "votes" for a regression co-efficient. Consider the regular formula for an unstandardized bivariate regression weight:

$$b = \frac{\sigma_{xy}}{\sigma_x^2} = \frac{\sum_{i=1}^{N}(X_i - \overline{X})(Y_i - \overline{Y})}{\sum_{i=1}^{N}(X_i - \overline{X})^2} \tag{6.1}$$

If we multiply both numerator and denominator of this equation by $\sum_{i=1}^{N}(X_i - \overline{X})$ and rearrange the terms of summation a bit, this gives us:

$$b = \sum_{i=1}^{N} \left(\frac{(X_i - \overline{X})^2}{\sum_{i=1}^{N} (X_i - \overline{X})^2} \right) \left(\frac{Y_i - \overline{Y}}{X_i - \overline{X}} \right) \tag{6.2}$$

Notice that this equation describes the overall regression weight in terms of contributions by each individual.

The first term in parentheses is a weight term which determines how much of a "vote" the individual has in determining the overall regression weight based solely on the individual's variation from the average X score. To make this a bit more explicit, some people write these two components to the lever in two parts as follows:

$$b = \sum_{i=1}^{N} h_i \left(\frac{Y_i - \overline{Y}}{X_i - \overline{X}} \right) \tag{6.3}$$

with h_i equal to

$$h_i = \frac{(X_i - \overline{X})^2}{\sum_{i=1}^{N} (X_i - \overline{X})^2} = \frac{(X_i - \overline{X})^2/N}{\sigma_X^2} \tag{6.4}$$

Notice that the fraction in parentheses in Eq. (6.3) is, in some sense, an estimate of an unstandardized regression weight in that it represents the amount of change in the criterion given change in the predictor. The weight that this vote receives, h_i, is a squared z-score representing the distance of each observation from the average predictor value.

6.4.3 DFFITS

DFFITS is a measure of how much influence one particular observation has on the fitted value ($\hat{Y}_i$). One easy way to remember this is to recall that the letters DF stand for the difference between the fitted value of $\hat{Y}_i$ for the ith case when all cases are used in fitting the regression function and the predicted value $\hat{Y}_{i(i)}$ is the predicted value obtained when the ith case is not used in fitting the regression function. The formula is:

$$(\text{DFFITS})_i = \frac{\hat{Y}_i - \hat{Y}_{i(i)}}{\sqrt{\text{MSE}_{(i)} h_{ii}}} \tag{6.5}$$

Notice that the denominator is the estimate of the standard deviation of $\hat{Y}_i$, but that it uses the mean square from the regression in which the ith case is omitted for estimating the variance. The numerator is therefore standardized by this denominator so that it represents the number of standard deviation of $\hat{Y}_i$ that the fitted value increases or decreases when the ith case is included in the regression model. Generally speaking, DFFITS that exceed 1 for small to medium data sets are thought to be influential, and values of $2\sqrt{\frac{p}{n}}$ are used for large data sets (where n is the size of the data set and p is the number of predictors).

6.4.4 DFBETA

DFBETA is a measure of how influential the ith case is on each of the regression co-efficients of a model. In form, it is similar to DFFITS except that DFFITS assesses the influence of an observation on the remaining predicted values, while DFBETA focuses on the influence associated with the regression weights. Analogous to the formula for DFFIT, the formula for DFBETA$_{(i)}$ is the difference between a given regression weight and the regression weight that would be observed if the ith observation is excluded (i.e. (DFBETA)$_{ij} = b_j - b_{j(i)}$ for each of the jth regression weights and for the ith observation). This DFBETA, however, is not scaled, making it difficult to compare across regression weights and across samples. The scaling of DFBETA is accomplished by dividing DFBETA by the estimated standard error of this difference. If we take c_{kk} to be the kth diagonal element of $(X'X)^{-1}$. The formula for DFBETA that results is:

$$\text{DFBETA}_{k(i)} = \frac{b_k - b_{k(i)}}{(\text{MSE}_{(i)}c_{kk})} \quad (\text{for } k = 0, 1, \ldots, p^{-1}) \tag{6.6}$$

6.4.5 Results from Influence Diagnostics

On one hand, the influence diagnostics for individual observations represent a variety of justifications for excluding atypical observations from the analyses which, it is expected, might result in identification of a less ambiguous assessment of directional dependence. Just as importantly, however, it is also possible that a reasonable skeptic could argue that a favored direction dependence conclusion is an artifact of the inclusion of influential observations for an alternative directional hypothesis. Eliminating all observations across all influence criteria risks dramatically reducing the power of the model comparison and more importantly, given that non-normality of the predictor represents the key element in direction dependence tests, excluding all observations risks invalidating the procedure. As such, it seems most reasonable to conduct two additional sets of direction

Table 6.2 Cross-tabulations of influential observations for target and alternate models.

		(Target) (Social → Enhancement)	
		Studentized Deleted Residual	
		Not influential	Influential
(Alternate)	Not influential	2536	8
(Enhancement → Social)	Influential	7	0
	Lever		
		Not influential	Influential
(Alternate)	Not influential	2448	100
(Enhancement → Social)	Influential	0	3
	DFFITS		
		Not influential	Influential
(Alternate)	Not influential	2478	37
(Enhancement → Social)	Influential	36	0
	DFBETA		
		Not influential	Influential
(Alternate)	Not influential	1284	404
(Enhancement → Social)	Influential	391	472
	Across all criteria		
		Not influential	Influential
(Alternate)	Not influential	1227	423
(Enhancement → Social)	Influential	395	506

dependence analyses in which one set eliminates possible influential observations from the target model and another eliminates influential observations from the alternate model.

Table 6.2 shown below shows the base-rates of influential observations for the target and alternate models for the precollege measures of Social and Enhancement Reasons for Drinking. Although the patterns of influential observations may be somewhat instructive for more general use, the patterns seen here are only illustrative and particular to the sample size and measures under consideration. As can be seen from Table 6.2, only a few observations were identified as influential using studentized deleted residuals. The observations so identified, however, were different individuals under the target and alternate models.

When leverage statistics are considered, 100 observations were identified as influential under the target model, however no observations were so identified under the alternate model. For the DFFITS statistic, a relatively modest number of individuals were identified, but as with the studentized deleted residuals, different observations were identified under the two competing models. DFBETA statistics, however, identified several influential observations using the suggested cutoff of 0.04 (876 and 863 individuals under the target and alternate models, respectively). Approximately only half of these observations, however, were identified as influential under both models. The last four rows of Table 6.2 show the pattern of influential observations across all influence diagnostics. A sizable number of potentially influential observations are so identified, largely due to base rates associated with the DFBETA test. Forty-six percent of individuals identified under the target model were not so identified under the alternate model while 44% of observations identified under the alternate model were not identified as influential under the target model.

Direction dependence analyses of the composite measures were then recalculated, deleting influential observations from the target model and then again deleting observations from the alternate model. The results of these analyses are shown in the bottom two thirds of Table 6.1. As can be seen from the table, under both scenarios the observed skew and kurtosis of the Social and Enhancement composites are quite different when influential observations are removed from the data. Both variables are positively, rather than negatively skewed. The Social composite is now platykurtic, rather than leptokurtic under both influence analyses and the Enhance composite is now leptokurtic of influential observations are removed based on the target model and substantially less platykurtic when observations are removed based on the alternate model.

These distributional changes appear to have a substantial effect on the estimation of directional dependence, with the results of most of the direction dependence tests now favoring the alternate model or producing non-significant results. Taken together, even though the general functional relationship between the two scales appears linear for both candidate regression models, analysis of the pre-college Reasons for Drinking scales suggest the analysis of all available data supporting the target model of Social Reasons for Drinking causing Enhancement Reasons could be an artifact of the presence of influential observations in the data. Although examination of this analysis of simple composite scores highlights the fact that influence diagnostics may temper an initial conclusion regarding direction of effect it does not take into account the possibilities that the observed distributions of scores may be unduly affected by the presence of measurement error, that measurement errors may be correlated for some items or the possibility that some items may actually be related to both constructs. In order to address these possibilities, the discussion now turns to an examination of direction dependence based on factor scores.

6.4.6 Directional Dependence Based on Factor Scores

Variation in residual scores in a particular regression model may be an artifact of some psychometric property of the dependent variable. As noted above, for this reason, von Eye and Wiedermann (2014) advised that researchers consider factor scores as an alternative to raw score composites in the candidate regression models. More broadly, within the context of longitudinal research, however, the use of factor scores can be extended to examine associations at a trait level across all measurement occasions but can also be additionally used to examine whether a differential pattern of causal effect is observed within each measurement occasion after adjusting the scores for the general trait. Factor scores based on an explicit measurement model allow the estimation of factor scores in the presence of more complex measurement models. The structural measurement model shown in Figure 6.1 represents a confirmatory factor model for the Enhancement and Social items of drinking motivations. Two small changes were made to a simple measurement model in which all items loaded on only their respective factors. The Enhancement item labeled "R" ("I drink because it's fun,") appears to have a small but significant standardized cross-loading ($r = 0.23$) on the Social factor. In addition, two of the Social items ("I drink to be sociable" and "I drink because it helps me enjoy a party") appeared to have small, but significant correlated residuals (0.29). The resulting confirmatory factor model appeared to fit well (CFI = 0.98; TLI = 0.97, $\chi^2_{40} = 424.84$; $p < 0.001$).

As with the analysis of composite variables, when directional dependence tests are conducted based on the factor scores from this model, it appeared to support the view that Social reasons for drinking cause Enhancement reasons for drinking as shown in the top third of Table 6.4. When influential observations are calculated based on the target and alternate models using factor scores obtained from the confirmatory factor model, a slightly smaller number of observations are identified as potentially influential, as shown in Table 6.3, but roughly equal numbers of influential observations are identified as influential under only one of the regression models. The general patterns of influential observations calculated separately for each influence diagnostic was roughly similar to that observed for the composite variables described above.

When the directional dependence analyses are rerun excluding influential observations the pattern of directional dependence continues to support the target model if influential observations are excluded based on the target model, with the Breusch–Pagan tests now showing a non-significant effect for the target model and the higher-order correlation tests also favoring the target model (Table 6.4).

Table 6.3 Frequencies of influential observations under target and alternate model based on factor scores.

		Target (Social → Enhancement)	
		Not influential	Influential
Alternate	Not influential	1307	897
(Enhancement → Social)	Influential	725	415

In contrast to the composite analysis, removal of influential observations based on the target model reduced but did not reverse the sign of the skewness of the Social factor scores. No significant directional effects were found when influential observations based on the alternate model except for the Breusch–Pagan tests, which now support the alternate model.

Taken together, although analyses based on the factor scores support the target model when based on all available data, the effect of removing influential observations on skew and kurtosis appears to have a substantial effect on the observed directional dependence. It appears that this effect is somewhat reduced when based on factor scores instead of simple composites. Thus, although analysis of all the available data appears to support the target model, a reasonable skeptic could still argue that this conclusion may be due to the inclusion of influential observations in the alternate model and that, had these been excluded or adjusted, the opposite conclusion would have been reached. Alternatively, however, a researcher who favors the target model could conclude that the statistical significance of the HSIC and Breusch–Pagan tests for the target model is due to the presence of influential observations identified as such under the target model.

6.5 Directional Dependence Based on Latent Difference Scores

One reason that the Enhancement and Social scales appear non-normally distributed is because of a failure to correctly adjust all inter-individual differences in the scores. Although use of sex as a covariate does reduce the non-normality of the distributions, other unincluded covariates exist (such as socio-economic status, school or community size, or access to alcohol outlets to name just a few) which may cause both observed distributions to depart from normality. Given that such inter-individual differences are likely legion, longitudinal data provides one

Table 6.4 Direction dependence analysis of factor scores.

	Predictor variable	Skewness	Kurtosis	BP-test	Robust BP-test	HSIC
All available data	Social	−0.58	0.15 ($p = 0.08$)	**8.67 ($p = 0.003$)**	**5.80 ($p = 0.016$)**	**4.07 ($p < 0.001$)**
	Enhance	−0.33	−0.40	148.05 ($p < 0.0001$)	99.54 ($p < 0.0001$)	8.09 ($p < 0.0001$)
	Difference (95% CI)	**0.25 (0.20–0.30)**	−0.25 (−0.45–0.00)			
	Higher-order correlation test (95% CI)	**0.07 (0.05–0.09)**	**1.41 (0.98–1.80)**			
Influential observations excluded based on Social → Enhancement	Social	−0.37	0.27	**1.62 ($p = 0.20$)**	**0.83 ($p = 0.36$)**	**3.74 ($p < 0.001$)**
	Enhance	−0.37	0.17	72.77 ($p < 0.0001$)	52.61 ($p < 0.0001$)	8.89 ($p < 0.0001$)
	Difference (95% CI)	0.00 (−0.08–0.08)	0.10 (−0.07–0.26)			
	Higher-order correlation test (95% CI)	**0.03 (0.01–0.05)**	**1.08 (0.48–1.71)**			
Influential observations excluded based on Enhancement → Social	Social	−0.08	−0.17	39.20 ($p < 0.0001$)	21.41 ($p < 0.0001$)	5.75 ($p < 0.001$)
	Enhance	−0.15	−0.22	**0.00 ($p = 0.98$)**	**0.98 ($p = 0.99$)**	6.57 ($p < 0.001$)
	Difference (95% CI)	−0.06 (−0.14–0.01)	−0.05 (−0.27–0.23)			
	Higher-order correlation test (95% CI)	−0.01 (−0.02–0.00)	0.52 (−0.15–1.07)			

way of adjusting scores. Use of a latent difference factor model, in which scores at a subsequent occasion are adjusted for the effects of a prior occasion can accomplish just such a general adjustment to the data. As pointed out earlier, a latent difference score model in which growth scores at the level of the latent variable factor scores are calculated may be thought of as a general covariate adjustment which takes into account all such interindividual differences which are stable across the two measurement occasion. To explore this, a latent difference model using the measurement model shown in Figure 6.1 is applied to the first two waves of assessment, the resulting structural model shown in Figure 6.2 is produced. The fit of the resulting model appears acceptable (CFI = 0.97; TLI = 0.94, χ^2_{210} = 1410.08; $p < 0.001$; RMSEA = 0.05; 95% CI (0.05–0.05)).

Although the estimates in Figure 6.2 are unstandardized in order to illustrate the time-invariant nature of the solution, it is worth noting that the correlation between the Social and Enhance latent difference scores is 0.46, a substantially different value than the 0.75 correlation observed between the factor measurement models based on only the pre-college assessment shown in Figure 6.1. This suggests that stable inter-individual differences may exist between these two constructs (Table 6.5).

As with the prior two analyses based on all available data, results appear to favor the target model. Specifically in this model, the skewness difference tests, the higher-order correlation based on skewness, the Breusch–Pagan and robust Breusch–Pagan all support the target model. When influential observations are calculated across both the target and alternative models a substantially larger number of observations are identified as influential across both regression models and a small number of influential observations are identified overall as being influential as shown in Table 6.6.

When the directional dependence analyses are re-conducted eliminating influential observations based on the target model, results from skewness and kurtosis difference, kurtosis higher-order correlation, BP-test and robust BP-test all favor the alternate model of Enhancement causing Social reasons for drinking. The HSIC, however, continues to support the target model. This effect is likely due to the fact that the observed skewness of the Social variable is now positively in the reduced data set, rather than negatively skewed as it was in the complete data set. When influential observations based on the alternate model are removed from the data, the results continue to favor the target model of Social causing Enhancement, with the skewness difference, kurtosis difference, BP-test, and robust BP-test all favoring the target model results of the HSIC test favor the alternate model, however.

Taken together, the results from the latent difference scores suggest a stronger case for the target model in that the complete data provide more support for the target model and, if influential observations are identified based on the alternate

Table 6.5 Directional analysis of latent difference.

	Predictor variable	Skewness	Kurtosis	BP-test	Robust BP-test	HSIC
	Social	−0.64	2.65	24.56 (p < 0.0001)	12.20 (p = 0.0004)	19.39 (p < 0.001)
	Enhance	−0.23	2.16	85.05 (p < 0.0001)	43.88 (p < 0.0001)	19.29 (p < 0.001)
All Available Data	Difference (95% CI)	0.40 (0.22–0.59)	0.49 (−0.02–1.01)			
	Higher-order correlation test (95% CI)	0.06 (0.03–0.12)	0.35 (−1.59–2.11)			
	Social	0.07	1.89	29.38 (p = 0.0001)	11.57 (p = 0.0006)	10.63 (p < 0.001)
	Enhance	−0.78	2.51	19.80 (p < 0.0001)	11.16 (p < 0.0008)	17.91 (p < 0.001)
Influential observations excluded based on Social → Enhancement	Difference (95% CI)	−0.71 (−0.97 to −0.47)	−0.62 (−1.44 to −0.03)			
	Higher-order correlation test (95% CI)	−0.02 (−0.07–0.02)	1.59 (0.58–2.62)			
	Social	−1.12	7.41	7.12 (p = 0.01)	3.95 (p = 0.05)	24.18 (p < 0.001)
	Enhance	0.06	1.86	32.82 (p < 0.0001)	6.41 (p = 0.01)	21.27 (p < 0.001)
Influential observations excluded based on Enhance-ment → Social	Difference (95% CI)	1.06 (0.61–1.78)	5.55 (3.33–10.28)			
	Higher-order correlation test (95% CI)	0.03 (−0.04 to 0.17)	−1.22 (−2.29 to 0.44)			

Table 6.6 Frequencies of influential observations under target and alternate model based on latent difference factor scores.

		Target (Social → Enhancement)	
		Not influential	Influential
Alternate	Not influential	1452	250
(Enhancement → Social)	Influential	258	957

model, a clearer pattern of support emerges for the target model. The apparent support for the alternate model which emerges when influence is defined based on the target model could be viewed as an artifact of the drastic change in skewness in the reduced data set which results.

6.6 Direction Dependence Based on State-Trait Models

State trait models may be thought of as a factor measurement model in which the individual measurement occasions represent the items of interest. The IMPACTS data set, for example, assessed reasons for drinking in a pre-college assessment and in the fall semester of the four subsequent years. As such, the general factor across all measurement occasions may be thought of as represent stable inter-individual differences in reasons for drinking. In addition, the residual scores specific to each measurement occasion may be thought of as state-level assessments of the constructs adjusted for these general inter-individual differences. When the structural model shown in Figure 6.3 is fit to the IMPACTS data, the resulting model appears to fit well (CFI = 0.98; TLI = 0.97, χ^2_{37} = 291.73; $p < 0.001$; RMSEA = 0.04; 95% CI (0.04–0.05)). The resulting estimated between the two constructs is quite high even after adjusting the scores for the effects of sex (0.82). Correlations between state-level associations within wave are also substantial, although more modest, ranging from 0.46 to 0.63.

The patterns of influential observations under the target and alternate model.

When directional dependence is examined for the trait factor score, the data appear to favor the target model of Social causing Enhancement reasons for drinking based on the skewness difference, higher-order correlations based on co-skewness and co-kurtosis, Breusch–Pagan test, and robust Breusch–Pagan tests as shown in the top third of Table 6.7.

When influential observations are calculated under the target and alternate models, the pattern of influential observations appears more similar to the composite or confirmatory factor model of Figure 6.1 with large numbers of

Table 6.7 Direction dependence of latent trait factor scores.

Predictor variable		Skewness	Kurtosis	BP-test	Robust BP-test	HSIC
	Social	−0.58	0.15	**8.67 (p = 0.003)**	**5.80 (p = 0.016)**	**4.06 (p < 0.001)**
	Enhance	−0.33	−0.40	148.05 (p < 0.0001)	99.54 (p < 0.0001)	8.05 (p < 0.001)
All Available Data	Difference (95% CI)	**0.25 (0.20–0.30)**	−0.25 (−0.45–0.00)			
	Higher-order correlation test (95% CI)	**0.07 (0.05–0.09)**	**1.41 (0.99–1.80)**			
	Social	−0.37	0.27	**1.62 (p = 0.20)**	**0.82 (p = 0.36)**	**3.69 (p < 0.001)**
	Enhance	−0.37	0.17	72.77 (p < 0.0001)	52.61 (p < 0.0001)	8.98 (p < 0.001)
Influential observations excluded based on Social → Enhancement	Difference (95% CI)	0.00 (−0.08 to 0.08)	0.10 (−0.07 to 0.26)			
	Higher-order correlation test (95% CI)	**0.03 (0.01–0.06)**	**1.08 (0.47–1.70)**			
	Social	−0.08	0.17	39.20 (p < 0.0001)	21.41 (p = 0.0001)	**5.75 (p < 0.001)**
	Enhance	−0.15	−0.22	**0.004 (p = 0.98)**	**0.002 (p = 0.99)**	6.64 (p < 0.001)
Influential observations excluded based on Enhancement → Social	Difference (95% CI)	−0.06 (−0.13 to 0.01)	−0.05 (−0.27 to 0.23)			
	Higher-order correlation test (95% CI)	−0.01 (−0.02 to 0.00)	0.52 (−0.13 to 1.07)			

Table 6.8 Frequencies of influential observations under target and alternate model for trait factor score of state-strait model.

		Target (Social → Enhancement)	
		Not influential	Influential
Alternate	Not influential	1307	897
(Enhancement → Social)	Influential	725	415

individuals identified as influential under one, but not the other regression model as shown in Table 6.8. When directional dependence influential observations are removed based on the target model, results continue to favor the target model based on the higher-order correlations based on co-skewness and co-kurtosis. Of note, the Breusch–Pagan and robust Breusch–Pagan tests show a non-significant χ^2 for the target model and highly significant effects for the alternate model when outliers are removed based on the target model. The HSIC as well also appears to favor target model. The difference in observed skewness and kurtosis measures were no longer significant, again, possibly due to the dramatic changes in these statistics when influential observations were removed from the data. When observations are removed from the data based on the alternate model, the Breusch–Pagan and robust Breusch–Pagan tests now favor the alternate model, however, with the remaining tests no longer showing differences between the target and alternate models. This pattern is again likely due to the fact that the observed skew and kurtosis of the reduced data set changes quite dramatically from the complete data set.

Direction dependence tests were then calculated for the residual scores associated with the individual waves of the study. These are shown in Tables 6.10–6.14. In contrast to the analyses removing influential observations from the trait model, removal of influential observations based on residual scores appears to result in only minor changes to the observed skewness and kurtosis values and the analyses appear to support the target model that social reasons for drinking cause enhancement reasons for drinking. Removal of influential observations based on the target model result in slight enhancement of the pattern of results favoring the target model except for Wave 7, where removal of influential observations based on the alternate model results in a pattern of results more in favor of the target model (Table 6.9).

Table 6.9 Frequencies of influential observations under target and alternate model based on for state trait model factor and error scores.

		(Target) (Social → Enhancement)	
		Trait	
		Not influential	Influential
(Alternate)	Not influential	1458	853
(Enhancement → Social)	Influential	682	351
		Wave 0	
		Not influential	Influential
(Alternate) (Enhancement → Social)	Not influential	1288	421
	Influential	331	508
		Wave 1	
		Not influential	Influential
(Alternate) (Enhancement → Social)	Not influential	1151	315
	Influential	195	493
		Wave 3	
		Not influential	Influential
(Alternate) (Enhancement → Social)	Not influential	553	335
	Influential	69	505
		Wave 5	
		Not influential	Influential
(Alternate) (Enhancement → Social)	Not influential	1017	243
	Influential	151	545
		Wave 7	
		Not influential	Influential
(Alternate) (Enhancement → Social)	Not influential	1097	278
	Influential	168	532

6.7 Discussion

This chapter has shown, at least in this real-world example that identification of influential observations depends considerably on whether the target or alternate model is considered. There is some evidence that the patterns of directional effects are somewhat more robust when factor scores are employed rather than simple

Table 6.10 Directional analysis of state residual scores for pre-college (wave 0).

	Predictor variable	Skewness	Kurtosis	BP-test	Robust BP-test	HSIC
	Social	−0.85	1.07	**5.07 ($p = 0.02$)**	**3.92 ($p = 0.05$)**	**1.47 ($p < 0.001$)**
	Enhance	−0.38	0.16	181.60 (**$p < 0.0001$**)	120.40 (**$p < 0.0001$**)	6.90
All Available Data	Difference (95% CI)	**0.47 (0.37–0.58)**	**0.91 (0.64–1.22)**			
	Higher-order correlation test (95% CI)	**0.14 (0.09–0.20)**	**1.90 (1.13–2.90)**			
	Social	−0.91	1.12	**5.50 ($p = 0.02$)**	**4.43 ($p = 0.04$)**	**1.34 ($p < 0.001$)**
	Enhance	−0.41	0.17	125.01 (**$p < 0.0001$**)	80.07 (**$p < 0.0001$**)	5.49 (**$p < 0.001$**)
Influential observations excluded based on Social → Enhancement	Difference (95% CI)	**0.50 (0.37–0.62)**	**0.95 (0.61–1.34)**			
	Higher-order correlation test (95% CI)	**0.16 (0.10–0.23)**	**1.84 (0.82–3.12)**			
	Social	−0.86	1.10	**3.07 ($p < 0.08$)**	**2.64 ($p = 0.10$)**	**1.15 ($p < 0.001$)**
	Enhance	−0.39	0.17	132.16 (**$p < 0.0001$**)	94.96 (**$p < 0.0001$**)	5.56 (**$p < 0.001$**)
Influential observations excluded based on Enhancement → Social	Difference (95% CI)	**0.48 (0.35–0.61)**	**0.93 (0.60–1.35)**			
	Higher-order correlation test (95% CI)	**0.15 (0.10–0.22)**	**2.02 (1.03–3.31)**			

Table 6.11 Directional analysis of residual scores for wave 1.

	Predictor variable	Skewness	Kurtosis	BP-test	Robust BP-test	HSIC
	Social	−0.68	1.67	9.62 (p = 0.0002)	7.03 (p = 0.0008)	1.98 (p < 0.001)
	Enhance	−0.21	0.89	54.64 (p < 0.0001)	26.85 (p < 0.0001)	3.58 (p < 0.001)
All Available Data	Difference (95% CI)	0.47 (0.30–0.66)	0.78 (0.30–1.36)			
	Higher-order correlation test (95% CI)	0.05 (0.02–0.11)	0.18 (−0.88 to 1.33)			
	Social	−0.71	1.66	6.90 (p = 0.01)	5.31 (p = 0.02)	1.33 (p < 0.001)
	Enhance	−0.19	0.83	35.27 (p < 0.0001)	17.10 (p < 0.0001)	2.73 (p < 0.001)
Influential observations excluded based on Social → Enhancement	Difference (95% CI)	0.52 (0.30–0.78)	0.83 (0.18–1.72)			
	Higher-order correlation test (95% CI)	0.05 (0.01–0.12)	0.07 (−1.18 to 1.48)			
	Social	−0.61	1.62	10.02 (p = 0.0002)	7.29 (p=0.0069)	1.59 (p<0.001)
	Enhance	−0.17	0.91	21.49 (p < 0.0001)	9.84 (p < 0.0002)	2.10 (p < 0.001)
Influential observations excluded based on Enhancement → Social	Difference (95% CI)	0.43 (0.22–0.67)	0.72 (0.13–1.50)			
	Higher-order correlation test (95% CI)	0.03 (0.00–0.08)	−0.43 (−1.59 to 0.62)			

Table 6.12 Directional analysis of residual scores for wave 3.

	Predictor variable	Skewness	Kurtosis	BP-test	Robust BP-test	HSIC
All Available Data	Social	−1.24	5.29	**14.35 (p = 0.0002)**	**9.65 (p = 0.002)**	**1.51 (p < 0.001)**
	Enhance	−0.22	1.03	**30.53 (p < 0.0002)**	**6.85 (p = 0.009)**	**2.71 (p < 0.001)**
	Difference (95% CI)	**1.02 (0.59–1.68)**	**4.26 (1.62–8.63)**			
	Higher-order correlation test (95% CI)	0.04 (−0.03 to 0.15)	−0.63 (−4.07 to 2.41)			
Influential observations excluded based on Social → Enhancement	Social	−1.16	5.16	6.75 (p = 0.009)	4.38 (p = 0.0036)	0.79 (p < 0.001)
	Enhance	0.00	0.95	**0.76 (p = 0.38)**	**0.14 (p = 0.71)**	**1.22 (p < 0.001)**
	Difference (95% CI)	**1.16 (0.56–2.41)**	**4.20 (0.02–12.14)**			
	Higher-order correlation test (95% CI)	**0.00 (−0.07 to 0.09)**	−2.72 (−9.56 to −1.08)			
Influential observations excluded based on Enhancement → Social	Social	−1.14	5.01	**10.50 (p = 0.001)**	7.03 (p = 0.008)	0.83 (p < 0.001)
	Enhance	−0.06	0.74	**3.03 (p = 0.08)**	**0.61 (p < 0.43)**	1.64 (p < 0.001)
	Difference (95% CI)	**1.08 (0.60–2.31)**	**4.27 (1.01–11.62)**			
	Higher-order correlation test (95% CI)	0.01 (−0.05 to 0.09)	−1.86 (−5.76 to 0.04)			

Table 6.13 Directional analysis of residual scores for wave 5.

	Predictor variable	Skewness	Kurtosis	BP-test	Robust BP-test	HSIC
	Social	−0.83	1.90	0.84 (p = 0.36)	0.57 (p = 0.45)	2.03 (p < 0.001)
	Enhance	−0.25	0.96	126.26 (p < 0.0001)	58.64 (p < 0.0001)	4.97 (p < 0.001)
All Available Data	Difference (95% CI)	0.58 (0.37–0.81)	0.94 (0.21–1.69)			
	Higher-order correlation test (95% CI)	0.11 (0.06–0.19)	0.07 (−1.54 to 1.10)			
	Social	−0.89	2.00	1.11 (p = 0.29)	0.72 (p = 0.39)	1.54 (p < 0.001)
	Enhance	−0.19	0.75	74.32 (p < 0.0001)	2.90 (p < 0.0089)	2.90 (p < 0.001)
Influential observations excluded based on Social → Enhancement	Difference (95% CI)	0.70 (0.44–1.00)	1.26 (0.41–2.25)			
	Higher-order correlation test (95% CI)	0.10 (0.03–0.21)	0.05 (−1.19 to 1.23)			
	Social	−0.84	2.00	2.34 (p = 0.13)	1.54 (p = 0.22)	1.27 (p < 0.001)
	Enhance	−0.21	0.87	94.55 (p < 0.0001)	46.55 (p < 0.0001)	2.77 (p < 0.001)
Influential observations excluded based on Enhancement → Social	Difference (95% CI)	0.63 (0.37–0.91)	1.12 (0.33–1.89)			
	Higher-order correlation test (95% CI)	0.14 (0.07–0.27)	1.00 (0.05–2.40)			

Table 6.14 Directional analysis of residual scores for wave 7.

	Predictor variable	Skewness	Kurtosis	BP-test	Robust BP-test	HSIC
	Social	−0.87	1.82	3.08 ($p = 0.08$)	2.06 ($p = 0.15$)	1.83 ($p < 0.001$)
	Enhance	−0.21	0.60	94.67 ($p < 0.0001$)	44.85 ($p < 0.0001$)	5.69 ($p < 0.001$)
All Available Data	Difference (CI)	0.67 (0.50–0.85)	1.21 (0.74–1.86)			
	Higher-order correlation test (95% CI)	0.10 (0.04–0.18)	1.23 (0.26–2.36)			
	Social	−0.94	1.96	4.88 ($p = 0.03$)	3.18 ($p = 0.07$)	0.86 ($p < 0.001$)
	Enhance	−0.23	0.75	54.54 ($p < 0.0001$)	25.41 ($p < 0.0001$)	3.19 ($p < 0.001$)
Influential observations excluded based on Social → Enhancement	Difference (95% CI)	0.71 (0.50–0.97)	1.21 (0.58–2.14)			
	Higher-order correlation test (95% CI)	0.11 (0.01–0.22)	1.19 (0.03–2.73)			
	Social	−0.91	2.04	1.86 ($p = 0.17$)	1.51 ($p = 0.22$)	1.17 ($p < 0.001$)
	Enhance	−0.31	0.67	80.91 ($p < 0.0001$)	39.44 ($p < 0.0001$)	3.69 ($p < 0.001$)
Influential observations excluded based on Enhancement → Social	Difference (95% CI)	0.60 (0.40–0.84)	1.36 (0.76–2.26)			
	Higher-order correlation test (95% CI)	0.11 (0.05–0.21)	0.91 (−0.24 to 2.43)			

composite scores (as seen by comparing Table 6.1 with Table 6.4, but that for the factor score, latent difference models as well as the trait factor score of the latent state-trait model, observations which were identified as influential significantly influenced the directional dependence analyses, depending on whether they were identified based on the target or alternative model (as seen from the results of Tables 6.4–6.7). Even though these observations do not appear to cause a nonlinear regression association based on a Loess regression, it appears that the association between regression assumptions and model choice is rather tightly bound in those situations where removal of influential observations results in a change in the observed skew and kurtosis of the reduced data set. The exception to this pattern appeared to be examination of directional associations between data gathered at an individual measurement occasion adjusted for trait-level effects (i.e. the analyses presented in Tables 6.10–6.14) in which removal of influential observations did not appear to appreciably affect the observed distributions of the variables under consideration.

Taken together, the analyses of the IMPACTS data set suggest that the evidence appears stronger for the target model of Social reasons for drinking causing Enhancement than for the converse but that the inter-individual differences in drinking motives as assessed in analyses based on composite variables, factor scores, latent difference scores or trait factor scores may be somewhat open to the counterargument that the simple linear regression model based on all available data does not appropriately apply. That said, however, examination of directional dependence across each individual measurement occasion, adjusted for inter-individual trait level effects appears to support the target model regardless of which observations are identified as influential.

It should also be noted that removal of influential observations consonant with a target model did not uniformly result in spurious improvement of directional hypotheses favoring the target model. Rather, the effect of removal of influential observations appears most influential when it changes the skewness and/or kurtosis of the variables under consideration. While this leaves the analyst of the original data in a somewhat difficult situation, alternatives to the simple removal of influential observations exist, which include winsorizing or a formal metric scaling analysis of the traits under consideration as options. More broadly, however, researchers may consider analysis of other types of experiments which would more precisely account for inter-individual differences in intra-individual variation. Such data sources would include longitudinally intensive data sets, for example, or experimental manipulations designed to affect one construct more than another. For example, interventions designed to educate students about social norms of alcohol consumption could concentrate on the fact that alcohol use does not enhance social situations or, alternatively, is not an objective enhancer of a given situation. Observations of the observed distributions of scores which result could more effectively discriminate between the target and alternative models.

More broadly, however, a reasonable skeptic could offer additional criticisms of any given claim of directional dependence based on a given data set. For example, a skeptic could claim that on observed distribution of scores could, in fact, represent a mixture of several latent subgroups which thereby produces a spurious pattern of skew and kurtosis. Although an intriguing possibility, such a criticism raises the rather familiar statistical issue that any observed distribution can be composed of mixtures of normally distributed variables (McLachlan & Peel, 2000). Although the question of whether qualitatively different subgroups of individuals exist in which the directionality of effect is different is intriguing, identification of such subgroups involves assumptions (such as conditional normality) which invalidate the comparison of the target and alternative models. If such a counterargument is advanced, however, such a counterargument could be addressed by considering longitudinal data which are based in which such inter-individual differences are removed from the data.

In sum, the identification of influential observations and functional form of the target and alternate models is an important adjunct to the interpretation of directional dependence tests, such examination is not a panacea for the resolution of directional effect, such analysis does help to identify atypical individuals who may spuriously produce a directional effect or may, alternatively, identify measurement models or data which may more accurately resolve a measurement model upon which a claim of causal direction may be made.

References

Cleveland, W. S. (1979). Robust locally weighted regression and smoothing scatterplots. *Journal of the American Statistical Association*, *74*(368), 829–836.

Cleveland, W. S., & Devlin, S. J. (1988). Locally weighted regression: An approach to regression analysis by local fitting. *Journal of the American Statistical Association*, *83*(403), 596–610.

Cooper, M. L. (1994). Motivations for alcohol use among adolescents: Development and validation of a four-factor model. *Psychological Assessment*, *6*, 117–128.

Crutzen, R., Kuntsche, E., & Schelleman-Offermans, K. (2013). Drinking motives and drinking behavior over time: A full cross-lagged panel study among adults. *Psychology of Addictive Behaviors*, *27*(1), 197–201.

Dodge, Y., & Rousson, V. (2000). Direction dependence in a regression line. *Communications in Statistics—Theory and Methods*, *29*, 9–10.

Dodge, Y., & Rousson, V. (2001). On asymmetric properties of the correlation coefficient in the regression setting. *The American Statistician*, *55*, 51–54.

Hamaker, E. L., Kuiper, R. M., & Grasman, R. P. P. P. (2015). A critique of the cross-lagged panel model. *Psychological Methods*, *20*(1), 102–116. doi:10.1037/a0038889

Kuntsche, E., Knibbe, R., Gmel, G., & Engels, R. (2005). Why do young people drink? A review of drinking motives. *Clinical Psychology Review*, *25*, 841–861. doi:10.1016/j.cpr.2005.06.002

Lannoy, L., Dormal, V., Billieux, J., & Maurage, P. (2019). Enhancement motivation to drink predicts binge drinking in adolescence: A longitudinal study in a community sample. *The American Journal of Drug and Alcohol Abuse*, *45*(3), 304–312. doi:10.1080/ 00952990.2018.1550089

McLachlan, G., & Peel, D. (2000). *Finite mixture models*. New York, NY: Wiley.

Pornprasertmanit, S., & Little, T. D. (2012). Determining directional dependency in causal associations. *International Journal of Behavioral Development*, *36*(4), 313–322. doi:10.1177/0165025412448944

Schmitt, M. J., & Steyer, R. (1993). A latent state-trait model (not only) for social desirability. *Personality and Individual Differences*, *14*(4), 519–529. doi:10.1016/0191-8869(93)90144-R

Sher, K. J., & Rutledge, P. (2007). Heavy drinking across the transition to college: Predicting first-semester heavy drinking from precollege variables. *Addictive Behaviors*, *32*(4), 819–835. doi:10.1016/j.addbeh.2006.06.024

Steyer, R., Mayer, A., Geiser, C., & Cole, D. A. (2015). A theory of states and traits—Revised. *Annual Review of Clinical Psychology*, *11*, 71–98. doi:10.1146/annurev-clinpsy-032813-153719

Steyer, R., Schwenkmezger, P., & Auer, A. (1990). The emotional and cognitive components of trait anxiety: A latent state-trait model. *Personality and Individual Differences*, *11*(2), 125–134. doi:10.1016/0191-8869(90)90004-B

Thoemmes, F. (2015). Empirical evaluation of directional dependence tests. *International Journal of Behavioral Development*, *39*, 560–569. doi:10.1177/0165025415582055

von Eye, A., & DeShon, R. P. (2012a). Directional dependence in developmental research. *International Journal of Behavioral Development*, *36*(4), 303–312.

von Eye, A., & DeShon, R. P. (2012b). Decisions concerning directional dependence. *International Journal of Behavioral Development*, *36*(4), 323–326.

von Eye, A., & Wiedermann, W. (2014). On direction of dependence in latent variable contexts. *Educational and Psychological Measurement*, *74*(1), 5–30. doi:10.1177/0013164413505863

von Eye, A., & Wiedermann, W. (2015). Manifest variable Granger causality models for developmental research: A taxonomy. *Applied Developmental Science*, *19*(4), 183–195. doi:10.1080/ 10888691.2014.1001512

Wiedermann, W., & Li, X. (2018). Direction dependence analysis: A framework to test the direction of effects in linear models with an implementation in SPSS. *Behavior Research Methods*, *50*, 1581–1601. doi:10.3758/s13428-018-1031-x

Wiedermann, W., & Li, X. (2019). *Direction dependence analysis in R*. Retrieved from http://www.r-project.com.

Wiedermann, W., Merkle, E. C., & von Eye, A. (2018). Direction of dependence in measurement error models. *British Journal of Mathematical and Statistical Psychology*, *71*, 117–145.

Williamson, J. (2006a). Causal pluralism versus epistemic causality. *Philosophica*, *77*, 69–96.

Williamson, J. (2006b). Dispositional versus epistemic causality. *Minds and Machines*, *16*, 259–276.

Williamson, J. (2009). Probabilistic theories. In H. Beebee, C. Hitchcock, & P. Menzies (Eds.), *The Oxford handbook of causation* (pp. 185–212). Oxford, UK: Oxford University Press.

7

Complete Dependence

A Survey

Santi Tasena

Department of Mathematics, Faculty of Science, Chiang Mai University, Chiang Mai, Thailand

The concept of complete dependence can probably be traced back to Rényi (1959). In 1959, Rényi defined a set of axioms for measuring association between (real-valued) random variables and one of these axioms is that the measure must reach the maximum value if one random variable is a (measurable) function of another, viz., one random variable is *completely dependent* on another. In a sense, a random variable Y is completely dependent on a random variable X means that we can always (theoretically) predict Y from X since Y itself is a function of X. In this case, we would have $Y = \mathbb{E}[Y \mid X]$ almost surely and hence recover the form of the function. Due to this reason, complete dependence is also called *functional dependence* and *regression dependence* by several others. Naturally, the concept of complete dependence can be extended to other types of random variables such as random vectors and random sets.

It should also be mentioned that complete dependence is not a symmetric relation. The simplest example would be the case $Y = X^2$ where X is a random variable symmetric around zero, for example, a standard normal random variable. In this case, each possible value of Y is associated with two values of X which implies X can never be a function of Y. Furthermore, it is also possible to construct a random variable Y that is a function of another random variable X in such a way that their joint distribution is closed to be independent (Boonmee & Tasena, 2016; Siburg & Stoimenov, 2015). See also Example 7.1. Thus, we may view complete dependence as a form of directional dependence.[1]

In general, a random variable Y might neither be completely dependent nor independent of X. For example, when $Y = X + Z$ where X and Z are independent.

---

[1] It is also possible to consider different notions for directional dependence, see for example, the work of Schweizer and Wolff (1981).

Direction Dependence in Statistical Modeling: Methods of Analysis, First Edition.
Edited by Wolfgang Wiedermann, Daeyoung Kim, Engin A. Sungur, and Alexander von Eye.
© 2021 John Wiley & Sons, Inc. Published 2021 by John Wiley & Sons, Inc.

So, how can we gauge the level that Y depends on X. This is where the concept of measures of complete dependence shine.

In the next section, we will review the concept of complete dependence and related properties. In Section 7.2, we will discuss basic properties of measures of complete dependence including some of Rényi's axioms (Rényi, 1959) and Lancaster's axioms (Lancaster, 1982). We will also provide definitions and basic properties of measures of complete dependence that we know of. In Section 7.3, we will provide example calculations of these measures in the case of the multivariate Gaussian distributions. We will also discuss some open problems and future works at the end of this chapter.

7.1 Basic Properties

Recall that the *(cumulative) distribution function* $F_{\vec{X}}$ of a d-dimensional random vector $\vec{X}$ is defined via

$$F_{\vec{X}}(\vec{x}) = \mathbb{P}(\vec{X} \leq \vec{x})$$

for all $\vec{x} \in \mathbb{R}^d$. Here the comparison is done component wise. Similarly, the *joint distribution function* $F_{\vec{X},\vec{Y}}$ of random vectors $\vec{X}$ and $\vec{Y}$ is defined via

$$F_{\vec{X},\vec{Y}}(\vec{x},\vec{y}) = \mathbb{P}(\vec{X} \leq \vec{x}, \vec{Y} \leq \vec{y})$$

for all vectors $\vec{x}$ and $\vec{y}$. By identifying $\mathbb{R}^d \times \mathbb{R}^k$ with $\mathbb{R}^{d+k}$, we may also view the joint distribution function $F_{\vec{X},\vec{Y}}$ as the distribution function $F_{(\vec{X},\vec{Y})}$ of the random vector $(\vec{X}, \vec{Y})$. It is well-known that the behavior of random vectors is governed by their distribution functions. Thus, it is natural to ask for an equivalent condition for complete dependence in term of distribution functions. It turns out that complete dependence is exactly happened when the conditional distribution function is degenerate. Recall that the *conditional distribution function* $F_{\vec{Y}|\vec{X}}$ of $\vec{Y}$ given $\vec{X}$ is defined by

$$F_{\vec{Y}|\vec{X}}(\vec{y} \mid \vec{x}) = \lim_{\vec{h} \searrow \vec{0}} \frac{\mathbb{P}(\vec{Y} \leq \vec{y}, \vec{x} - \vec{h} \leq \vec{X} \leq \vec{x} + \vec{h})}{\mathbb{P}(\vec{x} - \vec{h} \leq \vec{X} \leq \vec{x} + \vec{h})}$$

as long as the denominator inside the limit is never zero and $F_{\vec{Y}|\vec{X}}(\vec{y} \mid \vec{x}) = 0$ otherwise. Heuristically, $F_{\vec{Y}|\vec{X}}(\vec{y} \mid \vec{x})$ represents the probability that $\vec{Y}$ is at most $\vec{y}$ given that $\vec{X} = \vec{x}$, that is, $\mathbb{P}(\vec{Y} \leq \vec{y} \mid \vec{X} = \vec{x})$. Thus, the fact that $\vec{Y}$ is a function of $\vec{X}$ immediately implies this probability is trivial. Conversely, if this conditional distribution function is trivial, then we can define $f(\vec{x}) = \min\{\vec{y} \mid F_{\vec{Y}|\vec{X}}(\vec{y} \mid \vec{x}) = 1\}$ and get $f(\vec{X}) = \vec{Y}$ almost surely.

Similarly, we can also view complete dependence in term of conditional expectation. If $\vec{Y} = f(\vec{X})$, then we would have

$$\mathbb{E}[\vec{Y} \mid \vec{X} = \vec{x}] = \int \vec{y} \, dF_{\vec{Y}|\vec{X}}(\vec{y} \mid \vec{x}) = f(\vec{x}),$$

that is, $\mathbb{E}[\vec{Y} \mid \vec{X}] = f(\vec{X}) = \vec{Y}$ almost surely. Conversely, the fact that $\vec{Y} = \mathbb{E}[\vec{Y} \mid \vec{X}]$ immediately yields $\vec{Y}$ is a function of $\vec{X}$ since the latter is. Thus, we deduce the following facts.

Proposition 7.1 Let $\vec{X}$ and $\vec{Y}$ be random vectors. Then the following statements are equivalent.

1. $\vec{Y}$ is completely dependent on $\vec{X}$.
2. $F_{\vec{Y}|\vec{X}}$ is degenerate.
3. $\vec{Y} = f(\vec{X})$ almost surely for some measurable function f.
4. $\vec{Y} = \mathbb{E}[\vec{Y} \mid \vec{X}]$ almost surely.

Moreover, the first three statements remain equivalent for any pair of random variables defined in abstract spaces if we replace conditional distribution functions with conditional probability laws.

Another obvious property of complete dependence is *transitivity*, that is, if $\vec{Z}$ is completely dependent on $\vec{Y}$ and $\vec{Y}$ is completely dependent on $\vec{X}$, then $\vec{Z}$ is also completely dependent on $\vec{X}$. This result leads to other obvious facts, especially, those related to mutual complete dependence. A family of random vectors $\vec{X}_1, \ldots, \vec{X}_k$ is said to be *mutually completely dependent* on one another if $\vec{X}_j$ is completely dependent on $\vec{X}_i$ for all $i, j = 1, \ldots, n$. By transitivity property, the following statements hold.

1. Assume $\vec{X}$ and $\vec{Z}$ are mutually completely dependent. Then $\vec{Y}$ is completely dependent on $\vec{X}$ if and only if $\vec{Y}$ is completely dependent on $\vec{Z}$.
2. $(Y_1, \ldots, Y_d)$ is completely dependent on $\vec{X}$ if and only if each Y_i is completely dependent on $\vec{X}$.
3. $(X_1, \ldots, X_k)$ and $(F_{X_1}(X_1), \ldots, F_{X_k}(X_k))$ are mutually completely dependent.
4. $\vec{X}$ and $\Psi_{F_{\vec{X}}}(\vec{X})$ are mutually completely dependent where

$$\Psi_{F_{\vec{Y}}}(\vec{y}) = (F_{Y_1}(y_1), F_{Y_2|Y_1}(y_2 \mid y_1), \ldots, F_{Y_d|(Y_1, \ldots, Y_{d-1})}(y_d \mid (y_1, \ldots, y_{d-1})))$$

is the multivariate integral transformation on $\mathbb{R}^d$.

For a continuous random variable X, $F_X(X)$ has uniform distribution on the unit interval $\mathbb{I}$. Similarly, if $F_{\vec{Y}}$ has a density, then $\Psi_{F_{\vec{Y}}}(\vec{Y})$ has uniform distribution on $\mathbb{I}^d$. This leads naturally to only consider joint distribution functions with uniform marginals. When F_{Y_i} are continuous for all i, the restriction of the joint distribution function of $(F_{Y_1}(Y_1), \ldots, F_{Y_d}(Y_d))$ to $\mathbb{I}^d$ is called the *copula* associated with $(Y_1, \ldots, Y_d)$. It follows that a copula is a Lipschitz function and hence the first order partial derivative $\partial_u C$ of C with respected to the variable u exists almost everywhere. When both $F_{\overrightarrow{X}}$ and $F_{\overrightarrow{Y}}$ have densities, the restriction of the joint distribution function of $\Psi_{F_{\overrightarrow{X}}}(\overrightarrow{X})$ and $\Psi_{F_{\overrightarrow{Y}}}(\overrightarrow{Y})$ on $\mathbb{I}^k \times \mathbb{I}^d$ is called the *linkage* associated with $(\overrightarrow{X}, \overrightarrow{Y})$ (Li, Scarsini, & Shaked, 1996). In general, a linkage (a copula) associated with random vectors (variables) always exists but it is not unique unless the random vectors (variables) have densities. In addition, the joint distribution function of $\overrightarrow{Y} = (Y_1, \ldots, Y_d)$ can be recovered using their marginals and a copula $C_{\overrightarrow{Y}}$ associated with $\overrightarrow{Y}$ via

$$F_{\overrightarrow{Y}}(y_1, \ldots, y_d) = C_{\overrightarrow{Y}}(F_{Y_1}(y_1), \ldots, F_{Y_d}(y_d))$$

for all $y_i \in \mathbb{R}$. This is the so called *Sklar's Theorem*. Similarly, the joint distribution function of $\overrightarrow{X}$ and $\overrightarrow{Y}$ can be recovered using their associated linkage $L_{\overrightarrow{X},\overrightarrow{Y}}$ via

$$F_{\overrightarrow{X},\overrightarrow{Y}}(\overrightarrow{x}, \overrightarrow{y}) = L_{\overrightarrow{X},\overrightarrow{Y}}(\Psi_{F_{\overrightarrow{X}}}(\overrightarrow{x}), \Psi_{F_{\overrightarrow{Y}}}(\overrightarrow{y}))$$

for all vectors $\overrightarrow{x}$ and $\overrightarrow{y}$. For more information on linkage, see the work of Li et al. (1996).

In Section 7.2, we will review measures of complete dependence based on these concepts. For now, we will end this section with an example of a random variable V_n that is completely dependent on another random variable U and yet U barely depends on V_n when $n \to \infty$. See also the work of Siburg and Stoimenov (2015) for similar constructions.

Example 7.1 Let U be uniformly distributed on $\mathbb{I}$ and $V_n = nU - \lfloor nU \rfloor$ where $\lfloor x \rfloor$ is the floor of $x \in \mathbb{R}$. Clearly, V_n is completely dependent on U but U is not a function of V_n for all n. Now,

$$F_{U,V_n}(u, v) = \mathbb{P}(U \leq u, nU - \lfloor nU \rfloor \leq v)$$

$$= \sum_{i=0}^{\lfloor nu \rfloor} \mathbb{P}(U \leq u, nU - \lfloor nU \rfloor \leq v, \lfloor nU \rfloor = i)$$

$$= \sum_{i=0}^{\lfloor nu \rfloor} \mathbb{P}(i \leq nU \leq \min(i + v, nu))$$

$$= \lfloor nu \rfloor \frac{v}{n} + \frac{1}{n}\min(v, nu - \lfloor nu \rfloor)$$

for all $u, v \in \mathbb{I}$. Thus, V_n is uniformly distributed on $\mathbb{I}$ and $F_{U,V_n}(u, v) \to \Pi(u, v) = uv$ when $n \to \infty$.

7.2 Measure of Complete Dependence

In the previous section, we review complete dependence which we consider to be extremely opposite to independence. In practice, however, a random vector $\vec{Y}$ might be neither independent nor completely dependent on another random vector $\vec{X}$. How do we determine the dependency level in this case?

Denote $\mathcal{X}_d$ the set of d-dimensional random vectors. Basically, we want to order elements in $\mathcal{X}_d$ according to how much they depends on a specific random vector, say, $\vec{X}$. More generally, we might want to order elements of $\mathcal{X}_d \times \mathcal{X}_k$ according to how much the first random vector depends on the latter. This idea had been explored by Dette, Siburg, and Stoimenov (2013) using three stochastic orders on $\mathcal{X} = \mathcal{X}_1$ defined in term of their distribution functions. For any $X, Y \in \mathcal{X}$, define

$$X \leq_{cx} Y \Leftrightarrow F_X \leq_{cx} F_Y$$
$$\Leftrightarrow \mathbb{E}\phi(X) \leq \mathbb{E}\phi(Y) \text{ for all convex function } \phi,$$
$$X \leq_{dil} Y \Leftrightarrow F_X \leq_{dil} F_Y$$
$$\Leftrightarrow X - \mathbb{E}X \leq_{cx} Y - \mathbb{E}Y,$$
$$X \leq_{dis} Y \Leftrightarrow F_X \leq_{dis} F_Y$$
$$\Leftrightarrow F_X^-(b) - F_X^-(a) \leq F_Y^-(b) - F_Y^-(a) \text{ for all } a < b.$$

Here $F^-(t) = \inf\{x \in \mathbb{R} \mid F(x) \geq t\}$ is the quantile function associated with the distribution function F. Dette et al. define $(X, Y) \preccurlyeq_* (\widetilde{X}, \widetilde{Y})$ if and only if the set $\{u \in \mathbb{I} \mid \partial_u C_{(\widetilde{X}, \widetilde{Y})}(u, \cdot) \leq_* \partial_u C_{(X,Y)}(u, \cdot)\}$ has full measure where $\leq_*$ could be either one of $\leq_{cx}$, $\leq_{dil}$, and $\leq_{dis}$. They then show, when considering only continuous random variables, that minimal elements in $\preccurlyeq_*$ consist of those that are independent while maximal elements in $\preccurlyeq_*$ consist of the pair (X, Y) such that Y is completely dependent on X. In general, $(X, Y) \preccurlyeq_* (\widetilde{X}, \widetilde{Y})$ would be interpreted as the dependency level of Y given X is less than that of $\widetilde{Y}$ given $\widetilde{X}$. Since $\preccurlyeq_*$ is only a partial order, these pairs of random variables are not always comparable, however.

Another approach is to instead consider a linear ordering on $\mathcal{X}_d \times \mathcal{X}_k$, equivalently, a function $\xi : \mathcal{X}_d \times \mathcal{X}_k \to \mathbb{R}$ such that $\xi(\vec{Y} \mid \vec{X})$ quantify the level that the target random vector $\vec{Y}$ depends on the explanatory random vector $\vec{X}$. This is actually the idea behind classical statistics such as covariance and correlations. In this case, we would want $\xi(\vec{Y} \mid \vec{X})$ to be maximized when $\vec{Y}$ is completely dependent of $\vec{X}$ and be minimized when $\vec{X}$ and $\vec{Y}$ are independent. Of course, we may also scale ξ so that its value lying in $\mathbb{I}$. In summary, we would want ξ to satisfy the following properties.

(N) $0 \leq \xi(\vec{Y} \mid \vec{X}) \leq 1$.
(I) $\xi(\vec{Y} \mid \vec{X}) = 0$ if and only if $\vec{X}$ and $\vec{Y}$ are independent.

(C) $\xi(\vec{Y} \mid \vec{X}) = 1$ if and only if $\vec{Y}$ is completely dependent on $\vec{X}$.

The last two conditions can also be weaken to

(wI) $\xi(\vec{Y} \mid \vec{X}) = 0$ if $\vec{X}$ and $\vec{Y}$ are independent, and
(wC) $\xi(\vec{Y} \mid \vec{X}) = 1$ if $\vec{Y}$ is completely dependent of $\vec{X}$.

Other conditions considered here are the symmetric property:

(S) $\xi(\vec{Y} \mid \vec{X}) = \xi(\vec{X} \mid \vec{Y})$
and the invariant property:
(T) $\xi(\vec{Y} \mid f(\vec{X})) \leq \xi(\vec{Y} \mid \vec{X})$ for any measurable function f.

The conditions (N), (I), (S), and (wC) are parts of Rényi's axioms (Rényi, 1959) while conditions (N), (I), (S), and (C) are parts of Lancaster's axioms (Lancaster, 1982). Measures satisfy the symmetric condition (S) can also be constructed via symmetrical aggregation. For example, by defining

$$\hat{\xi}(\vec{X}, \vec{Y}) = \max(\xi(\vec{Y} \mid \vec{X}), \xi(\vec{X} \mid \vec{Y}))$$

or

$$\hat{\xi}(\vec{X}, \vec{Y}) = \frac{1}{2}(\xi(\vec{Y} \mid \vec{X}) + \xi(\vec{X} \mid \vec{Y})).$$

See also the works of Siburg and Stoimenov (2010), Tasena and Dhompongsa (2013), Boonmee and Tasena (2016), Kamnitui, Santiwipanont, and Sumetkijakan (2015) for this approach. In this case, $\hat{\xi}(\vec{X}, \vec{Y})$ will represent the dependency level that both random vectors depend on another.

Note that any measure of association satisfies (S) will lose its quantification as a measure of directional dependence, that is, it is meaningless to discuss which one is the source or the explanatory variable. For the quantification of directional dependence, it would be better for use asymmetric versions of these measures. For example, consider random variables U and V_n defined in Example 7.1. If a measure ξ satisfies (S), then $\xi(V_n \mid U) = \xi(U \mid V_n)$ which would be impossible to differentiate dependency level in each direction. If a measure ξ only satisfies (C), however, then $\xi(V_n \mid U) = 1$ while $\xi(U \mid V_n) < 1$ which states that two directions have different dependency levels.

It should also be mentioned that any measure satisfies (wI) and (wC) cannot be continuous with respect to the uniform convergence of distribution functions. This is equivalent to the continuity under distributional convergences of random vectors.

Proposition 7.2 There is no measure of association ξ satisfies (wI) and (wC) which is also continuous under the convergence in distribution.

Proof. Suppose such a measure ξ exists. Consider random variables V_n and U defined in Example 7.1 and define $\overrightarrow{Y}_n = (V_n, \dots, V_n) \in \mathcal{X}^d$ and $\overrightarrow{X} = (U, \dots, U) \in \mathcal{X}^k$. Then

$$F_{\overrightarrow{Y}_n, \overrightarrow{X}}(\overrightarrow{y}, \overrightarrow{x}) \to \Pi(\min(\overrightarrow{y}), \min(\overrightarrow{x}))$$

for all $\overrightarrow{y} \in \mathbb{I}^d$ and $\overrightarrow{x} \in \mathbb{I}^k$. Thus, $\xi(\overrightarrow{Y}_n \mid \overrightarrow{X}) \to 0$ which is impossible since $\xi(\overrightarrow{Y}_n \mid \overrightarrow{X}) = 1$ for all n. □

There is also another invariant property that can be considered together with (T) – the invariant under transformations of the target variable $\overrightarrow{Y}$. We might want $\xi(f(\overrightarrow{Y}) \mid \overrightarrow{X})$ to be comparable to $\xi(\overrightarrow{Y} \mid \overrightarrow{X})$ for all measurable function f. (See, for example, the work of Gebelein (1941), Ruankong, Santiwipanont, and Sumetkijakan (2013), and Kamnitui et al. (2015) for measures satisfy this property.) This is impossible to achieve under conditions (I) and (C), however. Consider, for example, independent random variables U, W, and B which are uniformly distributed on $\mathbb{I}$ and

$$V = \begin{cases} -U, & B \leq 0.5, \\ W, & B > 0.5. \end{cases}$$

If we define $f = 1_{[0, \infty)}$, then $f(V)$ is a function of B but it is independent of U while V itself is neither a function of B nor independent of U. Thus, $\xi(f(V) \mid U) = 0 < \xi(V \mid U)$ while $\xi(f(V) \mid B) = 1 > \xi(V \mid B)$ in this case. In practice, we may only hope to achieve the following property.

(M) If $Z_i = f_i(Y_i)$ almost surely and $Y_i = g_i(Z_i)$ almost surely for some nondecreasing functions $f_i, g_i : \mathbb{R} \to \mathbb{R}$, then $\xi((Y_1, \dots, Y_d) \mid \overrightarrow{X}) = \xi((Z_1, \dots, Z_d) \mid \overrightarrow{X})$.

Properties (T) and (M) lead naturally to the constructions of copula-based (linkage-based) measures of complete dependence for continuous random variables. Since in this case, we only need to consider, for example, either

$$\xi((F_{Y_1}(Y_1), \dots, F_{Y_d}(Y_d)) \mid (F_{X_1}(X_1), \dots, F_{X_k}(X_k)))$$

or

$$\xi(\Psi_{F_{\overrightarrow{Y}}}(\overrightarrow{Y}) \mid \Psi_{F_{\overrightarrow{X}}}(\overrightarrow{X})).$$

Dette et al. (2013) construct a measure $r(Y \mid X)$ between (continuous) random variables X and Y using the (modified) Sobolev L^2-distance. (See also the work of Siburg and Stoimenov (2010) for their symmetric version.) Around the same time, Trutschnig (2011) constructs another measure $\zeta(Y \mid X)$ based on the Sobolev L^1-distance. (See also the work of Trutschnig (2017).) Later, Li (2015a) defined a family of measures τ_φ via

$$\tau_\varphi(Y \mid X) = \int_{\mathbb{I}} \int_{\mathbb{I}} \varphi(\partial_u C_{X,Y}(u, v) - v) du dv$$

for each nonnegative convex function φ with $\varphi(0) = 0$. After normalization, this reduces to $\zeta(Y \mid X)$ when φ is the absolute function and this reduces to $r(Y \mid X)$ when φ is the squaring function:

$$\zeta(Y \mid X) = 3 \int_{\mathbb{I}} \int_{\mathbb{I}} |\partial_u C_{X,Y}(u,v) - v| du dv$$

and

$$r(Y \mid X) = 6 \int_{\mathbb{I}} \int_{\mathbb{I}} (\partial_u C_{X,Y}(u,v) - v)^2 du dv.$$

In general,

$$T_\varphi = \max_{X,Y} \tau_\varphi(Y \mid X) = \tau_\varphi(Y \mid Y) = \int_{\mathbb{I}} (\varphi(1-v)v + \varphi(-v)(1-v)) dv$$

so that the normalization of τ_φ is given by $\frac{1}{T_\varphi} \tau_\varphi$. These ideas also carry over to the case of random vectors. The case that the target variable is a random vector but the explanatory variable is a random variable is studied earlier by Tasena and Dhompongsa (2013). The case that the target is a continuous random variable but the explanatory is a continuous random vector is also studied by Li (2015b). In this case, Li defines

$$\tau_\varphi(Y \mid \vec{X}) = \int_{\mathbb{I}} \int_{\mathbb{I}^k} \varphi\left(\frac{\partial_{\vec{u}} C_{\vec{X},Y}(\vec{u},v)}{\partial_{\vec{u}} C_{\vec{X},Y}(\vec{u},1)} - v \right) d\vec{u} dv$$

where φ is again a nonnegative convex function φ with $\varphi(0) = 0$ and the measure $\tau_\varphi(Y \mid \vec{X})$ is only defined when F_Y is continuous and $F_{\vec{X}}$ has a density. Otherwise, $\partial_{\vec{u}} C_{\vec{X},Y}$ is undefined. In particular,

$$\tau_\varphi(Y \mid \vec{X}) = \int \int \varphi(F_{Y|\vec{X}}(y \mid \vec{x}) - F_Y(y)) dF_Y(y) dF_{\vec{X}}(\vec{x})$$

by the change of variable formula.

The case that both the target and the explanatory variables are continuous random vectors having densities is also studied by Boonmee and Tasena (2016) where they define

$$\zeta_p(\vec{Y} \mid \vec{X}) = \frac{1}{C_{p,d}} \left(\int_{\mathbb{I}^d} \int_{\mathbb{I}^k} |\partial_{\vec{u}} L_{\vec{X},\vec{Y}}(\vec{u},\vec{v}) - \Pi(\vec{v})|^p d\vec{u} d\vec{v} \right)^{1/p}$$

where $\Pi(\vec{v})$ is the product of components in $\vec{v}$, $L_{\vec{X},\vec{Y}}$ is the linkage associated with $\vec{X}$ and $\vec{Y}$, and

$$C_{p,d} = \left(\int_{\mathbb{I}^d} \Pi(\vec{v})(1 - \Pi(\vec{v}))^p + (\Pi(\vec{v}))^p (1 - \Pi(\vec{v})) d\vec{v} \right)^{1/p}.$$

Note that $\tau_\varphi(Y \mid \vec{X}) = \frac{2}{(p+1)(p+2)}\zeta_p^p(Y \mid \vec{X})$ when $\varphi(t) = |t|^p$ since both of them satisfy (T).

Several measures based directly on conditional distributions have also been constructed. For discrete random variables, a family of measures is constructed by Shan, Wongyang, Wang, and Tasena (2015). Assume that X and Y take values in countable sets Θ and $\Lambda \subseteq \mathbb{R}$, respectively. Shan et al. define

$$\omega_t^2(Y \mid X) = t \sum_{x \in \Theta} \sum_{y \in \Lambda} (\mathbb{P}(Y < y \mid X = x) - \mathbb{P}(Y < y))^2 \mathbb{P}(Y = y)\mathbb{P}(X = x)$$

$$+ (1 - t) \sum_{x \in \Theta} \sum_{y \in \Lambda} (\mathbb{P}(Y \le y \mid X = x) - \mathbb{P}(Y \le y))^2 \mathbb{P}(Y = y)\mathbb{P}(X = x)$$

which has the maximum value of

$$\omega_{t,\max}^2 = t \sum_{y \in \Lambda} \mathbb{P}(Y < y)(1 - \mathbb{P}(Y < y))\mathbb{P}(Y = y)$$

$$+ (1 - t) \sum_{y \in \Lambda} \mathbb{P}(Y \le y)(1 - \mathbb{P}(Y \le y))\mathbb{P}(Y = y).$$

For discrete random vectors, Li (2015c) defines a family of measures analogous to the continuous case as

$$\tau_\varphi(\vec{Y} \mid \vec{X}) = \sum_{\vec{x}} \sum_{\vec{y}} \varphi(\mathbb{P}(\vec{Y} \le \vec{y} \mid \vec{X} = \vec{x}) - \mathbb{P}(\vec{Y} \le \vec{y}))\mathbb{P}(\vec{Y} = \vec{y})\mathbb{P}(\vec{X} = \vec{x})$$

for each nonnegative convex function φ with $\varphi(0) = 0$. For discrete random variables X and Y, $\omega_0^2(Y \mid X) = 6\tau_\varphi(Y \mid X)$ when φ is the squaring function.

For general case, Tasena and Dhompongsa (2016) define

$$\overline{\omega}_p(\vec{Y} \mid \vec{X}) = \left(\frac{\iint \left|F_{\vec{Y} \mid \vec{X}}(\vec{y} \mid \vec{x}) - \frac{1}{2}\right|^p dF_{\vec{Y}}(\vec{y})dF_{\vec{X}}(\vec{x}) - \int \left|F_{\vec{Y}}(\vec{y}) - \frac{1}{2}\right|^p dF_{\vec{Y}}(\vec{y})}{\frac{1}{2^p} - \int \left|F_{\vec{Y}}(\vec{y}) - \frac{1}{2}\right|^p dF_{\vec{Y}}(\vec{y})}\right)^{1/p}$$

which reduces to

$$\overline{\omega}_2(\vec{Y} \mid \vec{X}) = \sqrt{\frac{\iint (F_{\vec{Y} \mid \vec{X}}(\vec{y} \mid \vec{x}) - F_{\vec{Y}}(\vec{y}))^2 dF_{\vec{Y}}(\vec{y})dF_{\vec{X}}(\vec{x})}{\int F_{\vec{Y}}(\vec{y})(1 - F_{\vec{Y}}(\vec{y}))dF_{\vec{Y}}(\vec{y})}}$$

when $p = 2$. Also,

$$\int \left|F_{\vec{Y}}(\vec{y}) - \frac{1}{2}\right|^p dF_{\vec{Y}}(\vec{y}) = \int_{[d} \left|\Pi(\vec{u}) - \frac{1}{2}\right|^p d\vec{u}$$

$$= \frac{(-1)^{d-1}}{(d-1)!}\int_{[} \left|u - \frac{1}{2}\right|^p \ln^{d-1} u \, du$$

when $\vec{Y}$ consists of independent continuous random variables. This value might be different in other cases, however. For example, when $\vec{Y} = (Y, \dots, Y)$ and F_Y is continuous,

$$\int \left| F_{\vec{Y}}(\vec{y}) - \frac{1}{2} \right|^p dF_{\vec{Y}}(\vec{y}) = \int \left| F_Y(y) - \frac{1}{2} \right|^p dF_Y(y)$$

$$= \int_{\mathbb{I}} \left| u - \frac{1}{2} \right|^p du$$

regardless of the dimension of $\vec{Y}$. When $\vec{X}$ is a continuous random variable, the measure $\overline{m}_2$ reduces to the one defined earlier by Tasena and Dhompongsa (2013). When F_Y is continuous and $F_{\vec{X}}$ has density, $\overline{\omega}_2(Y \mid \vec{X}) = \zeta_2(Y \mid \vec{X})$. These two measures are different when the target variable is a random vector, however.

Note that it is possible to define $\overline{\omega}_p(\vec{Y} \mid \vec{X})$ using different normalizations. For example, we could also define

$$\widetilde{\omega}_p(\vec{Y} \mid \vec{X}) = 1 - \frac{\frac{1}{2} - \left(\iint \left| F_{\vec{Y}|\vec{X}}(\vec{y} \mid \vec{x}) - \frac{1}{2} \right|^p dF_{\vec{Y}}(\vec{y}) dF_{\vec{X}}(\vec{x}) \right)^{1/p}}{\frac{1}{2} - \left(\int \left| F_{\vec{Y}}(\vec{y}) - \frac{1}{2} \right|^p dF_{\vec{Y}}(\vec{y}) \right)^{1/p}}$$

for any random vector $\vec{X}$ and $\vec{Y}$.

Note that all these measures excepted τ_φ satisfy properties (N), (I), (C), (T), and (M). The measure τ_φ also satisfies these properties after normalization.

Recently, another measure based on the conditional variance has also been constructed by Sungur (2005) and then generalized by Wei and Kim (2018). (See also Kamnitui et al. (2015) for related idea.) They define

$$\rho^2(Y \mid \vec{X}) = \rho^2_{(\vec{X} \to Y)} = \frac{\text{Var}(\mathbb{E}[Y \mid \vec{X}])}{\text{Var}(Y)} = \text{Corr}^2(Y, \mathbb{E}[Y \mid \vec{X}])$$

whenever the random variable Y has finite variance. This measure can also be rewritten in term of distribution functions as

$$\text{Var}(\mathbb{E}[Y \mid \vec{X}]) = \int \left(\int y \, dF_{Y|\vec{X}}(y \mid \vec{x}) - \int y \, dF_Y(y) \right)^2 dF_{\vec{X}}(\vec{x})$$

$$= \int \left(\int 1 - F_{Y|\vec{X}}(y \mid \vec{x}) dy - \int 1 - F_Y(y) dy \right)^2 dF_{\vec{X}}(\vec{x})$$

$$= \int \left(\int (F_{Y|\vec{X}}(y \mid \vec{x}) - F_Y(y)) dy \right)^2 dF_{\vec{X}}(\vec{x})$$

where the second equality follows from the Fubini Theorem. For the continuous case, a related measure has been defined earlier via

$$\widetilde{\rho}^2(Y \mid \vec{X}) = \rho^2(F_Y(Y) \mid (F_{X_1}(X_1), \dots, F_{X_k}(X_k)))$$

$$= 12 \int \left(\int (F_{Y|\vec{X}}(y \mid \vec{x}) - F_Y(y)) dF_Y(y) \right)^2 dF_{\vec{X}}(\vec{x})$$

$$= 12 \int \left(\int F_{Y|\vec{X}}(y \mid \vec{x}) \mathrm{d}F_Y(y) \right)^2 \mathrm{d}F_{\vec{X}}(\vec{x}) - 3$$

by Wei, Wang, and Kim (2016). Notice that $\widetilde{\rho}^2(Y \mid X) \leq 2r(Y \mid X)$ by Jensen inequality. These two measures satisfy (N), (wI), (C), and (T). The measure $\widetilde{\rho}^2$ also satisfies (M). For continuous random variables, Kamnitui et al. (2015) also studied measure $v(X, Y)$ which can be rewritten as

$$v(X, Y) = \sqrt{\frac{1}{2}(\widetilde{\rho}^2(Y \mid X) + \widetilde{\rho}^2(X \mid Y))}.$$

This can be considered as a symmetric version of $\widetilde{\rho}^2$.

7.3 Example Calculation

In this section, we calculate values of measures of complete dependence in case the target Y is a random variable given that the joint distribution of $(Y, \vec{X})$ is a (multivariate) Gaussian distribution. For the case of random vectors, see for example, the work of Tasena and Dhompongsa (2016) and the work of Boonmee and Tasena (2016). See also the work of Wei and Kim (2018) for the case of skew normal distributions.

Let $(Y, \vec{X})$ have a multivariate Gaussian distribution with the covariance matrix $\Sigma = \begin{bmatrix} \sigma^2 & \Gamma^t \\ \Gamma & \Sigma_{\vec{X}} \end{bmatrix}$. By properties (T) and (M), we may assume $(Y, \vec{X})$ has zero means. Given $\vec{X}$, the random variable Y has Gaussian distribution with mean $\Gamma^t \Sigma_{\vec{X}}^{-1} \vec{X}$ and variance $\sigma^2 - \Gamma^t \Sigma_{\vec{X}}^{-1} \Gamma$. Thus,

$$\rho^2(Y \mid \vec{X}) = \frac{\mathrm{Var}(\Gamma^t \Sigma_{\vec{X}}^{-1} \vec{X})}{\sigma^2} = \frac{1}{\sigma^2} \Gamma^t \Sigma_{\vec{X}}^{-1} \Gamma.$$

Notice that $\rho^2(Y \mid X)$ is the square of the correlation of Y and X in the case that X is a random variable and

$$\rho^2(Y \mid \vec{X}) = \sum_{i=1}^{k} \mathrm{Corr}^2(Y, Z_i) = \mathrm{Corr}^2(Y, \Gamma^t \Sigma_{\vec{X}}^{-1} \vec{X})$$

where $\vec{Z} = \Sigma_{\vec{X}}^{-1/2} \vec{X}$ in general. Therefore, $\rho^2(Y \mid \vec{X})$ can be thought of as the total correlation between Y and $\vec{X}$. Since the dependency structure between Gaussian random vectors only depends on their covariance matrix, we would expect that other measures of complete dependence provide results comparable to $\rho^2(Y \mid \vec{X})$. This will also serve as a benchmark for these measures. It turns out that all other measures can be written as a function of $\rho^2(Y \mid \vec{X})$ in this case.

Denote Φ_Σ the Gaussian distribution function with mean zero and covariance matrix Σ. Then $F_{Y|\vec{X}}(y \mid \vec{x}) = \Phi_{\sigma^2 - \Gamma^t \Sigma_{\vec{X}}^{-1} \Gamma}(y - \Gamma^t \Sigma_{\vec{X}}^{-1} \vec{x})$. We will use this formula to compute values of other measures of complete dependence.

First, consider $\tilde{\rho}^2(Y \mid \vec{X})$. In this case,

$$\tilde{\rho}^2(Y \mid \vec{X}) = 12 \int \left(\int \Phi_{\sigma^2 - \Gamma^t \Sigma_{\vec{X}}^{-1} \Gamma}(y - \Gamma^t \Sigma_{\vec{X}}^{-1} \vec{x}) d\Phi_{\sigma^2}(y) \right)^2 d\Phi_{\Sigma_{\vec{X}}}(\vec{x}) - 3$$

$$= 12 \int \left(\int \Phi_{\sigma^2 - \Gamma^t \Sigma_{\vec{X}}^{-1} \Gamma}(\sigma y - \Gamma^t \Sigma_{\vec{X}}^{-1/2} \vec{x}) d\Phi_1(y) \right)^2 d\Phi_I(\vec{x}) - 3$$

$$= 12 \int \left(\int \Phi_1 \left(\frac{\sigma y - \Gamma^t \Sigma_{\vec{X}}^{-1/2} \vec{x}}{\sqrt{\sigma^2 - \Gamma^t \Sigma_{\vec{X}}^{-1} \Gamma}} \right) d\Phi_1(y) \right)^2 d\Phi_I(\vec{x}) - 3$$

$$= 12 \int \left(\int \Phi_1 \left(\frac{\sigma y - x}{\sqrt{\sigma^2 - \Gamma^t \Sigma_{\vec{X}}^{-1} \Gamma}} \right) d\Phi_1(y) \right)^2 d\Phi_{\Gamma^t \Sigma_{\vec{X}}^{-1} \Gamma}(x) - 3$$

$$= 12 \int \left(\int \Phi_1 \left(\frac{\sigma y - x \sqrt{\Gamma^t \Sigma_{\vec{X}}^{-1} \Gamma}}{\sqrt{\sigma^2 - \Gamma^t \Sigma_{\vec{X}}^{-1} \Gamma}} \right) d\Phi_1(y) \right)^2 d\Phi_1(x) - 3$$

$$= 12\varrho(\sqrt{\rho^2(Y \mid \vec{X})}) - 3$$

where $\varrho(t) = \int \left(\int \Phi_1 \left(\frac{y - tx}{\sqrt{1 - t^2}} \right) d\Phi_1(y) \right)^2 d\Phi_1(x)$.

Next, we consider $\tau_\varphi(Y \mid \vec{X})$. Similar to $\tilde{\rho}^2(Y \mid \vec{X})$, we have

$$\tau_\varphi(Y \mid \vec{X}) = \int \int \varphi(\Phi_{\sigma^2 - \Gamma^t \Sigma_{\vec{X}}^{-1} \Gamma}(y - \Gamma^t \Sigma_{\vec{X}}^{-1} \vec{x}) - \Phi_{\sigma^2}(y)) d\Phi_{\sigma^2}(y) d\Phi_{\Sigma_{\vec{X}}}(\vec{x})$$

$$= \int \int \varphi(\Phi_{\sigma^2 - \Gamma^t \Sigma_{\vec{X}}^{-1} \Gamma}(\sigma y - \Gamma^t \Sigma_{\vec{X}}^{-1/2} \vec{x}) - \Phi_1(y)) d\Phi_1(y) d\Phi_I(\vec{x})$$

$$= \int \int \varphi \left(\Phi_1 \left(\frac{\sigma y - \Gamma^t \Sigma_{\vec{X}}^{-1/2} \vec{x}}{\sqrt{\sigma^2 - \Gamma^t \Sigma_{\vec{X}}^{-1} \Gamma}} \right) - \Phi_1(y) \right) d\Phi_1(y) d\Phi_I(\vec{x})$$

$$= \int \int \varphi \left(\Phi_1 \left(\frac{\sigma y - x}{\sqrt{\sigma^2 - \Gamma^t \Sigma_{\vec{X}}^{-1} \Gamma}} \right) - \Phi_1(y) \right) d\Phi_1(y) d\Phi_{\Gamma^t \Sigma_{\vec{X}}^{-1} \Gamma}(x)$$

$$= \int \int \varphi \left(\Phi_1 \left(\frac{\sigma y - x \sqrt{\Gamma^t \Sigma_{\vec{X}}^{-1} \Gamma}}{\sqrt{\sigma^2 - \Gamma^t \Sigma_{\vec{X}}^{-1} \Gamma}} \right) - \Phi_1(y) \right) d\Phi_1(y) d\Phi_1(x)$$

$$= \eta_\varphi(\sqrt{\rho^2(Y \mid \vec{X})})$$

where

$$\eta_\varphi(t) = \int\int \varphi\left(\Phi_1\left(\frac{y - tx}{\sqrt{1 - t^2}}\right) - \Phi_1(y)\right) d\Phi_1(y) d\Phi_1(x)$$

$$= \int\int \varphi(\Phi_1(z) - \Phi_1(y)) d\Phi_{\Lambda(t)}(y, z)$$

with $\Lambda(t) = \begin{bmatrix} 1 & \frac{1}{\sqrt{1-t^2}} \\ \frac{1}{\sqrt{1-t^2}} & \frac{1+t^2}{1-t^2} \end{bmatrix}$.

Last, we compute $\overline{\omega}_p(Y \mid \vec{X})$ and $\tilde{\omega}_p(Y \mid \vec{X})$. Consider

$$\int\int \left| F_{Y|\vec{X}}(y \mid \vec{x}) - \frac{1}{2} \right|^p dF_Y(y) dF_{\vec{X}}(\vec{x})$$

$$= \int\int \left| \Phi_{\sigma^2 - \Gamma^t\Sigma_{\vec{X}}^{-1}\Gamma}(y - \Gamma^t\Sigma_{\vec{X}}^{-1}\vec{x}) - \frac{1}{2} \right|^p d\Phi_{\sigma^2}(y) d\Phi_{\Sigma_{\vec{X}}}(\vec{x})$$

$$= \int\int \left| \Phi_{\sigma^2 - \Gamma^t\Sigma_{\vec{X}}^{-1}\Gamma}(z) - \frac{1}{2} \right|^p d\Phi_{\sigma^2 + \Gamma^t\Sigma_{\vec{X}}^{-1}\Gamma}(z)$$

$$= \int\int \left| \Phi_1\left(\frac{z}{\sqrt{\sigma^2 - \Gamma^t\Sigma_{\vec{X}}^{-1}\Gamma}}\right) - \frac{1}{2} \right|^p d\Phi_{\sigma^2 + \Gamma^t\Sigma_{\vec{X}}^{-1}\Gamma}(z)$$

$$= \int\int \left| \Phi_1\left(z\sqrt{\frac{\sigma^2 + \Gamma^t\Sigma_{\vec{X}}^{-1}\Gamma}{\sigma^2 - \Gamma^t\Sigma_{\vec{X}}^{-1}\Gamma}}\right) - \frac{1}{2} \right|^p d\Phi_1(z).$$

Denote $\varsigma_p(t) = \int\int \left| \Phi_1\left(z\sqrt{\frac{1+t}{1-t}}\right) - \frac{1}{2} \right|^p d\Phi_1(z)$. Then $\varsigma_p(0) = \frac{1}{2^p(p+1)}$ and

$$\overline{\omega}_p(Y \mid \vec{X}) = \left(\frac{\varsigma_p(\rho^2(Y \mid \vec{X})) - \varsigma_p(0)}{\frac{1}{2^p} - \varsigma_p(0)}\right)^{1/p}$$

$$= \left(\frac{2^p(p+1)}{p}\varsigma_p(\rho^2(Y \mid \vec{X})) - \frac{1}{p}\right)^{1/p}$$

while

$$\tilde{\omega}_p(Y \mid \vec{X}) = \frac{\varsigma_p^{1/p}(\rho^2(Y \mid \vec{X})) - \frac{1}{2(p+1)^{1/p}}}{\frac{1}{2} - \frac{1}{2(p+1)^{1/p}}}$$

$$= \frac{2(p+1)^{1/p}\varsigma_p^{1/p}(\rho^2(Y \mid \vec{X})) - 1}{(p+1)^{1/p} - 1}.$$

It should also be mentioned that these measures of complete dependence might not be functions of $\rho^2(Y \mid \vec{X})$ if $(Y, \vec{X})$ does not have multivariate Gaussian distribution. Otherwise, $\rho^2(Y \mid \vec{X})$ would satisfy the property (I).

7.4 Future Works and Open Problems

Even though there is a significant progress in the study of complete dependence, several questions remain unanswered and require further study. Example of these questions are the following.

1. Is it possible to construct measures of complete dependence for other types of random variables? In the case of random set, for instance, the capacities play the role of distribution functions. It is natural to ask whether we may replace conditional distribution functions with conditional capacities in the definition of these measures so that properties related to complete dependence remain intact.

2. Is it possible to study complete dependence of random vectors in term of stochastic ordering? Several stochastic orders for multivariate distributions have been study in the literature, see for example, the book of Shaked and Shanthikumar (2007). It would be natural to extend the work of Dette et al. (2013) in this direction. Also, would it still be possible to view measures of complete dependence as a linearization of these orders in this case?

3. Is it possible to provide a criteria for choosing measures of complete dependence? As we have seen, several measures have been proposed. A comparison has to be made to serve as a guideline for users. Among these measures, Wei and Kim (2018) provides some pros and cons between their measure $\rho^2(Y \mid X)$ and Dette et al.'s measure $r(Y \mid X)$. In our opinion, $\rho^2(Y \mid \vec{X})$ provides an interpretation in term of random variables which is better than other measures which only provide interpretation in term of distribution functions. The measure $\rho^2(Y \mid \vec{X})$ only satisfies (wI), however. This means its value could be near zero even though there are some relationship between $\vec{X}$ and Y. It is interesting to know whether a measure of association satisfying (I) with interpretation in term of random variables (vectors) can be constructed. Or, perhaps, such interpretation could be made for already defined measures. As for the interpretation in term of distribution functions, we feel that r, ζ, ζ_p, and τ_φ provide a more intuitive interpretation than $\overline{\omega}_p$ and $\tilde{\omega}_p$. The first four measures provide interpretation directly in term of distance between $F_{\vec{Y} \mid \vec{X}}$ and $F_{\vec{Y}}$.

4. Is it possible to construct good estimators for these measures? Wei and Kim (2018) provide an estimator for their measure $\rho^2(Y \mid \vec{X})$ in the case that $(Y, \vec{X})$

has skew normal distribution. Dette et al. (2013) also provide a kernel-based estimator for their measure under the condition that the associated copula is thrice differentiable. No estimators has been mentioned in other works. Using the fact that all Sobolev distance on the space of copulas are equivalent (Fernández Sánchez & Trutschnig, 2015), Dette et al.'s kernel estimator could also be used to estimate other measures of complete dependence in case of continuous random variables. The case of general random vectors remain opens, nevertheless.

References

Boonmee, T., & Tasena, S. (2016). Measure of complete dependence of random vectors. *Journal of Mathematical Analysis and Applications*, *443*(1), 585–595.

Dette, H., Siburg, K. F., & Stoimenov, P. A. (2013). A copula-based non-parametric measure of regression dependence. *Scandinavian Journal of Statistics*, *40*(1), 21–41.

Fernández Sánchez, J., & Trutschnig, W. (2015). Conditioning-based metrics on the space of multivariate copulas and their interrelation with uniform and levelwise convergence and iterated function systems. *Journal of Theoretical Probability*, *28*(4), 1311–1336.

Gebelein, H. (1941). Das statistische problem der korrelation als variations- und eigenwertproblem und sein zusammenhang mit der ausgleichsrechnung. *ZAMM—Journal of Applied Mathematics and Mechanics/Zeitschrift fÃŒr Angewandte Mathematik und Mechanik*, *21*(6), 364–379.

Kamnitui, N., Santiwipanont, T., & Sumetkijakan, S. (2015). Dependence measuring from conditional variances. *Dependence Modeling*, *3*(1), 98–112.

Lancaster, H. O. (1982). Measures and indices of dependence. In M. Kotz & N. L. Johnson (Eds.), *Encyclopedia of statistical sciences* (Vol. *2*, pp. 334–339). New York, NY: Wiley.

H. Li. Nonsymmetric dependence measures: The discrete case. *ArXiv e-prints*, December 2015a.

H. Li. A new class of nonsymmetric multivariate dependence measures. *ArXiv e-prints*, November 2015b.

H. Li. On nonsymmetric nonparametric measures of dependence. *ArXiv e-prints*, February 2015c.

Li, H., Scarsini, M., & Shaked, M. (1996). Linkages: A tool for the construction of multivariate distributions with given nonoverlapping multivariate marginals. *Journal of Multivariate Analysis*, *56*(1), 20–41.

Rényi, A. (1959). On measures of dependence. *Acta Mathematica Hungarica*, *10*(3–4), 441–451.

Ruankong, P., Santiwipanont, T., & Sumetkijakan, S. (2013). Shuffles of copulas and a new measure of dependence. *Journal of Mathematical Analysis and Applications*, *398*(1), 392–402.

Schweizer, B., & Wolff, E. F. (1981). On nonparametric measures of dependence for random variables. *The Annals of Statistics*, *9*(4), 879–885.

Shaked, M., & Shanthikumar, J. G. (2007). *Stochastic orders*. Springer Series in Statistics. New York, NY: Springer.

Shan, Q., Wongyang, T., Wang, T., & Tasena, S. (2015). A measure of mutual complete dependence in discrete variables through subcopula. *International Journal of Approximate Reasoning*, *65*, 11–23. SI: Modeling dependence in econometrics

Siburg, K. F., & Stoimenov, P. A. (2010). A measure of mutual complete dependence. *Metrika*, *71*(2), 239–251.

Siburg, K. F., & Stoimenov, P. A. (2015). Almost opposite regression dependence in bivariate distributions. *Statistical Papers*, *56*(4), 1033–1039.

Sungur, E. A. (2005). A note on directional dependence in regression setting. *Communications in Statistics - Theory and Methods*, *34*(9–10), 1957–1965.

Tasena, S., & Dhompongsa, S. (2013). A measure of multivariate mutual complete dependence. *International Journal of Approximate Reasoning*, *54*(6), 748–761.

Tasena, S., & Dhompongsa, S. (2016). Measures of the functional dependence of random vectors. *International Journal of Approximate Reasoning*, *68*, 15–26.

Trutschnig, W. (2011). On a strong metric on the space of copulas and its induced dependence measure. *Journal of Mathematical Analysis and Applications*, *384*(2), 690–705.

Trutschnig, W. (2017). Complete dependence everywhere? In M. Úbeda Flores, E. de Amo Artero, F. Durante, & J. Fernández Sánchez (Eds.), *Copulas and dependence models with applications* (pp. 225–240). New York, NY: Springer.

Wei, Z., & Kim, D. (2018). On multivariate asymmetric dependence using multivariate skew-normal copula-based regression. *International Journal of Approximate Reasoning*, *92*, 376–391.

Wei, Z., Wang, T., & Kim, D. (2016). *Multiple copula regression function and directional dependence under multivariate non-exchangeable copulas* (pp. 171–184). New York, NY: Springer.

Part III

Direction Dependence in Categorical Variables

8

Locating Direction Dependence Using Log-Linear Modeling, Configural Frequency Analysis, and Prediction Analysis

Alexander von Eye[1] and Wolfgang Wiedermann[2]

[1] Department of Psychology, Michigan State University, East Lansing, MI, USA
[2] Department of Educational, School, and Counseling Psychology, College of Education, & Missouri Prevention Science Institute, University of Missouri, Columbia, MO, USA

In this chapter, we resume the discussion of methods of analysis of direction dependence from an event-based, configural perspective (von Eye & Wiedermann, 2017a,b, 2018). In the contrasting, variable-oriented perspective, the effects of variables can be expressed in terms of changes in the distribution of the outcome variable (Dodge & Rousson, 2000, 2001; von Eye & DeShon, 2012; Wiedermann & von Eye, 2015; for an overview, see Wiedermann & Li, 2018, and Rosenström & García-Velázquez, 2020, this volume). The event-based perspective proceeds from the assumption that events (or causes) can be occurrences that did versus did not happen. For example, a car accident did versus did not occur, or a patient did versus did not take her medication. In addition to being suitable for this type of situation, the event-based perspective opens the doors to the analysis of both, effects that make an event occur and effects that prevent an event from occurring.

The extant literature on methods for the analysis of direction dependence, that is, the analysis of hypotheses that are compatible with direction dependence, started, in the domain of metric variables, with a univariate linear regression model (Dodge & Rousson, 2000, 2001; von Eye & DeShon, 2012; Wiedermann, Artner, & von Eye, 2017). This model requires the variable on the X-side of a model (the independent or causal variable) to be non-normal. The model is $Y = X\beta + \varepsilon$, where Y is the dependent, the outcome variable, X is the design matrix, and ε is the residual variable. In the original specification of this direction dependence model, there were only one variable each on the X- and Y-sides of the model, and, thus, X contained just the constant vector and the causal variable. When this model is applicable and X is non-normal, the outcome variable, Y, is closer to a normal distribution than X. This implies that X has a main effect on Y (i.e. $\beta \neq 0$).

In the domain of categorical variables, several approaches to analyzing direction dependence have been discussed. Examples include the approach by Eshima and Tabata, already proposed in 1999. This approach can be traced back to Goodman's (1973a, b) modified path analysis. It involves estimating path analysis-type log-linear models that represent but not explicitly test hypotheses of direction dependence.

A recent log-linear approach is based on von Eye and Wiedermann's (2016) *generalized direction of effect principle*. According to this principle, existing, that is, significant log-linear effects such as main effects and interactions can be removed from a fitting model when they are explained (caused) by other effects that are part of the model (cf. Wiedermann & von Eye, 2017).

These two approaches to direction dependence in categorical variables are clearly variable-oriented because they relate variables to each other, and effects are expressed in terms of main effects and interactions. A third approach is event-based and, thus, person-oriented (von Eye & Bergman, 2003; von Eye, Bergman, & Hsieh, 2015). Statements about the origin and outcome of effects are made in terms of categorical events instead of variables (von Eye & Wiedermann, 2018). A suitable method of analysis is CFA (Lienert & Krauth, 1975; von Eye, 2002; von Eye & Gutiérrez Peña, 2004), for which expected values are determined based on theoretical knowledge or estimated using log-linear models (the CFA base models; an explanation follows below).

The main characteristic of causal CFA is that configurations, that is, individual cells or groups of cells, emerge that are interpreted with a causal theory in mind. When such theory does not exist, CFA applications are exploratory in nature. That is, cells or groups of cells are searched for that reflect putative causal processes. In contrast to CFA, log-linear models considered here specify direction dependence, that is, directional hypotheses at the level of variable relations. Log-linear modeling is, usually, explanatory in nature. In addition, and just as in regression analysis-type direction dependence analysis, directional log-linear models are usually compared with competing models. Most of these are models that reverse the causal direction of effect. In CFA, competing models are not usually considered.

In this chapter, we pursue two goals. First, we discuss the *variable-oriented and the event-based approaches to direction dependence analysis in categorical variables*. We pursue an approach that has the following characteristics:

- it is asymmetric; that is, reversing the causal direction of effects can result in different goodness-of-fit statistics;
- it targets selected configurations and configuration patterns;

- it is explanatory; that is, targeted configurations or configuration patterns are specified a priori, that is, before data analysis; exploratory application is conceivable, but not discussed in detail in this chapter;
- it allows researchers to analyze directed effects that increase the probability of outcome events; it also allows one to analyze effects that decrease the probability of particular outcomes;
- it allows one to contrast configurations that are in support of a directional hypothesis with those that contradict this hypothesis, and
- it allows one to simultaneously test multiple hypotheses in the same data set.

The second aim pursued with this chapter is the *comparative application of three specific methods of analysis*: log-linear modeling, CFA, and Prediction Analysis (Froman & Hubert, 1980; Hildebrand, Laing, & Rosenthal, 1977; von Eye, 2005; von Eye & Brandtstädter, 1988a,b). The main goal of the comparison is to highlight characteristics of these three methods that are of importance in direction dependence analysis.

The remainder of this chapter is structured as follows. First, we explain how directional hypotheses can be specified for categorical variables. To this aim, we use statement calculus. Second, we explain how such hypotheses can be translated into specific hypotheses in log-linear models. Third, we present a data example and the comparative application of the three methods of analysis.

8.1 Specifying Directional Hypotheses in Categorical Variables

To be able to specify directional hypotheses in a way that possesses the above characteristics, we use two-valued statement calculus. Given that statements are either *true* (t) or *false* (f), statement calculus provides a set of rules that can be used to determine whether *composite statements* are true or false. Composite statements link two or more statements. Consider the two statements p and q. Classical links include the logical

- *or* ($\vee$): "*p or q*"; the meaning of this link is that at least one of $\{p, q\}$ must be true (must have been observed) for the composite statement $p \vee q$ to be true;
- *and* ($\wedge$): "*p and q*": the meaning of this link is that both of $\{p, q\}$ must be true (must have been observed) for the composite statement $p \wedge q$ to be true;
- *implication* ($\rightarrow$): "*if p then q*"; the meaning of this link is that "*p* is sufficient for q to occur," or, in other words, "*p* not without q," or "q is necessary for p;" and the

Table 8.1 Truth table for the two statements p and q and five links.

p	q	$p \vee q$	$p \wedge q$	$p \to q$	$p \leftrightarrow q$	$p \leftarrow q$
t[a)]	t	t	t	t	t	t
t	f	t	f	f	f	t
f[a)]	t	t	f	t	f	f
f	f	f	f	t	t	t

a) "t" indicates a true statement, and "f" indicates a false statement.

- *equivalence* ($\leftrightarrow$): "q *iff* p"; the meaning of this link is that "q if and only if p" or "p is necessary and sufficient for q" or "p and q must have the same truth value for the composite statement $p \leftrightarrow q$ to be true."

Table 8.1, a standard "truth table," summarizes the definition and the truth values for p, q, and the composite statements that use these four links. We also add the reverse implication, in the last column, the one from q to p, that can be used in the description of the methods discussed in this chapter and when competing, opposite-direction hypotheses are tested.

In the following sections, we extend the discussions by von Eye (2005) and von Eye and Wiedermann (2018) by focusing on the implication link, symmetry, and multivariate data situations. Before doing this, however, we illustrate how the logical links can be used in the analysis of direction dependence hypotheses.

Table 8.1 shows that the three links *or, and,* and *equivalence* are symmetric in the sense that reversing the order of the statements p and q will not change the truth of the composite statement. Because of this symmetry, these links are not of central importance for the analysis of directional hypotheses. In contrast, reversing the order of the statements in the *implication* can change the truth of the composite statement. This can be seen by comparing the fifth with the seventh column in Table 8.1. In the fifth column, the first *true* is given for the case in which p and q are both true. That is, an implication is true when both statements are true. In other words and in the context of direction dependence, an implication is true when both, the independent (the *causal*) and the dependent (the *outcome* or *effect*) events did occur. The sole *false* is given for the case in which p is true and q is false. That is, a false conclusion from a true statement cannot be considered a proper implication. In the following two cells, *true* is given because p is false, and, according to the rules, anything can be concluded when the premise is false.

In the seventh column of Table 8.1, the first true is also given again when p and q are both true (first row in columns 5 and 7). The sole false, however, is given in row 3, that is, when q is true and p is false. This pattern is true for the implication $p \to q$ but false for the reverse implication, $p \leftarrow q$. Therefore, the pattern {p = true,

q = false} can be used to distinguish $p \to q$ from $p \leftarrow q$, and is, thus of interest in the analysis of direction dependence (see also Peters, Janzing, & Schölkopf, 2017). This applies accordingly for the pattern {p = false, q = true} in the second row of Table 8.1.

Of note is that, for both, $p \to q$ and $p \leftarrow q$, the composite statement is true when the antecedent is false. In the context of observational research, this would mean that anything can be concluded when the antecedent was not observed. This kind of prediction is hard to justify in empirical, observational research. Therefore, we exclude, in the following discussion, patterns from consideration that are based on events that did not occur or were not observed. The question of how this can be reconciled with counterfactual causal theory is material for discussion in future work.

We now ask how to translate an implication from a truth table to a design matrix for contingency table analysis. We illustrate this translation using the case of two binary variables (von Eye & Wiedermann, 2018). Consider the two variables, X and Y, and the $X \times Y$ cross-classification. The cells ij of this cross-classification contain the joint probabilities of the categories of X and Y. The marginals represent the probabilities of the categories of X and of Y. Now, let the event-based direction of effect originate in the first category of X, x_1, and let it target the second category of Y, y_2, or, expressed differently, $x_1 \to y_2$. Then, Cell 1 2 contains those cases that confirm the hypothesis or, in other words, for which the implication comes true. For these cases, both, x_1 and y_2 were observed. Cell 1 1 contains those cases for which the implication is contradicted. For these cases, the antecedent, x_1, was observed, but not the consequence, y_2. Table 8.2 displays this scenario (cf. von Eye & Wiedermann, 2018).

When translating this scenario into a design matrix for a log-linear model (using effect coding), one obtains, when both main effects are considered, the matrix of a saturated model,

$$W = \begin{bmatrix} 1 & 1 & 1 & 1 \\ 1 & 1 & -1 & -1 \\ 1 & -1 & 1 & 0 \\ 1 & -1 & -1 & 0 \end{bmatrix}.$$

Table 8.2 Direction of effect for $x_1 \to y_2$.

Categories	y_1	y_2
x_1		
x_2		

In this matrix, the first column represents the constant. The second column represents the main effect of X. It contrasts the first row in Table 8.2 with the second. The third column contains the main effect of Y. It contrasts the first column in Table 8.2 with the second. The fourth column of W contrasts Cell 1 2 with Cell 1 1. These are the cells that confirm the implication $x_1 \rightarrow y_2$ (Cell 2 1) and those that disconfirm it (Cell 1 1).

It is worth discussing whether the third and the fourth vectors in this design matrix are contradictory. The fourth vector contrasts Cells 1 1 and 1 2. The third vector contrasts the same cells but, in addition, Cells 2 1 and 2 2. By using the same scores in the design matrix, the hypothesis proposed in the third vector proposes that Cells 1 1 and 2 1 are equiprobable, and that Cells 1 2 and 2 2 are equiprobable as well. It should be noted that the columns of W are not orthogonal. Thus, the two vectors are redundant, in part. The third vector is needed only if the second and third elements of the corresponding hypothesis are part of a model, that is, when Cells 1 2 and 2 2 need to be contrasted and when the statements of equiprobability need to be tested. When the third vector is part of the model, the redundancy results in a strong correlation between the main effect vector of Y and the vector for the implication. In the present example, this correlation is $r = 0.707$. In contrast, the correlation between the main effect vector of X and the vector for the implication is $r = 0$. This issue is taken up again later in this chapter. If, however, the directional hypothesis is hypothesized to substitute the main effect of the dependent variable, the third vector in the above design matrix is not needed, can be removed, and the matrix will be orthogonal.

To discuss the event-based perspective taken in this chapter more precisely, we need to define the term *event*. For the present purposes, we define *event* as a *pattern of categories* that are used to describe an observation. To test a directional hypothesis such as an implication, such a pattern needs to be specified both on the antecedent and the outcome sides. In the example in Table 8.2, x_1 is the event on the antecedent side and y_2 is the event on the outcome side. Both of these patterns contain just one variable category. The composite statement $x_1 \rightarrow y_2$ contains two variable categories, one from the independent and the other from the dependent variables.

To create *multivariate event patterns*, one uses multiple antecedents, that is, independent or causal variables, and/or multiple outcomes, that is, dependent variables. Let, for instance, A and B be on the antecedent side and D and E on the outcome side. Let A, D, and E be binary, and let B have three categories, that is, for example, person characteristics. Then, one implication could be $a_1 b_3 \rightarrow c_2 d_1$. Another could be $a_2 b_2 \rightarrow c_1 d_2$. Cells that contain cases that confirm composite statements, that is, in the present context, implications, are termed *hit cells*. Cells that contain cases that disconfirm composite statements are termed *error cells*. Table 8.3 depicts these two sample implications in a way analogous to Table 8.2.

Table 8.3 Hit cells for the implications $a_1 b_3 \rightarrow c_2 d_1$ and $a_2 b_2 \rightarrow c_1 d_2$.

Category patterns	Outcomes			
Antecedents	$c_1 d_1$	$c_1 d_2$	$c_2 d_1$	$c_2 d_2$
$a_1 b_1$				
$a_1 b_2$				
$a_1 b_3$	———————————————————→		◎	
$a_2 b_1$				
$a_2 b_2$	——————————→	◎		
$a_2 b_3$				

In Table 8.3, the two events under study for the first implication are $a_1 b_3$ and $c_2 d_1$. The cases in Cell 1 3 2 1 confirm the implication $a_1 b_3 \rightarrow c_2 d_1$. Cell 1 3 2 1 constitutes, thus, a *hit cell*. The cases in all other cells of this row disconfirm the implication. These are the Cells 1 3 1 1, 1 3 1 2, and 1 3 2 2. These cells, thus, constitute *error cells*. As was said above, the same cross-classification can be used to test multiple directional hypotheses. In the present example, the antecedent and the outcome events of the second implication are $a_2 b_2$ and $c_1 d_2$. The cases in Cell 2 2 1 2 confirm this implication, and the cases in Cells 2 2 1 1, 2 2 2 1, and 2 2 2 2 disconfirm it.

The present example illustrates two characteristics of the event-based approach to analyzing direction dependence hypotheses. First, both on the antecedent and the outcome sides, there can be multiple variables. The approach, thus, is applicable in multivariate settings, in a natural way. Second, multiple hypotheses can be tested in the same data matrix. Currently, when metric variables are analyzed, one assumes that the nature of direction dependence is the same across the entire range of possible scale values. This way, one cannot test multiple hypotheses. For multiple hypotheses, one would have to introduce specific interaction terms, specify moderator models (see, e.g. Li & Wiedermann, in press), or consider response surface models (for an introduction, see Kutner, Neter, Nachtsheim, & Li, 2005). Again, this is material for future work.

This example also illustrates what has been called *irrelevant cells* (Brandtstädter & von Eye, 1986). These cells are part of a cross-classification but they are neither hit nor error cells because they are not included in any directional hypothesis. One might be tempted to remove rows from a cross-classification that contain irrelevant cells. This must be done with great caution, however, because the frequencies in these cells can have an effect on the estimated expected frequencies in the other cells. In the following section, we discuss types of directional hypotheses.

8.2 Types of Directional Hypotheses

In this section, we discuss three types of directional hypotheses. The first involves two groups of events, one constituting the antecedent (or premise), the other constituting the consequence (or outcome). The second group of directional hypotheses has more than one premise for the same outcome, more than one outcome for the same premise, or multiple events on both the premise and the outcome sides. The third group of directional hypotheses covers moderator models. That is, this group covers models in which direction dependence varies over the categories of a variable, the moderator.

Examples of the first group of models were given already in the first section of this chapter (cf. von Eye & Wiedermann, 2018). Let $X_{i,j,...}$ be the antecedent event or premise, where subscript i indexes variables, and subscript j indexes categories, and let $Y_{i,j,...}$ be the event on the outcome side. Then, the implication that constitutes the directional hypothesis is $X_{i,j,...} \rightarrow Y_{i,j,...}$. The link used to specify both the antecedent and the outcome events in the first group of models is the logical $\wedge$, the *and* (see Table 8.1). Only events that exhibit all characteristics of a particular event can represent this event. More specifically, the above directional hypothesis can be written as $X_{i_1 j_1} \wedge X_{i_2 j_2} \wedge ... \rightarrow Y_{i_1 j_1} \wedge Y_{i_2 j_2} ...$

8.2.1 Multiple Premises and Outcomes

In contrast, in the second group of models, operators other than the logical $\wedge$ are also used to specify *sets of events*. Consider two events that are hypothesized to lead to the same outcome. This form or relation is known as the *wedge* (Mair & von Eye, 2007; von Eye & Brandtstädter, 1998), and the outcome is called *multicausal*. In the present context, this form can be represented using the logical $\wedge$ and the logical $\vee$, the *or*.

To give an example, consider the two binary predictors X_1 and X_2 and the binary outcome Y. Crossed, these three variables span a $2 \times 2 \times 2$ table. Let the multicausal directional hypothesis be $X_{1,1} \wedge X_{2,1} \vee X_{1,2} \wedge X_{2,1} \rightarrow Y_2$. Then, an illustration of this scenario could be, in analogy to Table 8.3, the one in Table 8.4 [1].

Now, consider an event that can have more than one outcome. This type of relation is known as the *fork* (Mair & von Eye, 2007; von Eye & Brandtstädter, 1998), and the corresponding antecedent is termed *multifinal*. Forks can also be represented by the logical $\wedge$ and the logical $\vee$. An example of a fork is the directional hypothesis $X_{i,j} \rightarrow Y_{i_1 j_1} \wedge Y_{i_2 j_2} ...$ In this example, outcome events exhibit Pattern $Y_{i_1 j_1}$, Pattern $Y_{i_2 j_2} \wedge ...$, or both. In either case, the antecedent event is X_{ij}.

---

1 Please note that the implication $X_{1,1} \wedge X_{2,1} \vee X_{1,2} \wedge X_{2,1} \rightarrow Y_2$ is redundant as far as X_1 is concerned. The hypothesis can also be read "regardless of the category that is observed for X, event $X_{2,1}$ leads to Y_2."

Table 8.4 Hit cells for the implication $X_{1,1} \wedge X_{2,1} \vee X_{1,2} \wedge X_{2,1} \to Y_2$.

	Outcome patterns	
Antecedent patterns	Y_1	Y_2
$X_{1,1}X_{2,1}$	$\longrightarrow$	◎
$X_{1,1}X_{2,2}$		
$X_{1,2}X_{2,1}$	$\longrightarrow$	◎
$X_{1,2}X_{2,2}$		

Table 8.5 Hit Cells for the implication $X_1 \to Y_{1,2} \wedge Y_{2,1} \vee Y_{1,2} \wedge Y_{2,2}$.

	Outcome patterns			
Antecedent patterns	$Y_{1,1}Y_{2,1}$	$Y_{1,1}Y_{2,2}$	$Y_{1,2}Y_{2,1}$	$Y_{1,2}Y_{2,2}$
X_1	$\longrightarrow$	◎		
X_2	$\longrightarrow$			◎

An example of a multifinal relation appears in Table 8.5. We use the binary antecedent X and the binary outcomes Y_1 and Y_2. The fork that is illustrated contains the joint hypotheses $X_1 \to Y_{1,2} \wedge Y_{2,1} \vee Y_{1,2} \wedge Y_{2,2}$.

In both forms, the wedge and the fork, additional logical links are conceivable. The same applies when both sides, the antecedent and the outcome, are constituted by multivariate events. An example of such a form is $X_{i_1j_1} \vee X_{i_2j_2} \wedge \ldots \to Y_{i_1j_1} \vee Y_{i_2j_2} \wedge \ldots$.

8.3 Analyzing Event-Based Directional Hypotheses

In this section, we review three approaches to analyzing event-based directional hypotheses. These approaches allow the researcher to answer different questions. The first approach is used to fit a log-linear model. The second approach is used to identify extreme cells in a cross-classification by way of CFA. The third approach contrasts hit cells with error cells by way of PA.

8.3.1 Log-Linear Models of Direction Dependence

Aims of this modeling approach include the specification of a model that allows researchers to make decisions concerning directional hypotheses in the context of a set of variables and their relations, and also concerning competing hypotheses, including those in which the direction of effect is reversed. A number of the effects included in the model will be unrelated to the hypothesized direction of effect. These effects must be included in the model nevertheless because the model must not fail because these effects were not considered. Examples of such effects include main effects. The directional relations are usually specified in the form of special effects. These effects are included in the model so that they replace standard hierarchical model terms such as interactions that could represent relations among antecedent and outcome variables. These effects cannot be cast in terms of a standard, hierarchical log-linear model or even a non-hierarchical model (e.g. Wiedermann & von Eye, 2017). Most models that are specified to test directional hypotheses are, thus, nonstandard (see Mair & von Eye, 2007). *Nonstandard models for directional hypotheses* select from the following six sets of effects:

1. Main effects of all antecedent variables, including possible covariates;
2. All possible interactions among antecedent variables, including possible covariates; the main effects and interactions among all antecedent variables must be included in directional models because directional hypotheses must not be rejected because these effects – they are unrelated to the directional hypotheses – were omitted from a model;
3. Main effects on the outcome side; it should be noted that main effects on the outcome side are not part of a model when special, directional effects of the kind exemplified in Tables 8.2–8.5 are hypothesized to cover (part of) the main effects; an example would be a directional hypothesis in which researchers propose that a main effect is carried by just a selection of categories;
4. Interactions among the variables on the outcome side; here again, interactions are not part of a model when special, directional effects are specified that are hypothesized to replace particular interactions; in addition, interactions are often not part of models for reasons of parsimony; this applies in particular to higher order interactions, and that even when these are not part of directional hypotheses;
5. Antecedent – outcome interactions; once more, these terms are not part of a model when they are hypothesized to be replaceable by special directional effects of the kind exemplified in the tables in the last section; the strive for parsimony can also have the effect that higher order antecedent – outcome interactions are deemed unnecessary for a fitting model;
6. Special effects that represent the directional hypotheses.

In the following example, we walk the reader through the process of specifying the design matrix for a log-linear model for the hypothesis $X_{1,1} \wedge X_{2,1} \vee X_{1,2} \wedge X_{2,1} \rightarrow Y_2$ from Table 8.4. First, we identify the patterns of interest and, then, we specify the design matrix.

Table 8.6 Truth table for the implication
$X_{1,1} \wedge X_{2,1} \vee X_{1,2} \wedge X_{2,1} \to Y_2$ (all variables binary).

Antecedents		Outcome	Predictive patterns observed – not observed	Implication
X_1	X_2	Y		
t	t	t	t	t
t	t	f	f	f
t	f	t	f	t
t	f	f	f	t
f	t	t	t	t
f	t	f	f	f
f	f	t	f	t
f	f	f	f	t

In the example, the two binary predictors X_1 and X_2 are crossed with the binary outcome Y, to span a $2 \times 2 \times 2$ table. To analyze the multicausal directional hypothesis illustrated in Table 8.4, we take an event-based perspective. Specifically, we note that both event patterns, $X_{1,1} \wedge X_{2,1}$ and $X_{1,2} \wedge X_{2,1}$, imply Y_2. For Y_2 to be observable, the antecedents are either $X_{1,1} \wedge X_{2,1}$, $X_{1,2} \wedge X_{2,1}$, or both. The truth table in Table 8.6 indicates the events that confirm versus disconfirm this implication.

There is an obvious difference between a standard truth table and Table 8.6. The link $\wedge$ does not yield the same results as the ones given in the fourth column in Table 8.6. Using the link $\wedge$, one asks whether two events co-occur that can be either true or false. In contrast, in the fourth column of Table 8.6, we ask whether the antecedent events $X_{1,1} \wedge X_{2,1}$ and $X_{1,2} \wedge X_{2,1}$ are observed with or without Y_2. We find the patterns of interest, that is, the patterns that confirm $X_{1,1} \wedge X_{2,1} \vee X_{1,2} \wedge X_{2,1} \to Y_2$, in rows 1 and 5 of Table 8.6 (shaded gray). Rows 2 and 6, shaded dark gray, disconfirm the implication.

It should be noted in addition, that the implication is true in many more instances than indicated by the shading in Table 8.6. In fact, with the exception of the disconfirming events in rows 2 and 6, all event patterns are in support of the implication. In the present context, we exclusively focus on the patterns t – t – t and f – t – t. The other events would confirm the implication as well. However, they do not describe the pattern of interest. For example, the fourth row represents a pattern for which the $\wedge$ would suggest that the joint event is false. Here, the joint event is true because the premise is false, and the outcome was not observed. As we said above, this kind of pattern is of lesser interest in the analysis of direction dependence because we ask whether two true events or a false and a true event were co-observed with a true outcome. We now derive the effect vectors to be included in the design matrix for a log-linear model.

Looking at the antecedent variables, we note that there are no constraints placed concerning their main effects and interactions. The directional hypothesis is posed regardless of the existence of these effects. Therefore, we estimate both the main effects of X_1 and X_2, and the $X_1 \times X_2$ interaction. Similarly, the directional hypothesis includes no statement about the main effect of the outcome variable, Y. Therefore, the main effect of Y will be part of the model as well.

The directional hypothesis itself consists of two parts. The first contrasts Pattern t – t – t (Cell 1 1 1) with Pattern t – t – f (Cell 1 1 2) the second part contrasts Pattern f – t – t (Cell 2 1 1) with Pattern f – t – f (Cell 2 1 2). When researchers hypothesize that these two effects can differ in magnitude, one effect vector each is specified for the two contrasts. When in contrast, the hypothesis is tested that the two effects are equal in size, one joint vector is specified. The design matrix for the first case is

$$
W = \begin{bmatrix}
1 & 1 & 1 & 1 & 1 & 1 & 0 \\
1 & 1 & 1 & -1 & 1 & -1 & 0 \\
1 & 1 & -1 & 1 & -1 & 0 & 0 \\
1 & 1 & -1 & -1 & -1 & 0 & 0 \\
1 & -1 & 1 & 1 & -1 & 0 & 1 \\
1 & -1 & 1 & -1 & -1 & 0 & -1 \\
1 & -1 & -1 & 1 & 1 & 0 & 0 \\
1 & -1 & -1 & -1 & 1 & 0 & 0
\end{bmatrix}.
$$

The first column in this design matrix represents the constant of the log-linear model. The second, third, and fourth columns represent the main effects of the variables that span the cross-classification of X_1, X_2, and Y. The fifth column represents the interaction between X_1 and X_2. The last two columns represent the two elements of the directional hypothesis. Specifically, in the sixth column, Cell 1 1 1 is contrasted with Cell 1 1 2, thus representing the first pair of events that are linked by the ∧ of the directional hypothesis $X_{1,1} \wedge X_{2,1} \vee X_{1,2} \wedge X_{2,1} \rightarrow Y_2$. The last column contrasts Cell 2 1 1 with Cell 2 1 2 and, thus, represents the second pair of events that are linked by an ∧.

When all of the vectors in this design matrix are used in the same model, this model can be represented by

$$
\log m = \lambda + \lambda^{X_1} + \lambda^{X_2} + \lambda^{X_1 \times X_2} + \lambda^{Y} + \lambda^{E_1} + \lambda^{E_2},
$$

Table 8.7 Correlation matrix of the design matrix for the hypothesis $X_{1,1} \wedge X_{2,1} \vee X_{1,2} \wedge X_{2,1} \rightarrow Y_2$.

	X_1	X_2	Y	$X_1 \times X_2$	E_1	E_2
X_1	1.0					
X_2	0.0	1.0				
Y	0.0	0.0	1.0			
$X_1 \times X_2$	0.0	0.0	0.0	1.0		
E_1	0.0	0.0	0.5	0.0	1.0	
E_2	0.0	0.0	0.5	0.0	0.0	1.0

where X_1 and X_2 represent the antecedent variables, Y represents the outcome variable, $X_1 \times X_2$ indicates the interaction between the two antecedents, and E_1 and E_2 represent the special contrasts of the directional hypothesis.

When the hypothesis is tested that the two effects are equal in size, the last two vectors in the design matrix are fused. One obtains $\{1 -1\ 0\ 0\ 1 -1\ 0\ 0\}$.

8.3.1.1 Identification Issues

For a clear-cut interpretation of parameters in log-linear models, it is required that the columns of the design matrix be orthogonal (Mair & von Eye, 2007; von Eye & Mun, 2013; Rindskopf, 1999). Evidently, this is not the case for the above example. The correlation matrix in Table 8.7 displays the correlations between the six effect coding vectors in the above design matrix, W.

The correlation matrix in Table 8.7 shows that those vectors of the log-linear model that are elements of a standard hierarchical model, that is, the vectors for the main effects of X_1, X_2, and Y, and the interaction between X_1 and X_2 are orthogonal and, thus, uncorrelated. In contrast, the two special effect vectors are correlated with the main effect of the outcome variable, Y.

When a model of this type is applied in real-world data analysis, researchers have three options. The first involves using all effects with the exception of the main effect of Y. Directional log-linear models of this type have been proposed by Wiedermann and von Eye (2017). This approach is preferred when the hypothesis implies that the special contrasts represent the main effect of Y entirely.

The second approach is a stepwise procedure. In the first step, the model is estimated without the special contrasts but including the main effect of Y. In the second step, the first of the two special effects is included in the model. In the third step, the second of the special effects is included. These three models are nested and can be compared statistically. When the difference in the overall goodness-of-fit values of these two models suggests that the second, third, or both are significantly better than the first, one can conclude that the directional hypothesis makes a significant contribution to the explanation of the joint frequency distribution of X_1, X_2, and Y. The inspection of the parameters for E_1 and E_2 can still be problematic because of the lack of orthogonality of the corresponding vectors in the design matrix. At the least, however, the difference test provides information about the viability of the directional hypothesis. This applies accordingly when the fused special effects vector is used.

The third approach involves using the entire design matrix given above, but performing a Schuster transformation before estimating the log-linear model (von Eye, Schuster, & Rogers, 1998). This approach has been discussed in the context of method development for direction dependence analysis by von Eye and Wiedermann (2017a, b) and will, therefore, not be described in detail in this chapter. In brief, this method transforms the original design matrix (e.g. the one created for the present example) such that parameter estimates can be interpreted as intended even when the original design matrix is not orthogonal. The only precondition that must be fulfilled for this transformation is that the inverse of $W'W$ exists, where W is the design matrix.

8.3.2 Confirmatory Configural Frequency Analysis (CFA) of Direction Dependence

CFA (Lienert & Krauth, 1975; von Eye, 2002; von Eye, Mair, & Mun, 2010) allows one to inspect individual cells or groups of cells in cross-classifications instead of variable relations. Cells stand out because they contain more cases than expected (CFA types) or fewer cases than expected (CFA antitypes). The expected cell frequencies are estimated by a base model, usually a log-linear model. This model includes all (and only) those effects that a researcher is *not* interested in. When, under this model, types or antitypes emerge, the effects of interest are bound to exist, and the types and antitypes indicate where, in the cross-classification, these effects are visible. The majority of CFA applications is exploratory in nature. In the present context, however, we discuss explanatory CFA.

For direction dependence analysis, a CFA base model contains the following effects (for a detailed discussion of CFA base models, see von Eye, 2004):

1. All main effects and all interactions among antecedent variables; in the above example, these would be the main effects of X_1 and X_2, and the interaction between X_1 and X_2;
2. All main effects and all interactions among the outcome variables; in the above example, this would entail the main effect of the sole outcome variable, Y.

To give an example of a CFA base model for the detection of elements of direction dependence, consider, again the example given in the last section. This example used the directional hypothesis $X_{1,1} \wedge X_{2,1} \vee X_{1,2} \wedge X_{2,1} \rightarrow Y_2$. For log-linear modeling, the model to be estimated was $\log m = \lambda + \lambda^{X_1} + \lambda^{X_2} + \lambda^{X_1 \times X_2} + \lambda^Y + \lambda^{E_1} + \lambda^{E_2}$. When this model fits, types and antitypes will not surface. For CFA, this model is estimated without the special effects. It, thus, becomes $\log m = \lambda + \lambda^{X_1} + \lambda^{X_2} + \lambda^{X_1 \times X_2} + \lambda^Y$. In configural analysis, the cells of interest are inspected. These are Cells 1 1 1, 1 1 2, 2 1 1, and 2 1 2. Of these, Cells 1 1 1 and 2 1 1 need to constitute types and Cells 1 1 2 and 2 1 2 need to constitute antitypes for the directional hypothesis to be confirmed.

It is important to note that the special effects that represent the directional hypotheses are not part of a CFA base model. In the above example, this would be the effects E_1 and E_2. The reason for this is that CFA focuses on individual cells or groups of cells, and asks whether they contain more (types) or fewer (antitypes) cases than expected under a base model. CFA does not contrast cells with each other. In the above example, the pairs of cells of interest are 1 1 1 versus 1 1 2, and 2 1 1 versus 2 1 2. To support the directional hypothesis and given the base model, Cell 1 1 1 should contain more cases than expected and, thus, constitute a CFA type. In addition and, also necessary to support the hypothesis, Cell 1 1 2 should contain fewer cases than expected and, thus, constitute an antitype. Correspondingly, Cell 2 1 1 should constitute a type and Cell 2 1 2 should constitute an antitype.

To test a directional hypothesis at the level of individual cells, researchers have two options. The first is to inspect each cell of interest individually. In the above example, this approach implies four CFA tests. The second approach views the type-constituting cells as a first composite and the antitype-constituting cells as a second composite. For each composite, one CFA test is performed.

When each cell is tested individually, most any test can be applied that has been proposed for CFA. This includes, when sampling is multinomial, for example, the binomial test, various χ^2-approximations, and the z-test. When sampling is product-multinomial, Lehmacher's (1981) or Lindner's (1984) hypergeometric tests become applicable.

Evidently, the approach proposed here is one of *confirmatory* CFA. In contrast to exploratory CFA, this approach comes with considerably more power. In all CFA applications, mutual dependence of multiple tests as well as capitalizing on chance are counteracted by adjusting the significance threshold with reference to the number of tests performed. In exploratory CFA, this number is t, the number of cells in a cross-classification (minus the number of fixed cells). In confirmatory CFA, this number is given by the number of cells of interest, t'. Usually, $t' < t$, which renders the adjusted significance threshold much less extreme or prohibitive.

Consider, again, the above example. The $X_1 \times X_2 \times Y$ cross-classification has eight cells. Let the nominal significance threshold be 0.05. Using the Bonferroni procedure, the protected significance threshold is $0.05/8 = 0.00\,625$. Confirmatory CFA specifies only four cells of interest. Using the Bonferroni procedure again, the protected threshold becomes $0.05/4 = 0.0125$, that is, a far less extreme threshold.

8.3.3 Prediction Analysis of Cross-Classifications

Originally proposed by Hildebrand, Laing, and Rosenthal (1974); Hildebrand et al. (1977); Froman & Hubert, 1980; cf. Goodman & Kruskal, 1974a, b; Hildebrand, Laing, & Rosenthal, 1973; Hildebrand et al., 1974; von Eye & Brandtstädter, 1988a, b), *Prediction Analysis of Cross-Classifications* (PA) is a method that allows one to evaluate the success of predictive, that is, directional hypotheses in cross-classifications of categorical variables. PA proceeds in three steps: (i) specification of prediction hypotheses, (ii) estimation of expected cell frequencies, and (iii) comparison of observed and expected cell frequencies in hit cells and error cells. In the following paragraphs, we explicate each of these steps.

(1) To specify *prediction hypotheses*, the authors used tools parallel to the ones described above, that is, statement calculus. These tools, developed specifically for PA, are termed *prediction logic*. They follow the same rules as statement calculus. However, they also integrate the context of statistical testing. Therefore, statements are probabilistic instead of deterministic. In the present chapter, we continue the use of standard statement calculus. The sets of hit and error cells that are identified are identical to the ones one would identify with prediction logic, and, in either case, statistical hypotheses are formulated using composite statements such as implications. The context of statistical testing is obvious. Therefore, we consider the introduction of prediction logic unnecessary for the present purposes and simplify matters by using just one method of hypothesis specification.

(2) For the *estimation of expected cell frequencies*, we use log-linear models. This is, again, in contrast to Hildebrand et al. (1977) work, who presented closed forms for the estimation process. These forms are, as far as we are aware, special cases of log-linear models. Specifically, these are models that Goodman (1979) called *simple models*. The iterative estimation methods used for all log-linear models converge for these models in just one iteration. This way, closed forms can be given.

Here, we use the more general log-linear formulation. This approach allows one to generalize PS and introduce more complex models such as, for example, models with special effects, structural zeros, or covariates. The models to be specified for PA group variables in two classes, those on the predictor side and those on the outcome side. The models are saturated on both sides and posit independence between predictors and outcome variables. Therefore, just as the log-linear models that are specified for log-linear and configural analysis of direction dependence, the models for PA contain the following terms.

1. All main effects of the variables on the predictor side;
2. all possible interactions among the variables on the predictor side;
3. all main effects of the variables on the outcome side; and
4. all interactions of the variables on the outcome side.

Neither effects that link predictor with outcome variables nor special effects are part of a PA model in the Hildebrand et al. (1977) tradition. The reason for this is that hit and error cells are compared based on the observed cell frequencies and with reference to the thus estimated expected cell frequencies. Please note that, in principle, covariates can be included in a PA model, and moderator hypotheses can be considered as well. To the best of our knowledge, however, neither of these options has been elaborated in the literature. We conclude, already at this point of the discussion, that PA is comparable to CFA in that it uses the same base model and inspects hit and error cells in a step after the estimation of expected cell frequencies, and that PA is comparable to log-linear modeling in that it contrasts hit with error cells.

To give an example of a PA model for the detection of elements of direction dependence, consider, again the example given in the last section. This example used the directional hypothesis $X_{1,1} \wedge X_{2,1} \vee X_{1,2} \wedge X_{2,1} \to Y_2$. For log-linear modeling, the model to be estimated was $\log m = \lambda + \lambda^{X_1} + \lambda^{X_2} + \lambda^{X_1 \times X_2} + \lambda^Y + \lambda^{E_1} + \lambda^{E_2}$. When this model fits, types and antitypes will not surface. For CFA, this model is $\log m = \lambda + \lambda^{X_1} + \lambda^{X_2} + \lambda^{X_1 \times X_2} + \lambda^Y$. In PA, the same model is used as in CFA. However, the cells of interest are inspected in a different way than in CFA. This is explained in the following paragraphs.

(3) *Comparison of observed and expected cell frequencies in hit and error cells.* PA compares the observed frequencies with respect to the expected frequencies in all hit cells and all error cells in just one test. In the following paragraphs, we describe the elements that go into this test, and the test itself.

Several tests have been proposed that can be used to make statistical decisions concerning directional hypotheses in PA. These tests include Hildebrand et al. (1977) z-approximation and von Eye and Brandtstädter's (1988b) binomial test. In this chapter, we apply the latter. To introduce the test, consider an $I \times J$ cross-classification in which the rows represent the I possible predictor patterns and the columns represent the J possible outcome patterns. The first element of this test is the sum of observed frequencies of those cases that confirm the directional hypothesis,

$$S_1 = \sum_{ij} o_{ij} \omega_{ij},$$

where $\omega_{ij} = 1$ if Cell ij contains cases that confirm the directional hypothesis, $\omega_{ij} = 0$ else, and o_{ij} is the observed frequency in Cell ij.

The second element of the test is the sum of expected frequencies in the same cells,

$$S_2 = \sum_{ij} e_{ij}\omega_{ij},$$

where e_{ij} is the expected frequency for Cell ij. This frequency was estimated using a log-linear model of the form discussed above. This model posits independence of predictors from outcome variables, i.e. no X–Y interactions are considered in the model.

Using S_1 and S_2, one can calculate the one-sided tail probability of the S_1 and extremer numbers of cases that confirm the directional hypothesis under the assumption of predictor – outcome independence as

$$P(S_1) = \sum_{S_1} \binom{n}{i} p^i (1-p)^{n-i},$$

where

$p = S_1/n$, with n being the sample size, is the relative frequency of a case confirming the directional hypothesis.

When $np > 10$, the binomial distribution sufficiently approximates the normal, and we can use the z-approximation

$$z = \frac{S_1 - S_2}{\sqrt{np(1-p)}}.$$

Both of these tests are meaningful in particular when more cases than expected confirm the hypotheses, or, when $S_1 - S_2 > 0$.

8.3.3.1 Descriptive Measures of Prediction Success

Based on S_1 and S_2, two descriptive measures have been derived that allow one to describe prediction success. The first (Hildebrand et al., 1977) is the *proportionate reduction in error* measure PRE. It is defined by

$$\text{PRE} = \frac{S_2 - S_1}{S_2} = 1 - \frac{S_1}{S_2}.$$

PRE can be interpreted as portion (or percentage, when multiplied by 100) of events that disconfirm the prediction hypothesis in the observed in comparison to the expected frequency distribution. For example, a value of PRE $= 0.35$ indicates that, in the observed frequency distribution, 35% fewer disconfirming cases are found in the observed than in the expected frequency distribution (for more detail on characteristics of PRE, see von Eye and Brandtstädter (1988a)).

Focusing on hit cells requires that parameter ω be redefined as
$\omega_{ij} = 1$ if Cell ij contains cases that confirm the directional hypothesis,

$\omega_{ij} = 0$ else.

Accordingly, $S_1 = \sum_{ij} o_{ij}\omega_{ij}$ and $S_2 = \sum_{ij} e_{ij}\omega_{ij}$ now describe hits instead of misses. Using this new definition, one can define

$$\mathrm{PIH}_1 = \frac{S_1 - S_2}{S_2} = \frac{S_1}{S_2} - 1$$

as a measure of *proportionate increase in hits* (von Eye & Brandtstädter, 1988a, b). The interpretation of this measure is that PIH_1 describes the relative increase in hits in the observed as compared to the expected frequency distribution. Given that the maximum increase is $\mathrm{PIH}_{1,\max} = \frac{n}{S_2} - 1$, one can express the proportionate increase in hits with reference to the maximally possible increase in hits as

$$\mathrm{PIH}_2 = \frac{\mathrm{PIH}_1}{\mathrm{PIH}_{1,\max}} = \frac{S_1 - S_2}{n - S_2}.$$

The range of possible values of PIH_2 is $-S_2/(n - S_2) \le \mathrm{PIH}_2 \le 1$. In the following section, we present a data example.

8.4 Data Example

In this section, we revisit the example given by von Eye and Brandtstädter (1988b; cf. von Eye, 2005). In a study on subjective control over development and perceptions of development in partnerships (Brandtstädter & von Eye, 1986), 950 respondents answered questions concerning *subjective developmental attainment* (SDA), *personal control over development* (PCD), *perceived marital support* (PMS), and *depressive outlook toward future development* (DEP). The first three of these variables were scored such that 1 indicates below median and 2 indicates above median scores. DEP had the three levels 1 = absence of depression, 2 = little depression, and 3 = strong depression.

von Eye (1991, 2005) showed that particular patterns of SDA, PCD, and PMS are predictive of particular levels of DEP. Here, we ask whether, taking an event-based perspective, DEP can be predicted from PCD and PMS alone. Table 8.8 presents the cross-classification of PCD and PMS with DEP. The table is arranged such that the four bivariate patterns of PCD and PMS constitute the rows and the three levels of DEP constitute the columns of the 4×3 [PCD, OMS] $\times$ [DEP] cross-classification (from von Eye, 1991, p. 109).

For the analysis of this frequency table, we put forth the same four directional hypotheses as von Eye (2005). However, instead of just using PA to analyze the

Table 8.8 [PCD, OMS] × [DEP] cross-classification.

PCD × PMS	Levels of depression		
	1	2	3
1 1	46	86	132
1 2	43	91	84
2 1	57	83	54
2 2	140	98	36

Hit cells are shaded (explanation follows below).

hypotheses, we also create results using log-linear models and CFA. The four hypotheses are:

1. Subjective lack of control over one's own development in tandem with lack of marital support results in depressed outlook concerning one's own future development. In terms of statement calculus, this hypothesis is specified as (PCD = 1) ∧ (PMS = 1) → (DEP = 3). The hit cell for this hypothesis is Cell 1 1 3. The corresponding error cells are 1 1 1 and 1 1 2.

2. Subjective lack of control over one's own development in tandem with above average marital support results in average or increased depressed outlook concerning one's own future development. In terms of statement calculus, this hypothesis is specified as (PCD = 1) ∧ (PMS = 2) → (DEP = 2) ∨ (DEP = 3). The hit cells for this hypothesis are 1 2 2 and 1 2 3. The corresponding error cell is 1 2 1.

3. Subjective control over one's own development in tandem with below average marital support results in absence of or little depressed outlook concerning one's own future development. In terms of statement calculus, this hypothesis is specified as (PCD = 2) ∧ (PMS = 1) → (DEP = 1) ∨ (DEP = 2). The hit cells for this hypothesis are 2 1 1 and 2 1 2. The corresponding error cell is 2 1 3.

4. Subjective control over one's own development in tandem with above average marital support results in absence of depressed outlook concerning one's own future development. In terms of statement calculus, this hypothesis is specified as (PCD = 2) ∧ (PMS = 2) → (DEP = 1). The hit cell for this hypothesis is 2 2 1. The corresponding error cells are 2 2 2 and 2 2 3.

In the following paragraphs, we re-analyze these four directional hypotheses using log-linear modeling, CFA, and PA.

8.4.1 Log-Linear Analysis

The following log-linear analysis is performed in two steps. The first involves estimating a hierarchical model in which all main effects and the two-way interaction between PCD and PMS are included. This model proposes independence between the two predictors, PCD and PMS, and the outcome variable, DEP. In other words, this model negates the directional dependence of DEP upon PCD and PMS. Specifically, this is the model

$$\log \hat{m} = \lambda + \lambda^{\text{PCD}} + \lambda^{\text{PMS}} + \lambda^{\text{DEP}} + \lambda^{\text{PCD}\times\text{PMS}}.$$

This model serves as base model in the same sense as CFA. If this model fails, we add special effect vectors for the four hypotheses to the design matrix. These are the vectors

- Hypothesis 1: $\{-0.5 \; -0.5 \; 1 \; 0 \; 0 \; 0 \; 0 \; 0 \; 0 \; 0 \; 0 \; 0\}$
- Hypothesis 2: $\{0 \; 0 \; 0 \; -1 \; 0.5 \; 0.5 \; 0 \; 0 \; 0 \; 0 \; 0 \; 0\}$
- Hypothesis 3: $\{0 \; 0 \; 0 \; 0 \; 0 \; 0 \; 0.5 \; 0.5 \; -1 \; 0 \; 0 \; 0\}$
- Hypothesis 4: $\{0 \; 0 \; 0 \; 0 \; 0 \; 0 \; 0 \; 0 \; 0 \; 1 \; -0.5 \; -0.5\}$.

Each of these vectors contrasts hit with error cells. There is one vector per row of the matrix in Table 8.8. The model is, thus

$$\log \hat{m} = \lambda + \lambda^{\text{PCD}} + \lambda^{\text{PMS}} + \lambda^{\text{DEP}} + \lambda^{\text{PCD}\times\text{PMS}} + \lambda^{H_1} + \lambda^{H_2} + \lambda^{H_3} + \lambda^{H_4},$$

where H_1 through H_4 label the four directional hypotheses. Table 8.9 summarizes overall goodness-of-fit information for these two models.

The two comparison models are hierarchically related. Therefore, they can be compared statistically. We obtain $\Delta X^2 = 111.90$ and $\Delta LR\text{-}X^2 = 113.38$, which, for $\Delta df = 4$, suggests that the model with the four special effects represents a significant improvement over the base model. Evidently, the directional hypotheses explain a significant portion of the variability in the $[\text{PCD}, \text{OMS}] \times [\text{DEP}]$ cross-classification. However, the results in Table 8.9 suggest that the total amount of variability that the model explains fails to result in acceptable fit. We, therefore, reject the model and conclude that, overall and in the context of a log-linear model, the directional hypotheses do not explain a sufficiently large portion of the variability of the frequency distribution in Table 8.8.

Table 8.9 Goodness of fit of two models for the analysis of the $[\text{PCD}, \text{OMS}] \times [\text{DEP}]$ cross-classification.

Model	df	X^2	$p(X^2)$	LR-X^2	$p(\text{LR-}X^2)$
Base model	6	127.16	<0.001	128.92	<0.001
Model with special effects	2	15.26	<0.001	15.55	<0.001

Without going into any detail, we note that the individual hypotheses could be evaluated individually. This could be achieved by estimating a series of models in which the vectors for individual hypotheses are added to the model, one after the other. This would result in a hierarchical model structure that is comparable to hierarchical regression models. In these structures, the contribution of the newly added term can be evaluated with reference to the terms already in the model. Considering, however, that the overall model does not represent the frequency distribution well, we refrain from performing these analyses. Another option involves estimating the model that includes the special terms after a Schuster transformation (von Eye et al., 1998). This transformation allows one to interpret the special effect parameters as intended. This step, however is not needed here either because the model does not fit. We, therefore, proceed to the configural analysis of the data in Table 8.8.

8.4.2 Configural Analysis

Using CFA, an event-based perspective is taken in which the observed frequencies of individual patterns of events are compared with the corresponding expected patterns. If types and antitypes emerge, it is asked whether they speak to the directional hypotheses. In the present example, the base model is the same as for the log-linear approach. In the CFA, we use the z-test and the Holland–Copenhaver procedure for the protection of the significance threshold which had been set to 0.05. Overall goodness-of-fit is the same as in the first result row of Table 8.9. Table 8.10 presents the CFA results.

The results in Table 8.10 suggest that three antitypes and two types emerged. We now ask whether these types and antitypes can be related to the directional hypotheses formulated above.

The first characteristic of the type–antitype pattern in Table 8.10 is that types and antitypes emerge for Hypotheses 1, 2, and 4, but not for Hypothesis 3. Specifically, we find for

- Hypothesis 1
  - Pattern 1 1 1 constitutes an antitype. This antitype suggests that, with reference to the base model, fewer respondents than expected exhibit below median personal control over their own development, below median marital support, and absence of depression. In the context of Hypothesis 1, this antitype can be interpreted such that it is less likely than chance, that is, less likely than one would expect with reference to the base model, that below median personal control and marital support cause absence of depression.
  - Pattern 1 1 3 constitutes a type. This type can be viewed as the counterpoint to the first antitype. It can be interpreted such that it is more likely than chance

Table 8.10 CFA of the [PCD, OMS] × [DEP] cross-classification.

Configuration	m	$\hat{m}$	Statistic	p	Type/antitype decision
111	46.00	79.478	−3.7552	0.000 087	Antitype
112	86.00	99.486	−1.3521	0.088 170	
113	132.00	85.036	5.0929	0.000 000	Type
121	43.00	65.629	−2.7933	0.002 608	Antitype
122	91.00	82.152	0.9762	0.164 472	
123	84.00	70.219	1.6446	0.050 028	
211	57.00	58.404	−0.1837	0.427 108	
212	83.00	73.107	1.1570	0.123 637	
213	54.00	62.488	−1.0738	0.141 454	
221	140.00	82.488	6.3323	0.000 000	Type
222	98.00	103.255	−0.5171	0.302 534	
223	36.00	88.257	−5.5625	0.000 000	Antitype

that below median levels of both personal control and martial support cause strong depression.
- Hypothesis 2
  - Pattern 1 2 1 constitutes an antitype. This antitype corresponds to antitype under Hypothesis 1. It suggests that it is less likely than chance that below median personal control in tandem with above median marital support cause absence of depression.
- Hypothesis 4
  - Pattern 2 2 1 constitutes a type. This type suggests that it is more likely than chance that above median personal control and marital support cause absence of depression.
  - Pattern 2 2 3 constitutes an antitype. This antitype suggests that it is less likely than chance that above median personal control and marital support cause strong depression.

The CFA results highlight the difference between CFA and log-linear modeling. In the latter, researchers attempt to specify a model that incorporates their hypotheses and describes the entire body of data. In CFA, individual cell frequencies are compared with their expectancies and statements are derived that characterize those cases that are found in individual cells. Types and antitypes indicate where significant deviations from the base model can be found.

Not surprisingly, the results from applying log-linear modeling and CFA to the present example are quite different. The log-linear models specified under the present four hypotheses fail to describe the frequency distribution in Table 8.10. Therefore, individual parameters cannot be interpreted (von Eye & Mun, 2013). In contrast, CFA did identify five patterns that contradict the base model (Table 8.10). Each of these patterns lends partial support to one of the hypotheses.

The type and antitype found for Hypothesis 1 reflect the basic idea of this hypothesis rather well. Below median control and marital support do lead to strong depression but not to absence of depression. Similarly, for Hypothesis 2, we find that above median control and below median marital support do not lead to absence of depression. This, again, can be interpreted as partially supporting Hypothesis 2. The other two patterns for predictor configuration 1 2 are not extreme enough to constitute types. They do describe, however, slightly more cases than expected. For Hypothesis 3, there is no support at all. No type or antitype emerged. For Hypothesis 4, support is as strong as for Hypothesis 1.

In all, and in contrast to the log-linear models, the results from CFA

1. do lend partial support to the four hypotheses of the data example, and, in addition,
2. tell the researcher where, in the table, hypotheses are supported and where not.

8.4.3 Prediction Analysis

PA proceeds in the following two steps. First, hit and error cells are specified. This was done in the context of Table 8.8, above. Second, the descriptive measures PRE and PIH are calculated. Third, significance tests are performed. In the present example, we obtain the following results (cf. von Eye, 2005).

The proportionate reduction in errors is 0.272. This corresponds to a 27.2% reduction in prediction errors when the prediction hypothesis is used as compared to the model of independence of predictors, PCD and PMS, and the outcome variable, DEP. Accordingly, using the prediction hypothesis results in a 30.04% increase in hits. The significance statistic $z = 8.81$ suggests that the prediction hypothesis is successful ($p < 0.001$). In all, we conclude that PA indicates that the prediction hypothesis finds significant support.

Now, one of the most important elements of the analysis of directional hypotheses and causal effects is directionality itself. Before retaining a directional hypothesis, one must establish direction or, in other words, one must establish the status of variables as predictors and outcome variables, or as cause and effect variables. To accomplish this in the present example, we now reverse the direction of effect by treating the two predictors, PCD and PMS, as outcome variables and DEP as the predictor.

8.5 Reversing Direction of Effect

The four hypotheses analyzed so far were

$(\text{PCD} = 1) \wedge (\text{PMS} = 1) \rightarrow (\text{DEP} = 3)$;

$(\text{PCD} = 1) \wedge (\text{PMS} = 2) \rightarrow (\text{DEP} = 2) \vee (\text{DEP} = 3)$;

$(\text{PCD} = 2) \wedge (\text{PMS} = 1) \rightarrow (\text{DEP} = 1) \vee (\text{DEP} = 2)$; and

$(\text{PCD} = 2) \wedge (\text{PMS} = 2) \rightarrow (\text{DEP} = 1)$.

When reversing direction results in the new specification, we realize that the original outcome variable, DEP, has three categories, and the combined original predictor variables, PCD and PMS, have four categories. We, thus have to arrive at the same hit cells coming from three categories instead of four. The number of hypotheses reduces from four to three. Specifically, the new prediction hypotheses are

- Hypothesis 1: $(\text{DEP} = 1) \rightarrow (\text{PCD} = 2) \wedge (\text{PMS} = 1) \vee (\text{PCD} = 2) \wedge (\text{PMS} = 2)$;
- Hypothesis 2: $(\text{DEP} = 2) \rightarrow (\text{PCD} = 1) \wedge (\text{PMS} = 2) \vee (\text{PCD} = 2) \wedge (\text{PMS} = 1)$; and
- Hypothesis 3: $(\text{DEP} = 3) \rightarrow (\text{PCD} = 1) \wedge (\text{PMS} = 1) \vee (\text{PCD} = 1) \wedge (\text{PMS} = 2)$.

The hit cells for the new first hypothesis are Cell 2 1 1 and Cell 2 2 1. The error cells are now Cells 1 1 1, and 1 2 1. The hit cells for the second hypothesis are Cells 1 2 2 and 2 1 2. The corresponding error cells are Cells 1 1 2 and 2 2 2. The hit cells for the third hypothesis are 1 1 3 and 1 2 3. The error cells for the third hypothesis are 2 1 3 and 2 3 3. Looking at Table 8.8, we find that, overall, the same cells are hit and error cells as before. However, for each individual hypothesis, the error cells are now in the columns of Table 8.8 instead of the rows. We now proceed to analyze the new hypotheses using log-linear modeling, CFA, and PA.

8.5.1 Log-Linear Modeling of the Re-Specified Hypotheses

As in the first analysis, two log-linear models are estimated to test the re-specified hypotheses. The first of these is the base model, unchanged from the first analysis. The second includes three special effect vectors, one each for the three hypotheses. Specifically, the three vectors are for

- Hypothesis 1: $\{-0.5, 0, 0, -0.5, 0, 0, 0.5, 0, 0, 0.5, 0, 0\}$;
- Hypothesis 2: $\{0, -0.5, 0, 0, -0.5, 0, 0, 0.5, 0, 0, 0.5, 0\}$, and
- Hypothesis 3: $\{0, 0, 0.5, 0, 0, 0.5, 0, 0, -0.5, 0, 0, -0.5\}$.

Table 8.11 Goodness of fit of base model and reverse-direction model for the analysis of the [PCD, OMS] × [DEP] cross-classification.

Model	df	X^2	$p(X^2)$	LR-X^2	$p($LR-$X^2)$
Base model	6	127.16	<0.001	128.92	<0.001
Model with special effects	3	54.25	<0.001	59.03	<0.001

Table 8.11 presents the overall goodness-of-fit results for the base model (from Table 8.9) and the model in which the direction of effect was reversed.

Clearly, the results in Table 8.11 suggest that the model in which the direction of effect was reversed fails to explain the frequency distribution in Table 8.8. The model is significantly better than the base model but far from being satisfactory.

Unfortunately, the reverse-direction and the original-direction models are not hierarchically related. Therefore, they cannot be statistically compared. Not even the numerically greater X^2-values of the reverse-direction model can be interpreted. Therefore, we conclude that the log-linear approach does not allow us to retain either directional model, in the present example.

8.5.2 CFA of the Re-Specified Hypotheses

In a configural analysis, the effects of interest are *not* part of the base model. Therefore, in the present context, the base models for the original-direction and the reverse-direction models are the same (cases where causally competing base models differ from each other have been discussed by von Eye and Wiedermann (2018)). In addition, the configurations of interest are the same. Therefore, the standard base model of the configural approach does not allow one to distinguish between two models in which opposite directional hypotheses are posited. However, application of von Eye and Mair's (2008a,b) *functional CFA* in combination with the *generalized principle of directional effects* (von Eye & Wiedermann, 2016) may help us shed light on the direction of effect in the present example.

Specifically, functional CFA is a search procedure in which patterns of types and antitypes are inspected that surface or disappear when particular effects are removed or included in a model. In the present context, the effects of interest are those that involve directional hypotheses. The generalized direction of effect principle posits that effects in a model disappear (become non-significant) when those effects are included in the model that are seen as causes of the disappearing effects. In CFA, this principle implies that types and antitypes disappear when the effects are added to the base model that explain these types and antitypes or are their

Table 8.12 CFA of the [PCD, OMS] × [DEP] cross-classification under two directional hypotheses.

	Original directional Hypothesis			Reversed directional Hypothesis		
Configuration	m	$\hat{m}$	Statistic	$\hat{m}$	Statistic	
111	46.00	49.99	−0.564	51.91	−0.820	
112	86.00	82.01	0.441	106.67	−2.001	
113	132.00	132.00	0.000	105.42	2.588	Type
121	43.00	42.99	0.000	37.09	0.970	
122	91.00	107.05	−1.551	105.58	−1.419	
123	84.00	67.96	1.946	75.33	0.999	
211	57.00	53.02	0.547	76.77	−2.257	Antitype
212	83.00	86.98	−0.427	68.42	1.763	
213	54.00	54.00	−0.000	48.81	0.743	
221	140.00	139.99	0.000	120.23	1.803	
222	98.00	81.97	1.771	77.36	2.350	Type
223	36.00	52.04	−2.223	76.44	−4.625	Antitype

causes. In the present example, these are the effects that were included in the log-linear models estimated in this chapter.

As was explained in the sections on log-linear modeling, the two sets of special effects were specified to represent the original and the opposite directions of effect. Included in the base model, these effects may or may not make the types and antitypes in Table 8.10 disappear. We, therefore, now add the four special effects (H_1–H_4) that led to the results in Table 8.9 to a first run of CFA and ask whether the type/antitype pattern remains unchanged. In a second run of CFA, we ask the same question for the three special effects (H_1–H_3) that led to the results in Table 8.11. The results of these two runs appear in Table 8.12.

The overall goodness-of-fit values for these two base models are the same as in the section on log-linear modeling. Both models are rejected. Therefore, types and antitypes are possible. For the model that represents the original directional hypothesis (left panel of Table 8.12), not a single type or antitype emerges. Not even the last configuration comes with a test statistic extreme enough to constitute a type (the p-value for −2.223 is $p = 0.0076$, a value larger than the Holland and Copenhaver-adjusted critical value of $p = 0.0042$).

In contrast, CFA of the reversed-direction hypothesis shows two types and two antitypes. One of these types (Cell 1 1 3) and one of these antitypes (Cell 2 2 3) had emerged already in Table 8.9, that is, when the special effect vectors had not been included. The antitype in Cell 2 1 1 and the type in Cell 2 2 3 had not emerged before.

The conclusion we draw from these results is clear. The special effect vectors for the original directional hypotheses make the types and antitypes completely disappear that supported the causal assumption. We, therefore, retain the causal assumption in the analysis at the event level.

8.5.3 PA of the Re-Specified Hypotheses

Virtually all cross-classifications to be analyzed with PA can be arranged in a row-by-column format (Hildebrand et al., 1977; von Eye & Brandtstädter, 1988a). In this format, the predictor patterns constitute the rows and the outcome patterns constitute the columns. Therefore, reversing the direction of a prediction hypothesis amounts to transposing the original cross-classification. The base model is unchanged, as it posits independence of predictors and outcome variables. The pattern of hit and error cells is unchanged as well. Accordingly, the overall goodness-of-fit of the reverse-direction cross-classification will be the same, and so will be the overall PRE and PIH statistics.

Therefore, PA does not allow one to distinguish between directional hypotheses that go in opposite directions. This applies even when PRE and PIH statistics are calculated row- versus column-wise. Their values are bound to differ. However, there are no significance tests for the row-wise statistics and there are, to the best of our knowledge, no approaches that would allow one to compare these row-wise statistics.

In all, PA is useful in particular when opposite-direction hypotheses can be excluded a priori. This is the case, for instance, when logic, factual knowledge, or temporal arrangement of predictors and outcome variables preclude an inversion of the directional hypotheses. In those instances in which such an inversion is possible, PA can be set aside and log-linear models or CFA might be preferable.

8.6 Discussion

In this chapter, we attempt to compare, via theoretical discussion and a data example, log-linear modeling, CFA, and PA in their approaches to the analysis of directional hypotheses. The conclusions of this comparison are as follows:

1. All three methods are capable of testing directional hypotheses. The models are comparable in that they all allow one (i) to analyze directional hypotheses in categorical variables and (ii) use the same base models. However, they go about testing the directional hypotheses in different ways.

(i) Log-linear modeling attempts to fit models that fit the entire cross-classification in a satisfactory way, that is, without significant local model-data deviations.[2] This approach is in accordance with the usual model fitting in the metric variables domain. The downside to this approach is that a model may fail not because the directional hypotheses are unrelated to reality but because the model is unable to describe the data in other sectors of the table. This may have been the case in the data example in the last sections.

(ii) CFA inspects individual cells or groups of cells in a cross-classification. The interpretation of extreme cells is usually done with reference to theory. In the present context, this is a theory that allows one to derive directional or causal hypotheses (cf. von Eye & Brandtstädter, 1997). Both, configural analysis and interpretation operate at the level of events. Variable relations, that is, the focus of log-linear modeling, are part of a CFA model in the sense that the relations of interest are excluded from the base model. Types and antitypes must, then, be the result of the existence of these relations.

(iii) PA identifies the sets of hit and error cells. These are the same cells that log-linear modeling and CFA of directional hypotheses focus on. Specific to PA is that all hit cells as a group are set in relation to all error cells as a group, with reference to a base model. The statistical test tells the researcher whether, overall, the directional hypotheses can be retained. This is possible even when individual hypotheses are strongly contradicted. Log-linear modeling and CFA would identify such contradictory hypotheses.[3] In contrast to CFA, individual cells are not inspected. In contrast to log-linear modeling, variable relations are not part of the modeling effort.

2. Among the most important characteristics of methods of analysis of directional hypotheses is their capability of distinguishing between hypotheses that reverse the direction of effect. In this respect, the three methods compared here differ fundamentally.

(i) Log-linear modeling methods allow one to design models that are specific for particular directional hypotheses. In practically all cases, these models can be estimated even when special care needs to be

---

2 Please notice that we did not discuss the options of (i) excluding cells from a model by declaring them structural cells and (ii) fitting local models (see Hand & Viniciotti, 2003). These options have also been discussed in the context of CFA (von Eye et al., 2010), but not in the context of PA. Therefore, these options were not included in this chapter.

3 Please remember that PA does allow one to inspect individual hypotheses by calculating descriptive measures row- or column-wise (when the reverse-direction hypothesis is tested). For the reasons elaborated above, this option was not employed in this chapter.

taken of non-orthogonality in design matrices. However, models of opposite-direction hypotheses are rarely hierarchically related. Therefore, they cannot be compared statistically, and the decision for or against a particular model can be based only on the goodness-of-fit of individual models or on individual parameters (general practice is to select the model with the lower objective function). Still, models do reflect the hypothesized direction of effect.

(ii) In CFA, the design matrix of the base model will not include vectors for the directional effects. Therefore, types and antitypes can be assumed to reflect such effects. This assumption can be defended in particular when including such directional effects makes types or antitypes disappear. Based on the patterns of presence/absence of types and antitypes, decisions are possible about the direction of effect. Thus, whenever reversion of direction of effect is possible, log-linear modeling and CFA will allow one to do justice to direction of effect.

(iii) This is in contrast to PA. Reversing the direction of effect will not alter the base model. This is unchanged from log-linear modeling and CFA. However, when the hit cell pattern remains unchanged, overall measures of prediction success will be exactly the same for the two opposite-direction models. PA can, therefore, be the method of choice only when reversing the direction of effect is implausible or logically impossible.

3. The questions that can be answered with the three comparison methods all concern direction of effect. However, they differ in important respects.

(i) Log-linear analysis of direction of effect is performed in the context of a model. This model can contain hypotheses in addition to those that are directional. For example, interaction hypotheses can be tested, covariates can be taken into account, or moderator variables can be used to compare groups of respondents.

(ii) In CFA, model fit is not targeted. To the contrary, it is asked where in the space of the cross-classification deviations from a base model can be observed. When the base model contains all effects except the directional ones, types and antitypes indicate where effects can be observed. Specifically, CFA types indicate where effects result in an increased occurrence of observations, and antitypes indicates where effects prevent events from occurring (or being observed). Interactions and covariates can be considered as well.

(iii) In PA, all cells that are not hit cells are considered error cells. It is in comparison to these error cells that hit cells are evaluated. It is of note that, although all hit cells are evaluated with respect to all error cells, hit cells specifically indicate where, in the space of the cross-classification, effects are expected. In contrast to CFA, the focus of analysis is the possible increase in observations. Decreases are not considered.

In sum, the three comparison methods discussed in this chapter, log-linear modeling, CFA, and PA, are all suitable for the analysis of directional hypotheses in cross-classifications of categorical variables. Considering that the characteristics of the three methods are quite different, data analysts must select the method that best matches their theories.

References

Brandtstädter, J., & von Eye, A. (1986). Hypothesenevaluation in der multivariaten Prädiktionsanalyse. Methodische Erweiterungen für multivariate Anwendungen. *Psychologische Beiträge, 28*, 400–424.

Dodge, Y., & Rousson, V. (2000). Direction dependence in a regression line. *Communications in Statistics: Theory and Methods, 32*, 2053–2057.

Dodge, Y., & Rousson, V. (2001). On asymmetric properties of the correlation coefficient in the regression setting. *The American Statistician, 55*, 51–54.

Eshima, N., & Tabata, M. (1999). Effect analysis in loglinear model approach to path analysis of categorical variables. *Behaviormetrika, 26*, 221–233.

Froman, T., & Hubert, L. J. (1980). Application of prediction analysis to developmental priority. *Psychological Bulletin, 87*, 136–146.

Goodman, L. A. (1973a). Causal analysis of data from panel studies and other kinds of surveys. *American Journal of Sociology, 78*, 1135–1191.

Goodman, L. A. (1973b). The analysis of multidimensional contingency tables when some variables are posterior to others: A modified path analysis approach. *Biometrika, 60*, 179–192.

Goodman, L. A. (1979). Simple models for the analysis of association in cross-classifications having ordered categories. *Journal of the American Statistical Association, 74*, 537–552.

Goodman, L. A., & Kruskal, W. (1974a). Measures of association for cross-classifications. *Journal of the American Statistical Association, 65*, 226–256.

Goodman, L. A., & Kruskal, W. (1974b). Empirical evaluation of formal theory. *Journal of Mathematical Sociology, 3*, 483–494.

Hildebrand, D., Laing, J., & Rosenthal, H. (1977). *Prediction analysis of cross-classifications.* New York, NY: Wiley.

Hildebrand, D. K., Laing, J. D., & Rosenthal, H. (1973). Prediciton logic: A method for empirical evaluation of formal theory. *Journal of Mathematical Sociology, 3*, 163–185.

Hildebrand, D. K., Laing, J. D., & Rosenthal, H. (1974). Prediction logic and quasi-independence in empirical evaluation of formal theory. *Journal of Mathematical Sociology, 3*, 197–209.

Kutner, M. H., Neter, J., Nachtsheim, C. J., & Li, W. (2005). *Applied linear statistical models* (5th ed.). New York, NY: McGraw Hill Irwin.

Lehmacher, W. (1981). A more powerful simultaneous test procedure in Configural Frequency Analysis. *Biometrical Journal, 23*, 429–436.

Li, X., & Wiedermann, W. (in press). Conditional direction dependence analysis: Testing the causal direction of linear models with interaction terms. *Multivariate Behavioral Research*. doi:10.1080/00273171.2019.1687276

Lienert, G. A., & Krauth, J. (1975). Configural Frequency Analysis as a statistical tool for defining types. *Educational and Psychological Measurement, 35*, 231–238.

Lindner, K. (1984). Eine exakte Auswertungsmethode zur Konfigurationsfrequenzanalyse. *Psychologische Beiträge, 26*, 393–415.

Mair, P., & von Eye, A. (2007). Application scenarios for nonstandard log-linear models. *Psychological Methods, 12*, 139–156.

Peters, J., Janzing, D., & Schölkopf, B. (2017). *Elements of causal inference*. Cambridge, MA: MIT Press.

Rindskopf, D. (1999). Some hazards of using nonstandard log-linear models, and how to avoid them. *Psychological Methods, 4*, 339–347.

Rosenström, T., & García-Velázquez, R. (2020). Distribution-based causal inference: A review and practical guidance for epidemiologists. In W. Wiedermann, D. Kim, E. A. Sungur, & A. von Eye (Eds.), *Direction dependence in statistical modeling: Methods and analysis*. Wiley. this volume

von Eye, A. (1991). Einführung in die Prädiktionsanalyse. In A. von Eye (Ed.), *Prädiktionsanalyse. Vorhersagen mit kategorialen Variablen* (pp. 45–155). Weinheim, Germany: Psychologie Verlags Union.

von Eye, A. (2002). *Configural Frequency Analysis—Methods, models, and applications*. Mahwah, NJ: Lawrence Erlbaum.

von Eye, A. (2004). Base models for Configural Frequency Analysis. *Psychology Science, 46*, 150–170.

von Eye, A. (2005). The odds of prediction success. In W. Greve, K. Rothermund, & D. Wentura (Eds.), *The adaptive self: Personal continuity and intentional self-development* (pp. 71–93). Göttingen, Germany: Hogrefe.

von Eye, A., & Bergman, L. R. (2003). Research strategies in developmental psychopathology: Dimensional identity and the person-oriented approach. *Development and Psychopathology, 15*, 553–580.

von Eye, A., Bergman, L. R., & Hsieh, C.-A. (2015). Person-oriented methodological approaches. In W. F. Overton & P. C. M. Molenaar (Eds.), *Handbook of child psychology and developmental science—Theory and methods* (pp. 789–841). New York, NY: Wiley.

von Eye, A., & Brandtstädter, J. (1988a). Application of prediction analysis to cross-classifications of ordinal data. *Biometrical Journal, 30,* 651–655.

von Eye, A., & Brandtstädter, J. (1988b). Formulating and testing developmental hypotheses using statement calculus and non-parametric statistics. In P. B. Baltes, D. Featherman, & R. M. Lerner (Eds.), *Life-span development and behavior* (Vol. *8,* pp. 61–97). Hillsdale, NJ: Erlbaum.

von Eye, A., & Brandtstädter, J. (1997). Configural Frequency Analysis as a searching device for possible causal relationships. *Methods of Psychological Research Online, 2*(2), 1–23.

von Eye, A., & Brandtstädter, J. (1998). The wedge, the fork, and the chain—modeling dependency concepts using manifest categorical variables. *Psychological Methods, 3,* 169–185.

von Eye, A., & DeShon, R. P. (2012). Directional dependency in developmental research. *International Journal of Behavioral Development, 36,* 303–312.

von Eye, A., & Gutiérrez Peña, E. (2004). Configural Frequency Analysis—The search for extreme cells. *Journal of Applied Statistics, 31,* 981–997.

von Eye, A., & Mair, P. (2008a). Functional Configural Frequency Analysis: Explaining types and antitypes. *Bulletin de la Société des Sciences Médicales, Luxembourg, 144,* 35–52.

von Eye, A., & Mair, P. (2008b). A functional approach to Configural Frequency Analysis. *Austrian Journal of Statistics, 37,* 161–173.

von Eye, A., Mair, P., & Mun, E.-Y. (2010). *Advances in configural frequency analysis.* New York, NY: Guilford Press.

von Eye, A., & Mun, E.-Y. (2013). *Log-linear modeling—Concepts, interpretation and applications.* New York, NY: Wiley.

von Eye, A., Schuster, C., & Rogers, W. M. (1998). Modeling synergy using manifest categorical variables. *International Journal of Behavioral Development, 22,* 537–557.

von Eye, A., & Wiedermann, W. (2016). Direction of effects in categorical variables—A structural perspective. In W. Wiedermann & A. von Eye (Eds.), *Statistics and causality: Methods for applied empirical research* (pp. 107–130). Hoboken, NJ: Wiley.

von Eye, A., & Wiedermann, W. (2017a). Testing event-based forms of causality. *Integrative Psychological and Behavioral Science, 51,* 324–344.

von Eye, A., & Wiedermann, W. (2017b). Direction of effects in categorical variables—Looking inside the table. *Journal of Person-Oriented Research, 3,* 11–27.

von Eye, A., & Wiedermann, W. (2018). Locating event-based causal effects: A configural perspective. *Integrative Psychological and Behavioral Science, 52,* 307–330.

Wiedermann, W., Artner, R., & von Eye, A. (2017). Heteroscedasticity as basis of direction dependence in reversible linear regression models. *Multivariate Behavioral Research, 52,* 222–241.

Wiedermann, W., & Li, X. (2018). Direction dependence analysis: A framework to test the direction of effects in linear models with implementation in SPSS. *Behavior Research Methods, 50,* 1581–1601.

Wiedermann, W., & von Eye, A. (2015). Direction of effects in multiple linear regression models. *Multivariate Behavioral Research, 50,* 23–40.

Wiedermann, W., & von Eye, A. (2017). Log-linear models to evaluate direction of effect in binary variables. *Statistical Papers,* 1–30. doi:10.1007/s00362-017-0936-2

Hand, D. J., & Viniciotti, V. (2003). Local versus global models for classification problems: Finding models where it matters. *The American Statistician, 57,* 124–131.

9

Recent Developments on Asymmetric Association Measures for Contingency Tables

Xiaonan Zhu[1,3], Zheng Wei[2], and Tonghui Wang[3]

[1]*Department of Mathematics, University of North Alabama, Florence, AL, USA*
[2]*Department of Mathematics and Statistics, University of Maine, Orono, ME, USA*
[3]*Department of Mathematical Sciences, New Mexico State University, Las Cruces, NM, USA*

9.1 Introduction

Contingency tables are heavily used in survey research, business intelligence, engineering and scientific research, because they provide a basic picture of the interrelation of variables and describe their interactions. A crucial problem of multivariate statistics is finding the directional dependence structure underlying the variables contained in high-dimensional contingency tables.

Recently, many measures of directional dependence or functional dependence of random variables or random vectors have been defined and studied by many researchers, e.g. Boonmee and Tasena (2016), Shan, Wongyang, Wang, and Tasena (2015), Siburg and Stoimenov (2010), Tasena and Dhompongsa (2013), Tasena and Dhompongsa (2016), Wei and Kim (2017), Zhang and Song (2013) and Zhu, Wang, Choy, and Autchariyapanitkul (2018). By directional dependence or the functional dependence, we mean the relationship between two variables is that one variable can be considered as a dependent (response) variable and the other as an independent variable (predictor) in a functional relation. Among those works, three measures are important for contingency tables with discrete random variables, (i) the functional chi-square statistic defined by Zhang and Song (2013), (ii) the subcopula-based mutually complete dependence (MCD) measure mu defined by Shan et al. (2015) (MCD here means that two variables are functions of each other almost surely), and (iii) the subcopula-based association measure ρ^2 defined by Wei and Wei and Kim (2017).

The basic idea of the functional chi-square statistic is to compute the difference between the sum of the Pearson's chi-square statistic of Y conditioned on X and the Pearson's chi-square statistic of Y, and so one can assess the deviation of Y

Direction Dependence in Statistical Modeling: Methods of Analysis, First Edition.
Edited by Wolfgang Wiedermann, Daeyoung Kim, Engin A. Sungur, and Alexander von Eye.
© 2021 John Wiley & Sons, Inc. Published 2021 by John Wiley & Sons, Inc.

from a uniform distribution explainable by X (Zhang & Song, 2013). Note that Wei and Kim (2017) showed that the standardized functional chi-square statistic is identical to the Goodman–Kruskal tau defined in Goodman and Kruskal (1954), therefore, the functional chi-square statistics can be applied for tables with nominal X and Y.

The subcopula-based mutually complete dependence measure μ defined by Shan et al. (2015) is the standardized distance between the conditional distribution of Y given X and marginal distribution of Y, and between the conditional distribution of X given Y and marginal distribution of X. Their measure is the discrete version of the MCD measure omega defined by Siburg and Stoimenov (2010) for two continuous random variables,

$$\omega(X, Y) = (3 \, \|C\|^2 - 2)^{\frac{1}{2}},$$

where X and Y are continuous random variables with the copula C and $\| \cdot \|$ is the Sobolev norm of bivariate copulas given by

$$\|C\| = \left(\int\int |\nabla C(u, v)|^2 \mathrm{d}u \mathrm{d}v \right)^{\frac{1}{2}},$$

where $\nabla C(u, v) = \left(\frac{\partial C(u,v)}{\partial u}, \frac{\partial C(u,v)}{\partial v} \right)$ is the gradient of $C(u, v)$.

The subcopula-based association measure $\rho_{X \rightarrow Y}^2$ defined by Wei and Kim (2017) can be interpreted as the proportions of total variation of the distribution function for a putative response variable Y that is explained by the subcopula regression based on a putative predictor X. Note that their measure and Shan et al.'s measure are free of the supports of X and Y, which is useful in applications, as the marginal supports of X and Y are categorical, not numerical.

This chapter is organized as follows. In Section 9.2, three measures on two way contingency tables and their properties are briefly reviewed. In Section 9.3, two measures of complete dependence on three-way contingency tables are defined, and the subcopula-based association measure for three-way contingency tables provided by Wei and Kim (2017) and Kim and Wei (2020, this volume) are reviewed. Their nonparametric estimators are obtained. In Section 9.4, simulation on three-way contingency tables shows that two measures defined in Section 9.3 are insensitive to the number of categories in response variables. Lastly, two real data sets are used to illustrate our results in Section 9.5.

9.2 Measures on Two-Way Contingency Tables

9.2.1 Functional Chi-Square Statistic

Zhang and Song (2013) defined functional chi-square statistic for two-way contingency tables. Let the $I \times J$ matrix $[n_{ij}]$ be an observed contingency table of discrete random variables X, a putative independent variable, and Y, a putative dependent variable, where n_{ij} is the number of observations in the cell of the ith row and jth column, $i = 1, ..., I$, and $j = 1, ..., J$. The functional chi-square statistic of X and Y is defined by

$$\chi^2(f : X \to Y) = \sum_{i=1}^{I} \sum_{j=1}^{J} \frac{(n_{ij} - n_{i.}/J)^2}{n_{i.}/J} - \sum_{j=1}^{J} \frac{(n_{.j} - n/J)^2}{n/J}, \tag{9.1}$$

where n is the total number of observations, and $n_{i.}$ and $n_{.j}$ are sums of numbers of observations in ith row and jth column, respectively.

Theorem 9.1 (Zhang & Song, 2013) For the functional chi-square defined above, the following properties can be obtained:

(i) If X and Y are empirically independent, then $\chi^2(f : X \to Y) = 0$.
(ii) $\chi^2(f : X \to Y) \geq 0$ for any contingency table.
(iii) The functional chi-square is asymmetric, that is, $\chi^2(f : X \to Y)$ does not necessarily equal to $\chi^2(f : Y \to X)$ for a given contingency table.
(iv) $\chi^2(f : X \to Y)$ is asymptotically chi-square distributed with $(I - 1)(J - 1)$ degrees of freedom under the null hypothesis that Y is uniformly distributed conditioned on X.
(v) $\chi^2(f : X \to Y)$ attains maximum if and only if the column variable Y is a function of the row variable X in the case that a contingency table is feasible. Moreover, the maximum of the functional chi-square is given by $nJ\left(1 - \sum_j (n_{.j}/n)^2\right)$.

Wongyang (2015) proved that the functional chi-square statistic has following additional property.

Proposition 9.1 For any injective function $\phi : \text{supp}(X) \to \mathbb{R}$ and $\psi : \text{supp}(Y) \to \mathbb{R}$,

$$\chi^2(f : \phi(X) \to Y) = \chi^2(f : X \to Y) \quad \text{and}$$
$$\chi^2(f : X \to \psi(Y)) = \chi^2(f : X \to Y),$$

where $\text{supp}(\cdot)$ is the support of the random variable.

From Proposition 9.1, the functional chi-square statistics are invariant with respect to the permutation over the support sets of both X and Y in a contingency table. Therefore, the functional chi-square statistics can be applied for the contingency tables with nominal variables.

9.2.2 Measures of Complete Dependence

Let X and Y be random vectors. Y is *completely dependent* (CD) on X if Y is a measurable function of X almost surely, i.e. there is a measurable function ϕ such that $P(Y = \phi(X)) = 1$. X and Y are said to be *mutually completely dependent* (MCD) if X and Y are completely dependent on each other. In 2015, Shan et al. (2015) defined measures of CD and MCD for discrete random variables through subcopulas. Here, we rewrite their measures for two-way contingency tables.

Let $P = \{p_{ij}\}$ be the matrix of the joint cell proportions in the $I \times J$ contingency table of X and Y, where $i = 1, \ldots, I$ and $j = 1, \ldots, J$. Let $p_{i\cdot}$ and $p_{\cdot j}$ be the ith row and jth column marginal proportions, respectively, and $p_{j|i} = \frac{p_{ij}}{p_{i\cdot}}$ and $p_{i|j} = \frac{p_{ij}}{p_{\cdot j}}$ be the conditional proportions of Y given X and X given Y, respectively. A measure $\mu(Y \mid X)$ for Y completely depending on X and a MCD measure $\mu(X, Y)$ of X and Y are, respectively, defined by

$$
\mu(Y \mid X) = \left[\frac{\omega^2(Y \mid X)}{\omega_{\max}^2(Y \mid X)} \right]^{\frac{1}{2}} = \left[\frac{\displaystyle\sum_{j=1}^{J}\sum_{i=1}^{I}\left(\sum_{j' \leq j}(p_{j'|i} - p_{\cdot j'})\right)^2 p_{i\cdot} p_{\cdot j}}{\displaystyle\sum_{j=1}^{J}\left[\sum_{j' \leq j}p_{\cdot j'} - \left(\sum_{j' \leq j}p_{\cdot j'}\right)^2\right]p_{\cdot j}} \right]^{\frac{1}{2}}
\tag{9.2}
$$

and

$$
\mu(X, Y) = \left[\frac{\omega^2(Y \mid X) + \omega^2(X \mid Y)}{\omega_{\max}^2(Y \mid X) + \omega_{\max}^2(X \mid Y)} \right]^{\frac{1}{2}},
\tag{9.3}
$$

where $\omega^2(X \mid Y)$ and $\omega_{\max}^2(X \mid Y)$ are similarly defined as $\omega^2(Y \mid X)$ and $\omega_{\max}^2(Y \mid X)$ by interchanging X and Y.

Note that the numerator of Eq. (9.2) can be understood as the *distance between the conditional distribution of Y given X and marginal distribution of Y*. Shan et al. (2015) showed that the denominator of Eq. (9.2) is the maximum distance which can be reached if and only if Y is a function of X.

Theorem 9.2 (Shan et al., 2015) For any discrete random variables X and Y, measures $\mu(Y \mid X)$, $\mu(X \mid Y)$ and $\mu(X, Y)$ have the following properties:

(i)

$$0 \leq \mu(Y \mid X), \mu(X \mid Y), \mu(X, Y) \leq 1.$$

(ii)

$$\mu(X, Y) = \mu(Y, X).$$

(iii) $\mu(Y \mid X) = \mu(X \mid Y) = \mu(X, Y) = 0$ if and only if X and Y are independent.

(iv) $\mu(X, Y) = 1$ if and only if X and Y are MCD.

(v) $\mu(Y \mid X) = 1$ if and only if Y is completely dependent on X.

(vi) $\mu(X \mid Y) = 1$ if and only if X is completely dependent on Y.

As will be shown in Section 9.3, the complete dependence measure $\mu(Y \mid X)$ is invariant with respect to the permutation over the support set on the putative independent variable X. Furthermore, the complete dependence measure $\mu(Y \mid X)$ is also invariant when the putative response variable Y is dichotomous. In other words, the complete dependence measure $\mu(Y \mid X)$ can be utilized for contingency tables that the putative independent variable X is nominal or ordinal and the putative response variable Y is ordinal or dichotomous variable.

9.2.3 A Measure of Asymmetric Association Using Subcopula-Based Regression

In 2017, Wei and Kim (2017) defined a *measure of subcopula-based asymmetric association* of discrete random variables. Let X and Y be two discrete random variables with I and J categories having the supports S_0 and S_1, where $S_0 = \{x_1, x_2, \ldots, x_I\}$, and $S_1 = \{y_1, y_2, \ldots, y_J\}$, respectively. Denote the marginal distributions of X and Y to be $F(x)$, $G(y)$, and the joint distribution of (X, Y) be $H(x, y)$, respectively.

Let $P = \{p_{ij}\}$ be the matrix of the joint cell proportions in the $I \times J$ contingency table of X and Y, where $i = 1, \ldots, I$ and $j = 1, \ldots, J$. Set $U = F(X)$ and $V = G(Y)$, then the supports of U and V are $D_0 = F(S_0) = \{u_1, u_2, \ldots, u_I\}$ and $D_1 = G(S_1) = \{v_1, v_2, \ldots, v_J\}$, respectively. The elements in supports D_0 and D_1 can be defined by

$$u_i = \sum_{s=1}^{i} p_{s \cdot} \quad \text{and} \quad v_j = \sum_{t=1}^{j} p_{\cdot t}. \tag{9.4}$$

By Sklar's theorem, there exists an unique function C, named the subcopula (Sklar, 1959), defined on $D_0 \times D_1$ such that $C(F(x), G(y)) = H(x, y)$.

In order to define the asymmetric dependence measure to quantitatively measure the asymmetric dependence in an $I \times J$ contingency table, the subcopula regression functions associated with U and V in an $I \times J$ ordinal contingency table are utilized.

Definition 9.1 (Wei & Kim, 2017) For discrete random variables X and Y with the corresponding subcopula $C(u, v)$, the **subcopula regression functions** of V on U and U on V are defined as follows, respectively:

$$r_{U|V}^{C}(u_i) \equiv E_C(V \mid U = u_i) = \sum_{v_j \in D_1} v_j c_{V|U}(v_j \mid u_i) = \sum_{j=1}^{J} v_j p_{j|i}, \quad \text{for} \quad i = 1, \ldots, I.,$$

$$\tag{9.5}$$

$$r_{U|V}^{C}(v_j) \equiv E_C(U \mid V = v_j) = \sum_{u_i \in D_0} u_i c_{U|V}(u_i \mid v_j) = \sum_{i=1}^{I} u_i p_{i|j}, \text{ for } j = 1, \ldots, J.$$
(9.6)

Based on the subcopula based regressions given in above definitions, Wei and Kim (2017) defined a *measure of subcopula-based asymmetric association* to quantify the variation of the distribution function of a putative response variable that captured by the subcopula regression.

Definition 9.2 (Wei & Kim, 2017) For the two random variables X and Y with subcopula $C(u, v)$ in an $I \times J$ contingency table, let $U = F(X)$ and $V = G(Y)$, then, a **measure of subcopula-based asymmetric association** of Y on X and X on Y are defined as follows, respectively:

$$
\rho_{(X \to Y)}^2 \equiv \frac{\text{Var}(r_{V|U}^C(U))}{\text{Var}(V)} = \frac{E[(r_{V|U}^C(U) - E(V))^2]}{\text{Var}(V)}
$$

$$
= \frac{\sum_{i=1}^{I} \left(\sum_{j=1}^{J} v_j p_{j|i} - \sum_{j=1}^{J} v_j p_{\cdot j} \right)^2 p_{i\cdot}}{\sum_{j=1}^{J} \left(v_j - \sum_{j=1}^{J} v_j p_{\cdot j} \right)^2 p_{\cdot j}},
$$
(9.7)

$$
\rho_{(Y \to X)}^2 \equiv \frac{\text{Var}(r_{U|V}^C(V))}{\text{Var}(U)} = \frac{E[(r_{U|V}^C(V) - E(U))^2]}{\text{Var}(U)}
$$

$$
= \frac{\sum_{j=1}^{J} \left(\sum_{i=1}^{I} u_i p_{i|j} - \sum_{i=1}^{I} u_i p_{i\cdot} \right)^2 p_{\cdot j}}{\sum_{i=1}^{I} \left(u_i - \sum_{i=1}^{I} u_i p_{i\cdot} \right)^2 p_{i\cdot}}.
$$
(9.8)

The properties of subcopula-based asymmetric association are investigated and summarized by Theorem 9.3.

Theorem 9.3 (Wei & Kim, 2017) *Let X and Y be two variables with subcopula $C(u, v)$ in an $I \times J$ contingency table, and let $U = F(X)$ and $V = G(Y)$. Then*

(i) $0 \leq \rho_{X \to Y}^2 \leq 1$.

(ii) If X and Y are independent, then $\rho_{X \to Y}^2 = 0$; Furthermore, if $\rho_{X \to Y}^2 = 0$, then the correlation of U and V is 0.

(iii) $\rho_{X \to Y}^2 = 1$ if and only if $Y = g(X)$ almost surely for some measurable function g.

(iv) If $X_1 = g_1(X)$, where g_1 is an injective function of X, then $\rho^2_{X_1 \to Y} = \rho^2_{X \to Y}$.

(v) If X and Y are both dichotomous variables with only 2 categories, then $\rho^2_{X \to Y} = \rho^2_{Y \to X}$.

From the Properties (iv) and (v), we see that measures of subcopula-based asymmetric association Eqs. (9.7) and (9.8) can be utilized for contingency tables that the putative independent variable is nominal or ordinal and the putative response variable is ordinal or dichotomous variable.

9.3 Asymmetric Measures of Three-Way Contingency Tables

9.3.1 Measures of Complete Dependence for Three Way Contingency Table

In 2018, Zhu et al. (2018) defined a subcopula-based measure of mutually complete dependence for discrete random vectors.

Definition 9.3 *(Zhu et al., 2018)* Let X and Y be discrete random vectors of dimension p and q. The measure of Y completely dependent on X is defined by

$$\mu^2(Y \mid X) = \frac{\sum_y \sum_x [P(Y \leq y \mid X = x) - P(Y < y)]^2 P(X = x) P(Y = y)}{\sum_y [P(Y \leq y) - (P(Y \leq y)^2] P(Y = y)}.$$

$$(9.9)$$

Based on their work, we define asymmetric measures of complete dependence for three-way contingency tables with ordinal response variable(s) and ordinal/nominal independent variable(s). First, we define a complete dependence measure for the scenario when there is one ordinal response variable.

Definition 9.4 Let X, Y, and Z be three discrete random variables. A measure of complete dependence for Z on X and Y is defined by

$$\mu(Z \mid X, Y) = \left[\frac{\sum_z \sum_y \sum_x [P(Z \leq z \mid X = x, Y = y) - P(Z \leq z)]^2 P(X = x, Y = y) P(Z = z)}{\sum_z [P(Z \leq z) - P(Z \leq z)^2] P(Z = z)} \right]^{\frac{1}{2}}. \quad (9.10)$$

Similarly, measures of complete dependence for X on Y and Z and, Y on X and Z are, respectively, defined by

$$\mu(X \mid Y, Z) = \left[\frac{\sum_z \sum_y \sum_x \frac{[P(X \le x \mid Y = y, Z = z) - P(X \le x)]^2}{P(X = x)P(Y = y, Z = z)}}{\sum_x [P(X \le x) - P(X \le x)^2]P(X = x)} \right]^{\frac{1}{2}}. \qquad (9.11)$$

and

$$\mu(Y \mid X, Z) = \left[\frac{\sum_z \sum_y \sum_x \frac{[P(Y \le y \mid X = x, Z = z) - P(Y \le y)]^2}{P(X = x, Z = z)P(Y = y)}}{\sum_Y [P(Y \le y) - P(Y \le y)^2]P(Y = y)} \right]^{\frac{1}{2}}. \qquad (9.12)$$

We further define a complete dependence for the scenario when there are two ordinal response variables.

Definition 9.5 Let X, Y, and Z be three discrete random variables. A measure of complete dependence for Y and Z on X is defined by

$$\mu(Y, Z \mid X) = \left[\frac{\sum_z \sum_y \sum_x \frac{[P(Y \le y, Z \le z \mid X = x) - P(Y \le y, Z \le z)]^2}{P(X = x)P(Y = y, Z = z)}}{\sum_y \sum_z [P(Y \le y, Z \le z) - P(Y \le y, Z \le z)^2]P(Y = y, Z = z)} \right]^{\frac{1}{2}}.$$

$$(9.13)$$

Similarly, measures of complete dependence for X and Z on Y, and X and Y on Z are, respectively, defined by

$$\mu(X, Z \mid Y) = \left[\frac{\sum_z \sum_y \sum_x \frac{[P(X \le x, Z \le z \mid Y = y) - P(X \le x, Z \le z)]^2}{P(X = x, Z = z)P(Y = y)}}{\sum_x \sum_z [P(X \le x, Z \le z) - P(X \le x, Z \le z)^2]P(X = x, Z = z)} \right]^{\frac{1}{2}}.$$

$$(9.14)$$

and

$$\mu(X, Y \mid Z) = \left[\frac{\sum_z \sum_y \sum_x \frac{[P(X \le x, Y \le y \mid Z = z) - P(X \le x, Y \le y)]^2}{P(X = x, Y = y)P(Z = z)}}{\sum_x \sum_y [P(X \le x, Y \le y) - P(X \le x, Y \le y)^2]P(X = x, Y = y)} \right]^{\frac{1}{2}}.$$

$$(9.15)$$

Note that above measures can also be understood as standardized distance between a conditional distribution and a marginal distribution like the two-dimensional measure given by Eq. (9.2).

Theorem 9.4 For any three-way contingency table of variables X, Y, and Z, measures $\mu(Z\,|\,X, Y)$, $\mu(Y\,|\,X, Z)$, and $\mu(X\,|\,Y, Z)$ have the following properties:

(i) $0 \le \mu(Z\,|\,X, Y)$, $\mu(Y\,|\,X, Z)$, $\mu(X\,|\,Y, Z) \le 1$.

(ii) $\mu(Z\,|\,X, Y) = 0$ if and only if (X, Y) and Z are independent.
  $\mu(Y\,|\,X, Z) = 0$ if and only if (X, Z) and Y are independent.
  $\mu(X\,|\,Y, Z) = 0$ if and only if (Y, Z) and X are independent.

(iii) $\mu(Z\,|\,X, Y) = 1$ if and only if Z is a function of (X, Y).
  $\mu(Y\,|\,X, Z) = 1$ if and only if Y is a function of (X, Z).
  $\mu(X\,|\,Y, Z) = 1$ if and only if X is a function of (Y, Z).

(iv) $\mu(Z\,|\,X, Y)$, $\mu(Y\,|\,X, Z)$ and $\mu(X\,|\,Y, Z)$ are invariant under strictly increasing transformations of X, Y and Z, i.e. if ψ_1, ψ_2, and ψ_3 are strictly increasing functions defined on ranges of X, Y, and Z, respectively, then $\mu(\psi_3(Z)\,|\,\psi_1(X), \psi_2(Y)) = \mu(Z\,|\,X, Y)$, $\mu(\psi_2(Y)\,|\,\psi_1(X), \psi_3(Z)) = \mu(Y\,|\,X, Z)$, and $\mu(\psi_1(X)\,|\,\psi_2(Y), \psi_3(Z)) = \mu(X\,|\,Y, Z)$.

Proof.
The detailed proofs of properties (i)–(iv) for a more general case can be found in lemma 1, lemma 2, and theorem 1 in Zhu et al. (2018). $\qquad\square$

Theorem 9.5 For any three-way contingency table of variables X, Y, and Z, measures $\mu(Y, Z\,|\,X)$, $\mu(X, Z\,|\,Y)$, and $\mu(X, Y\,|\,Z)$ have the following properties:

(i) $0 \le \mu(Y, Z\,|\,X)$, $\mu(X, Z\,|\,Y)$, $\mu(X, Y\,|\,Z) \le 1$.

(ii) $\mu(Y, Z\,|\,X) = 0$ if and only if (Y, Z) and X are independent.
  $\mu(X, Z\,|\,Y) = 0$ if and only if (X, Z) and Y are independent.
  $\mu(X, Y\,|\,Z) = 0$ if and only if (X, Y) and Z are independent.

(iii) $\mu(Y, Z\,|\,X) = 1$ if and only if (Y, Z) is a function of X.
  $\mu(X, Z\,|\,Y) = 1$ if and only if (X, Z) is a function of Y.
  $\mu(X, Y\,|\,Z) = 1$ if and only if (X, Y) is a function of Z.

(iv) $\mu(Y, Z\,|\,X)$, $\mu(X, Z\,|\,Y)$, and $\mu(X, Y\,|\,Z)$ are invariant under strictly increasing transformations of X, Y, and Z, i.e. if ψ_1, ψ_2, and ψ_3 are strictly increasing functions defined on ranges of X, Y, and Z, respectively, then $\mu(\psi_2(Y)$, $\psi_3(Z)\,|\,\psi_1(X)) = \mu(Y, Z\,|\,X)$, $\mu(\psi_1(X)$, $\psi_3(Z)\,|\,\psi_1(Y)) = \mu(X, Z\,|\,Y)$, and $\mu(\psi_1(X), \psi_2(Y)\,|\,\psi_3(Z)) = \mu(X, Y\,|\,Z)$.

The proofs are similar to those of theorem 1 in Zhu et al. (2018).

Table 9.1 Three-way contingency table of X, Y, and Z with $P = \{p_{ijk}\}$.

Z	-2			0			2		
X	-1	0	1	-1	0	1	-1	0	1
Y									
-2	0	0	2/18	0	1/18	0	2/18	0	0
0	0	0	0	6/18	2/18	1/18	0	0	0
2	1/18	0	0	0	2/18	0	0	0	1/18

The following example shows the calculation of the proposed measures for scenario with one response variable in Definition 9.4 and the scenario with two response variables in Definition 9.5. We will use it as a running example.

Example 9.1 Table 9.1 shows a three-way contingency table for a three-dimensional ordinal random vector (X, Y, Z) with the joint p.m.f. $P = \{p_{i,j,k}\}$ where $i = j = k = 1,2,3$. Note that the three variables have the functional relationship, $Z = X \times Y$ with probability 1. The supports of X, Y, and Z are $S_0 = \{-2, 0, 2\}$, $S_1 = \{-1, 0, 1\}$, and $S_2 = \{-2, 0, 2\}$, respectively, and the corresponding marginal distributions for X, Y, and Z are obtained by $P(X = x) = \{p_i \cdot \cdot\} = \{5/18, 9/18, 4/18\}$, $P(Y = y) = \{p_{\cdot j \cdot}\} = \{5/18, 9/18, 4/18\}$, and $P(Z = z) = \{p_{\cdot \cdot k}\} = \{3/18, 12/18, 3/18\}$.

By calculation, it can be shown that $\mu(Z|X, Y) = 1$, $\mu(Y|X, Z) = 0.7743$ and $\mu(X|Y, Z) = 0.8238$, $\mu(Y, Z|X) = 0.5618$, $\mu(X, Z|Y) = 0.2725$ and $\mu(X, Y|Z) = 0.4946$. Specifically, $\mu(Z|X, Y) = 1$ indicates that there is a prefect functional relation from (X, Y) to Z. There is a relatively weak functional dependence from Y to (X, Z) since $\mu(X, Z|Y) = 0.2725$.

The three two-way marginal p.m.f.s for (Y, Z), (X, Z), and (X, Y) are given in Table 9.2(a), (b),and (c), respectively.

By calculation, it can be shown that $\mu(Y|X) = 0.3355$, $\mu(X|Y) = 0.2845$, $\mu(Z|Y) = 0.3333$, $\mu(Y|Z) = 0.3218$, $\mu(Z|X) = 0.4690$, and $\mu(X|Z) = 0.4117$.

The following theorem shows invariant properties for the complete dependence measures with one dependent response variable and two independent variables given in Definition 9.4.

Theorem 9.6 For any three-way contingency table of variables X, Y, and Z, measures $\mu(Z|X, Y)$, $\mu(Y|X, Z)$, and $\mu(X|Y, Z)$ have the following invariant properties:

(i) $\mu(Z|X, Y)$, $\mu(Y|X, Z)$, and $\mu(X|Y, Z)$ are invariant under injective transformations of predictor variables, i.e. if ψ_1, ψ_2, and ψ_3 are injective

Table 9.2 (a) Joint p.m.f of Y and Z; (b) joint p.m.f of X and Z; (c) joint p.m.f of X and Y.

(a) $\{p_{ij\cdot}\}$

Y	-2	0	2
Z			
-1	1/18	6/18	2/18
0	0	5/18	0
1	2/18	1/18	1/18

(b) $\{p_{i\cdot k}\}$

X	-2	0	2
Z			
-2	2/18	1/18	2/18
0	0	9/18	0
2	1/18	2/18	1/18

(c) $\{p_{ij\cdot}\}$

X	-1	0	1
Y			
-2	2/18	1/18	2/18
0	6/18	2/18	1/18
2	1/18	2/18	1/18

functions defined on ranges of X, Y, and Z, respectively, then $\mu(Z\,|\,\psi_1(X),$ $\psi_2(Y)) = \mu(Z\,|\,X,\ Y)$, $\mu(Y\,|\,\psi_1(X),\ \psi_3(Z)) = \mu(Y\,|\,X,\ Z)$, and $\mu(X\,|\,\psi_2(Y),$ $\psi_3(Z)) = \mu(X\,|\,Y,Z)$.

(ii) $\mu(Z\,|\,X,\ Y)$, $\mu(Y\,|\,X,\ Z)$, and $\mu(X\,|\,Y,\ Z)$ are invariant under injective transformations of response variables when response variables are dichotomous with only two categories, i.e. if ψ_1, ψ_2, and ψ_3 are injective functions defined on ranges of X, Y, and Z, respectively, then $\mu(\psi_3(Z)\,|\,X,\ Y) = \mu(Z\,|\,X,\ Y)$, $\mu(\psi_2(Y)\,|\,X,Z) = \mu(Y\,|\,X,\ Z)$, and $\mu(\psi_1(X)\,|\,Y,Z) = \mu(X\,|\,Y,\ Z)$.

Proof.
(i) Note that the denominator of $\mu(Z\,|\,X,\ Y)$ is free of X and Y, so it is invariant under injective transformations of X and Y. On the other hand, for the numerator part of $\mu(Z\,|\,X,\ Y)$, $\sum_y\sum_x[P(Z \leq z \,|\, X = x, Y = y) - P(Z \leq z)]^2 P(X = x, Y = y)$ is invariant under injective transformations of X and Y for each fixed z. Thus, Property (i) holds.

(ii). Suppose that Z is dichotomous with only two categories z_1 and z_2. Let's assume that z_1 is the first category and z_2 is the second category. Since $P(Z = z_1) + P(Z = z_2) = 1$ and $P(Z = z_1 \mid X = x, Y = y) + P(Z = z_2 \mid X = x, Y = y) = 1$, it implies $[P(Z \leq z_2 \mid X = x, Y = y) - P(Z \leq z_2)]^2 P(X = x, Y = y)P(Z = z_2) = 0$, for any x, y. Then, we have

$$\frac{\sum_z \sum_y \sum_x [P(Z \leq z \mid X = x, Y = y) - P(Z \leq z)]^2 P(X = x, Y = y)P(Z = z)}{\sum_z [P(Z \leq z) - P(Z \leq z)^2]P(Z = z)}$$

$$= \frac{\sum_y \sum_x [P(Z = z_1 \mid X = x, Y = y) - P(Z = z_1)]^2 P(X = x, Y = y)P(Z = z_1)}{[P(Z = z_1) - P(Z = z_1)^2]P(Z = z_1)}$$

$$= \frac{\sum_y \sum_x [P(Z = z_1 \mid X = x, Y = y)^2 - 2P(Z = z_1 \mid X = x, Y = y)P(Z = z_1) + P(Z = z_1)^2]P(X = x, Y = y)P(Z = z_1)}{P(Z = z_1)(1 - P(Z = z_1))P(Z = z_1)}$$

$$= \frac{\sum_y \sum_x \left[\frac{P(X=x,Y=y,Z=z_1)^2}{P(X=x,Y=y)} - 2P(X = x, Y = y, Z = z_1)P(Z = z_1) + P(Z = z_1)^2 P(X = x, Y = y) \right] P(Z = z_1)}{P(Z = z_1)P(Z = z_2)P(Z = z_1)}$$

$$= \frac{\sum_y \sum_x \frac{P(X=x,Y=y,Z=z_1)^2}{P(X=x,Y=y)} - P(Z = z_1)^2}{P(Z = z_1)P(Z = z_2)}.$$

Similarly, when we assume that z_2 is the first category and z_1 is the second category, i.e. $z_2 < z_1$, we have

$$\frac{\sum_z \sum_y \sum_x [P(Z \leq z \mid X = x, Y = y) - P(Z \leq z)]^2 P(X = x, Y = y)P(Z = z)}{\sum_z [P(Z \leq z) - P(Z \leq z)^2]P(Z = z)}$$

$$= \frac{\sum_y \sum_x \frac{P(X=x,Y=y,Z=z_2)^2}{P(X=x,Y=y)} - P(Z = z_2)^2}{P(Z = z_1)P(Z = z_2)}.$$

Thus, we just need to show that

$$\sum_y \sum_x \frac{P(X = x, Y = y, Z = z_1)^2}{P(X = x, Y = y)} - P(Z = z_1)^2$$

$$= \sum_y \sum_x \frac{P(X = x, Y = y, Z = z_2)^2}{P(X = x, Y = y)} - P(Z = z_2)^2.$$

Indeed,

$$\sum_{y}\sum_{x} \frac{P(X=x, Y=y, Z=z_1)^2}{P(X=x, Y=y)} - P(Z=z_1)^2$$

$$= \sum_{y}\sum_{x} \frac{[P(X=x, Y=y)-P(X=x, Y=y, Z=z_2)]^2}{P(X=x, Y=y)} - [1 - P(Z=z_2)]^2$$

$$= \sum_{y}\sum_{x} \frac{\begin{array}{c}[P(X=x, Y=y)^2 - 2P(X=x, Y=y) \\ P(X=x, Y=y, Z=z_2) + P(X=x, Y=y, Z=z_2)^2]\end{array}}{P(X=x, Y=y)}$$

$$- [1 - 2P(Z=z_2) + P(Z=z_2)^2]$$

$$= \sum_{y}\sum_{x} \Big[P(X=x, Y=y) - 2P\left(X=x, Y=y, Z=z_2\right)$$

$$+ \frac{P(X=x, Y=y, Z=z_2)^2}{P(X=x, Y=y)} \Big]$$

$$- [1 - 2P(Z=z_2) + P(Z=z_2)^2]$$

$$= 1 - 2P(Z=z_2) + \sum_{y}\sum_{x} \frac{P(X=x, Y=y, Z=z_2)^2}{P(X=x, Y=y)}$$

$$- [1 - 2P(Z=z_2) + P(Z=z_2)^2]$$

$$= \sum_{y}\sum_{x} \frac{P(X=x, Y=y, Z=z_2)^2}{P(X=x, Y=y)} - P(Z=z_2)^2.$$

Thus, Property (ii) holds. □

The following theorem shows the invariant property of the complete dependence measures with two dependent response variables and one independent variable given in Definition 9.5. The proof is similar to those of Theorem 9.6.

Theorem 9.7 For any three-way contingency table of variables X, Y, and Z, $\mu(Y, Z|X)$, $\mu(X, Z|Y)$, and $\mu(X, Y|Z)$ are invariant under injective transformations of predictor variables, i.e. if ψ_1, ψ_2, and ψ_3 are injective functions defined on ranges of X, Y, and Z, respectively, then $\mu(Y, Z|\psi_1(X)) = \mu(Y, Z|X)$, $\mu(X, Z|\psi_2(Y)) = \mu(X, Z|Y)$, and $\mu(X, Y|\psi_3(Z)) = \mu(X, Y|Z)$.

Remark 9.1 From the Theorems 9.6 and 9.7, we know that the complete dependence measures are invariant with respect to the permutation over the support sets on independent variables. In other words, the complete dependence measures given in Definitions 9.4 and 9.5 can be utilized for contingency tables that the putative independent variables are nominal or ordinal.

In addition, from the Property (ii) in Theorem 9.6, the complete dependence measures given in Definition 9.4 with one dependent variable is also invariant

Table 9.3 Three-way contingency tables of dichotomous variables.

(a) X, Y, and Z				
Z	1		2	
X	1	2	1	2
Y				
1	1	3	5	7
2	2	4	6	8
(b) X, Y, and Z'				
Z'	2		1	
X	1	2	1	2
Y				
1	5	7	1	3
2	6	8	2	4

when the putative response variable is dichotomous. Note that, as shown in the following example, this invariant property does not hold for the complete dependence measures given in Definition 9.5 with two dependent variables. Furthermore, it can also be shown that the complete dependence measures with one response variable given in Definition 9.4 is not invariant with respect to the permutation over the support on response variable when the response variable is not dichotomous.

Example 9.2 Let X, Y, and Z be dichotomous variables. Assume Z' is the variable of Z by switching the order of categories in Z. Suppose that we have the following contingency tables given by Table 9.3. Assume the variables X and Z (or Z') are putative dependent variables. It can be show that $\mu(X, Z \mid Y) = 0.0684$, but $\mu(X, Z' \mid Y) = 0.0795$. When there is only one response variable Z, $\mu(Z \mid X, Y) = \mu(Z' \mid X, Y) = 0.1301$.

9.3.2 Subcopula Based Measure for Three Way Contingency Table

To introduce the subcopula-based association measure for three-way contingency table provided by Wei and Kim (2017) and by Kim and Wei (2020, this volume), we first start with the introduction of the three dimensional subcopula.

Definition 9.6 A three-dimensional **subcopula** is a function $C : D_0 \times D_1 \times D_2 \to [0, 1]$ such that

1. $D_0 \times D_1 \times D_2$ is the domain of C where D_0, D_1, and D_2 are subsets of $[0, 1]$ containing 0 and 1;
2. $C(u, v, w) = 0$ if at least one of u, v, w is zero; For every $u \in D_0$, $v \in D_1$, and $w \in D_2$, $C(u,1,1) = u$, $C(1, v, 1) = v$, and $C(1, 1, w) = w$;
3. $C(u', v', w') - C(u', v', w) - C(u', v, w') - C(u, v', w') + C(u, v', w) + C(u, v', w) + C(u, v, w') - C(u, v, w) \geq 0$ where $[u, u'] \times [v, v'] \times [w, w']$ is a rectangle with all vertices in $D_0 \times D_1 \times D_2$.

Given the joint p.m.f. $P = \{p_{ijk}\}$ of the $I \times J \times K$ ordinal contingency table for X, Y, and Z, we denote the marginal distributions of X, Y, and Z to be $F_0(x)$, $F_1(y)$, and $F_2(z)$, respectively, and let $U = F_0(X)$, $V = F_1(Y)$, and $W = F_2(Z)$. Then, we define the supports for the subcopula C of U, V, and W to be $D_0 = \{u_0 = 0, u_1, \ldots, u_I = 1\}$, $D_1 = \{v_0 = 0, v_1, \ldots, v_J = 1\}$, and $D_2 = \{w_0 = 0, w_1, \ldots, w_K = 1\}$, respectively, where

$$u_i = \sum_{s=1}^{i} p_{\cdot\cdot s}, \quad v_j = \sum_{t=1}^{j} p_{\cdot t \cdot}, \quad \text{and} \quad w_k = \sum_{r=1}^{k} p_{\cdot\cdot r}. \tag{9.16}$$

For the joint distribution H of ordinal variables X, Y, and Z, there exists a three dimensional copula C on the domain $D_0 \times D_1 \times D_2$, such that

$$H(x_i, y_j, z_k) = \sum_{s=1}^{i} \sum_{t=1}^{j} \sum_{r=1}^{k} p_{str} = C(u_i, v_j, w_k).$$

In addition, the joint p.m.f. of C, the conditional p.m.f.s of U given (V, W), V given (U, W), and W given (U, V), are

$$c(u_i, v_j, w_k) = p_{ijk}, \quad c_{U|V,W}(u_i \mid v_j, w_k) = p_{i|jk} = p_{ijk}/p_{\cdot jk},$$

$$c_{V|U,W}(v_j \mid u_i, w_k) = p_{j|ik} = p_{ijk}/p_{i\cdot k}, \quad c_{W|U,V}(w_k \mid u_i, v_j) = p_{k|ij} = p_{ijk}/p_{ij\cdot}.$$

Wei and Kim (2017) and Kim and Wei (2020, this volume) provide the formulas for the trivariate subcopula regression based measures to quantify the asymmetric association among ordinal variables X, Y, and Z in an $I \times J \times K$ contingency table.

Definition 9.7 *(Kim & Wei, 2020, this volume; Wei & Kim, 2017)* For the ordinal random variables X, Y, and Z with the trivariate subcopula $C(u, v, w)$ in an $I \times J \times K$ contingency table, let $U = F_0(X)$, $V = F_1(Y)$, and $W = F_2(Z)$. Then, a **measure of subcopula-based asymmetric association measure** of X on Y

and Z is,

$$\rho^2_{(Y,Z\to X)} \equiv \frac{\text{Var}(r^C_{U|V,W}(V,W))}{\text{Var}(U)} = \frac{E[(r^C_{U|V,W}(V,W)) - E(U))^2]}{\text{Var}(U)}$$

$$= \frac{\sum_{k=1}^{K}\sum_{j=1}^{J}\left(\sum_{i=1}^{I} u_i p_{i|jk} - \sum_{i=1}^{I} u_i p_{i\cdot\cdot}\right)^2 p_{\cdot jk}}{\sum_{i=1}^{I}\left(u_i - \sum_{i=1}^{I} u_i p_{i\cdot\cdot}\right)^2 p_{i\cdot\cdot}}.$$

Similarly, the **measures of subcopula-based asymmetric association measure** of Y on X and Z, and of Z on X and Y are

$$\rho^2_{(X,Z\to Y)} = \frac{\sum_{k=1}^{K}\sum_{i=1}^{I}\left(\sum_{j=1}^{J} v_j p_{j|ik} - \sum_{j=1}^{J} v_j p_{\cdot j\cdot}\right)^2 p_{i\cdot k}}{\sum_{j=1}^{J}\left(v_j - \sum_{j=1}^{J} v_j p_{\cdot j\cdot}\right)^2 p_{\cdot j\cdot}},$$

$$\rho^2_{(X,Y\to Z)} = \frac{\sum_{j=1}^{J}\sum_{i=1}^{I}\left(\sum_{k=1}^{K} w_k p_{k|ij} - \sum_{k=1}^{K} w_k p_{\cdot\cdot k}\right)^2 p_{ij\cdot}}{\sum_{k=1}^{K}\left(w_k - \sum_{k=1}^{K} w_k p_{\cdot\cdot k}\right)^2 p_{\cdot\cdot k}}.$$

The proposed measures of subcopula-based asymmetric association measure given in Definition 9.7 can be interpreted as the average proportion of variance for the distribution function of the (putative) response variable explained by the sub-copula copula regression of the distribution functions for the (putative) predictors (Kim & Wei, 2020, this volume).

For better understanding of the trivariate subcopula-based measures, we illustrate the utilization of trivariate subcopula based asymmetric measures using the three way table given in Example 9.1.

Example 9.3 From the Table 9.1 given in Example 9.1, we can find the supports for U, V, and W in Eq. (9.16) are $D_0 = (u_0, u_1, u_2, u_3) = (0, 5/18, 14/18, 1)$, $D_1 = (v_0, v_1, v_2, v_3) = (0, 9/18, 14/18, 1)$, and $D_2 = (w_0, w_1, w_2, w_3) = (0, 3/18, 15/18, 1)$. The joint p.m.f. of the subcopula C is obtained in Table 9.4.

Using the computed values of u_i, v_j, and w_k, and the joint p.m.f. of (U, V, W), we calculate the proposed measures in Definition 9.7: $\rho^2_{(Y, Z \to X)} = 0.733$, $\rho^2_{(X, Z \to Y)} = 0.625$, and $\rho^2_{(X, Y \to Z)} = 1$. This result supports the true relationship between X, Y and Z: Z is a function of X and Y (i.e. $Z = XY$ with probability 1), but not vice versa. Recall that $\mu(Z \mid X, Y) = 1$. Both subcopula-based asymmetric

Table 9.4 Supports of U, V, W and the joint p.m.f. of C.

Z		3/18			5/18			1	
Y	9/18	14/18	1	9/18	14/18	1	9/18	14/18	1
X									
5/18	0	0	2/18	0	1/18	0	2/18	0	0
14/18	0	0	0	6/18	2/18	1/18	0	0	0
1	1/18	0	0	0	2/18	0	0	0	1/18

measure $\rho^2_{(X, Y \to Z)}$ and complete dependence measure $\mu(Z \mid X, Y)$ can be used as a measure of functional relationship among variables.

Furthermore, we can also find the bivariate subcopula-based asymmetric measures for three two-way marginal using the p.m.f.s for (Y, Z), (X, Z), and (X, Y) are given in Table 9.2(a), (b), and (c). For (X, Y), we have $\rho^2_{(Y \to X)} = 0.057$, and $\rho^2_{(Y \to X)} = 0.121$. For (Y, Z), we have $\rho^2_{(Y \to Z)} = 0.176$, and $\rho^2_{(Z \to Y)} = 0.101$. And for (X, Z), we have $\rho^2_{(X \to Z)} = 0.106$, and $\rho^2_{(Z \to X)} = 0.199$.

9.3.3 Estimation

In this section, we provide the estimation methods for measures given in Section 9.3.1 and 9.3.2. Let $\{n_{ijk}\}$, $i = 1, ..., I$, $j = 1, ..., J$, and $k = 1, ..., K$ denote the cell counts in an $I \times J \times K$ ordinal contingency table obtained by classifying $n \left(= \sum_{i=1}^{I} \sum_{j=1}^{J} \sum_{k=1}^{K} n_{ijk} \right)$ cases according to the I categories of X, J categories of Y, and K categories of Z. The row sums, column sums, and tube sums (i.e. the marginal frequencies) are denoted as $n_{i \cdot \cdot} = \sum_{j=1}^{J} \sum_{k=1}^{K} n_{ijk}$, $n_{\cdot j \cdot} = \sum_{i=1}^{I} \sum_{k=1}^{K} n_{ijk}$, and $n_{\cdot \cdot k} = \sum_{i=1}^{I} \sum_{j=1}^{J} n_{ijk}$, respectively. We then define the estimators for p_{ijk}, $p_{i \cdot \cdot}, p_{\cdot j \cdot}, p_{\cdot \cdot k}$, and $p_{i|jk} = \frac{p_{ijk}}{p_{\cdot jk}}$, $p_{j|ik} = \frac{p_{ijk}}{p_{i \cdot k}}$ and $p_{k|ij} = \frac{p_{ijk}}{p_{ij \cdot}}$ by

$$\hat{p}_{ijk} = \frac{n_{ijk}}{n}, \hat{p}_{i \cdot \cdot} = \frac{n_{i \cdot \cdot}}{n}, \hat{p}_{\cdot j \cdot} = \frac{n_{\cdot j \cdot}}{n}, \hat{p}_{\cdot \cdot k} = \frac{n_{\cdot \cdot k}}{n}, \hat{p}_{i|jk} = \frac{n_{ijk}}{n_{\cdot jk}}, \hat{p}_{j|ik}$$

$$= \frac{n_{ijk}}{n_{i \cdot k}}, \hat{p}_{k|ij} = \frac{n_{ijk}}{n_{ij \cdot}}.$$

The measures in Definitions 9.4 and 9.5 can be estimated by,

$$\hat{\mu}(Z \mid X, Y) = \left[\frac{\sum_{k=1}^{K} \sum_{j=1}^{J} \sum_{i=1}^{I} \left(\sum_{k' \leq k} (\hat{p}_{k'|ij} - \hat{p}_{\cdot \cdot k'}) \right)^2 \hat{p}_{ij \cdot} \hat{p}_{\cdot \cdot k}}{\sum_{k=1}^{K} \left(\sum_{k' \leq k} \hat{p}_{\cdot \cdot k'} - \left(\sum_{k' \leq k} \hat{p}_{\cdot \cdot k'} \right)^2 \right) \hat{p}_{\cdot \cdot k}} \right]^{\frac{1}{2}},$$

$$\hat{\mu}(Y \mid X, Z) = \left[\frac{\sum_{k=1}^{K}\sum_{j=1}^{J}\sum_{i=1}^{I}\left(\sum_{j'\le j}(\hat{p}_{j'|ik} - \hat{p}_{\cdot j'\cdot})\right)^2 \hat{p}_{i\cdot k}\hat{p}_{\cdot j\cdot}}{\sum_{j=1}^{J}\left(\sum_{j'\le j}\hat{p}_{\cdot j'\cdot} - \left(\sum_{j'\le j}\hat{p}_{\cdot j'\cdot}\right)^2\right)\hat{p}_{\cdot j\cdot}} \right]^{\frac{1}{2}},$$

$$\hat{\mu}(X \mid Y, Z) = \left[\frac{\sum_{k=1}^{K}\sum_{j=1}^{J}\sum_{i=1}^{I}\left(\sum_{i'\le i}(\hat{p}_{i'|jk} - \hat{p}_{i'\cdot\cdot})\right)^2 \hat{p}_{\cdot jk}\hat{p}_{i\cdot\cdot}}{\sum_{i=1}^{I}\left(\sum_{i'\le i}\hat{p}_{i'\cdot\cdot} - \left(\sum_{i'\le i}\hat{p}_{i'\cdot\cdot}\right)^2\right)\hat{p}_{i\cdot\cdot}} \right]^{\frac{1}{2}},$$

$$\hat{\mu}(Y, Z \mid X) = \left[\frac{\sum_{k=1}^{K}\sum_{j=1}^{J}\sum_{i=1}^{I}\left(\sum_{k'\le k}\sum_{j'\le j}(\hat{p}_{j'k'|i} - \hat{p}_{\cdot j'k'})\right)^2 \hat{p}_{i\cdot\cdot}\hat{p}_{\cdot jk}}{\sum_{k=1}^{K}\sum_{j=1}^{J}\left(\sum_{k'\le k}\sum_{j'\le j}\hat{p}_{\cdot j'k'} - \left(\sum_{k'\le k}\sum_{j'\le j}\hat{p}_{\cdot j'k'}\right)^2\right)\hat{p}_{\cdot jk}} \right]^{\frac{1}{2}},$$

$$\hat{\mu}(X, Z \mid Y) = \left[\frac{\sum_{k=1}^{K}\sum_{j=1}^{J}\sum_{i=1}^{I}\left(\sum_{k'\le k}\sum_{i'\le i}(\hat{p}_{i'k'|j} - \hat{p}_{i'\cdot k'})\right)^2 \hat{p}_{\cdot j\cdot}\hat{p}_{i\cdot k}}{\sum_{k=1}^{K}\sum_{i=1}^{I}\left(\sum_{k'\le k}\sum_{i'\le i}\hat{p}_{i'\cdot k'} - \left(\sum_{k'\le k}\sum_{i'\le i}\hat{p}_{i'\cdot k'}\right)^2\right)\hat{p}_{i\cdot k}} \right]^{\frac{1}{2}},$$

$$\hat{\mu}(X, Y \mid Z) = \left[\frac{\sum_{k=1}^{K}\sum_{j=1}^{J}\sum_{i=1}^{I}\left(\sum_{i'\le i}\sum_{j'\le j}(\hat{p}_{i'j'|k} - \hat{p}_{\cdot i'j'})\right)^2 \hat{p}_{\cdot\cdot k}\hat{p}_{ij\cdot}}{\sum_{i=1}^{i}\sum_{j=1}^{J}\left(\sum_{i'\le i}\sum_{j'\le j}\hat{p}_{i'j'\cdot} - \left(\sum_{i'\le i}\sum_{j'\le j}\hat{p}_{i'j'\cdot}\right)^2\right)\hat{p}_{ij\cdot}} \right]^{\frac{1}{2}}.$$

For subcopula based asymmetric measures given in Definition 9.7, Kim and Wei (2020, this volume) first estimate the ranges of the marginal distributions of X, Y and Z, D_0, D_1 and D_2 in Eq. ((9.16)),

$$\hat{u}_i = \sum_{s=1}^{i} \hat{p}_{s\cdot\cdot}, \quad \hat{v}_j = \sum_{t=1}^{j} \hat{p}_{\cdot t\cdot}, \quad \text{and} \quad \hat{w}_k = \sum_{r=1}^{k} \hat{p}_{\cdot\cdot r}.$$

Using the estimators obtained above, they provide estimators for subcopula-based asymmetric association measures $\rho^2_{(Y, Z \to X)}$, $\rho^2_{(X, Z \to Y)}$, and $\rho^2_{(X, Y \to Z)}$ given in Definition 9.7:

$$\hat{\rho}^2_{(Y, Z \to X)} = \frac{\sum_{k=1}^{K}\sum_{j=1}^{J}\left(\sum_{i=1}^{I}\hat{u}_i\hat{p}_{i|jk} - \sum_{i=1}^{I}\hat{u}_i\hat{p}_{i\cdot\cdot}\right)^2 \hat{p}_{\cdot jk}}{\sum_{i=1}^{I}\left(\hat{u}_i - \sum_{i=1}^{I}\hat{u}_i\hat{p}_{i\cdot\cdot}\right)^2 \hat{p}_{i\cdot\cdot}},$$

$$\hat{\rho}^2_{(X,Z \to Y)} = \frac{\sum_{k=1}^{K} \sum_{i=1}^{I} \left(\sum_{j=1}^{J} \hat{v}_j \hat{p}_{j|ik} - \sum_{j=1}^{J} \hat{v}_j \hat{p}_{\cdot j \cdot} \right)^2 \hat{p}_{i \cdot k}}{\sum_{j=1}^{J} \left(\hat{v}_j - \sum_{j=1}^{J} \hat{v}_j \hat{p}_{\cdot j \cdot} \right)^2 \hat{p}_{\cdot j \cdot}},$$

$$\hat{\rho}^2_{(X,Y \to Z)} = \frac{\sum_{j=1}^{J} \sum_{i=1}^{I} \left(\sum_{k=1}^{K} \hat{w}_k \hat{p}_{k|ij} - \sum_{k=1}^{K} \hat{w}_k \hat{p}_{\cdot \cdot k} \right)^2 \hat{p}_{ij \cdot}}{\sum_{k=1}^{K} \left(\hat{w}_k - \sum_{k=1}^{K} \hat{w}_k \hat{p}_{\cdot \cdot k} \right)^2 \hat{p}_{\cdot \cdot k}}. \tag{9.17}$$

9.4 Simulation of Three-Way Contingency Tables

It is known that the asymmetric measure Goodman–Kruskal tau for two-way contingency table is sensitivity to the number of categories of the response variable (Agresti & Kateri, 2011; Beh & Lombardo, 2014). Thus, the small magnitude of the tau index does not necessarily mean a low asymmetric association between the two variables (Wei & Kim, 2017). Wei and Kim (2017) showed by simulation that the subcopula-based association measure for two-way ordinal contingency tables is insensitive to the number of categories in the response variable. In this section, we examine the sensitivity of the complete dependence measure on the three-way contingency table with two response variables and one independent variable given in Section 9.1.

We consider the simulations with three-way contingency table with two response variables (Y, Z), and the one independent variable X. For each simulated three-way contingency table, we calculate the complete dependence measure $\hat{\mu}(Y, Z \mid X)$. We consider the magnitude of association among three categorical variables (X, Y, Z) with I categories in each variable, respectively, the sample size, the number of categories I.

For the magnitude of association, we applied the three values for the exchangeable correlation matrix (i.e. all three correlation parameters are identical in the correlation matrix) in Gaussian copula, $\theta = (0.8, 0.5, 0.2)$. As to the number of categories, 14 values are considered: $I = 2, 3, \ldots, 15$. We simulate data of size $n = 500, 1000$, and 2000 from the trivariate Gaussian copulas. Then we create an $I \times I \times I$-contingency table by classifying n data points by the three discrete variables, each with the equal-width categories in (X, Y, Z). And for each simulated table, the complete dependence measure $\hat{\mu}(Y, Z \mid X)$ is calculated.

Figure 9.1 shows the values of the complete dependence measure $\hat{\mu}(Y, Z \mid X)$ with respect to the number of the categories for different sample sizes and the

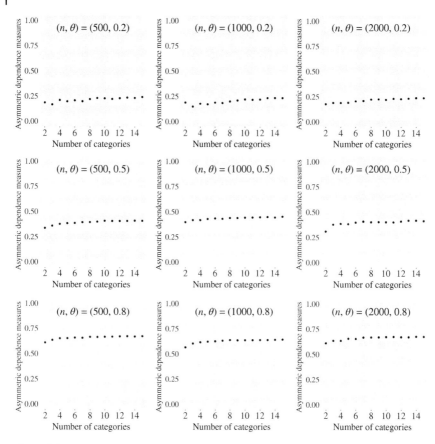

Figure 9.1 The complete dependence measure $\mu(Y, Z \mid X)$ with respect to the number of the categories for the sample size $n = 500, 1000, 2000$ (from left column to right column) and the correlation parameter values in Gaussian copula with $\theta = 0.8, 0.5, 0.2$ (from top panel to bottom panel).

correlation parameter values in Gaussian copula. We observe that the complete dependence measure for the case with two response variables Y, Z and one independent variable X, $\hat{\mu}(Y, Z \mid X)$, tends to take slightly larger values and then stabilize as the number of categories increases. This shows that the complete dependence measure $\mu(Y, Z \mid X)$ is insensitive to the number of categories in response variables Y and Z.

9.5   Real Data of Three-Way Contingency Tables

In this section, two real data sets are considered to illustrate our result. The first three-way contingency table (Table 9.5) has two independent variables and one response variable. The study was aimed at testing the effect of urbanization (U) and location (L) on the preference (P) for black olives using a six-point ordinal scale. The response is given by the variable preference for black olives of Armed Forces personnel with six growing ordered levels: A, B, C, D, E, F. The nominal-scale factors are location with three levels (NE, NW, and SW), and urbanization with two levels (urban, rural) (Agresti, 1990).

By calculation, the estimation of the complete dependence measure of the effect of urbanization (U) and location (L) on the preference (P) via the Definition 9.4 is $\hat{\mu}(P \mid L, U) = 0.1913$. There are relatively weak relations of complete dependence of U and L on P.

The second three-way contingency table (Table 9.6) has one independent variable and two response variables. It was a study of job satisfaction of workers in social enterprises of Caserta (survey of the Economics Faculty of the Second University of Naples, 2010). The two responses represent the satisfaction with respect to organizational aspects (O) and the satisfaction to relationships with users (R). The predictor is given by the status of workers (partner or non-partner) (S). The aim is to study the importance of the worker status (partner or non-partner) on the

Table 9.5   Black Olive preference (P) by location (L) and urbanization (U).

U	Urban			Rural		
L	NW	NE	SW	NW	NE	SW
P						
A	20	18	12	30	23	11
B	15	17	9	22	18	9
C	12	18	23	21	20	26
D	17	18	21	17	18	19
E	16	6	19	8	10	17
F	28	25	30	12	15	24

Table 9.6 Worker satisfaction for organizational aspects (O) and satisfaction for relationships with users (R) by worker status (S).

S	Partner			
R	Unsatisfied	Little satisfied	Satisfied	Very satisfied
O				
Unsatisfied	3	7	4	8
Little satisfied	5	9	10	12
Satisfied	1	9	22	30
Very satisfied	3	5	59	73

S	Non-partner			
R	Unsatisfied	Little satisfied	Satisfied	Very satisfied
O				
Unsatisfied	15	4	2	4
Little satisfied	4	8	14	11
Satisfied	2	5	22	53
Very satisfied	2	4	14	39

final satisfaction of 426 workers with respect to organizational aspects and relationships with users. We observe the job satisfaction with four levels (unsatisfied, little satisfied, satisfied, very satisfied) and worker status with two levels (partner, non-partner).

By calculation, estimates of complete dependence measures defined in Definitions 9.4 and 9.5 are given by $\hat{\mu}(R \mid O, S) = 0.334$ and $\hat{\mu}(O \mid R, S) = 0.395$, and $\hat{\mu}(O, R \mid S) = 0.157$.

References

Agresti, A. (1990). *Analysis of categorical data*. New York, NY: Wiley & Sons.

Agresti, A., & Kateri, M. (2011). Categorical data analysis. In *International encyclopedia of statistical science* (pp. 206–208). New York, NY: Springer.

Beh, E. J., & Lombardo, R. (2014). *Correspondence analysis: Theory, practice and new strategies*. New York, NY: Wiley & Sons.

Boonmee, T., & Tasena, S. (2016). Measure of complete dependence of random vectors. *Journal of Mathematical Analysis and Applications, 443*(1), 585–595.

Goodman, L. A., & Kruskal, W. H. (1954). Measures of association for cross classifications. *Journal of the American Statistical Association, 49*(268), 732–764.

D. Kim and Z. Wei (2020). Analysis of asymmetric dependence for three-way contingency tables using subcopula approach. In Wolfgang Wiedermann, Daeyoung Kim, Engin A. Sungur, Alexander von Eye *Direction dependence in statistical models: Methods of analysis* Wiley, this volume.

Shan, Q., Wongyang, T., Wang, T., & Tasena, S. (2015). A measure of mutual complete dependence in discrete variables through subcopula. *International Journal of Approximate Reasoning, 65*, 11–23.

Siburg, K. F., & Stoimenov, P. A. (2010). A measure of mutual complete dependence. *Metrika, 71*(2), 239–251.

Sklar, A. (1959). Fonctions de répartition à n dimensions et leurs marges. *Publications de l'Institut de statistique de l'Université de Paris, 8*, 229–231.

Tasena, S., & Dhompongsa, S. (2013). A measure of multivariate mutual complete dependence. *International Journal of Approximate Reasoning, 54*(6), 748–761.

Tasena, S., & Dhompongsa, S. (2016). Measures of the functional dependence of random vectors. *International Journal of Approximate Reasoning, 68*, 15–26.

Wei, Z., & Kim, D. (2017). Subcopula-based measure of asymmetric association for contingency tables. *Statistics in Medicine, 36*(24), 3875–3894.

Wongyang, T. (2015). *Copula and measures of dependence*. Las Cruces, NM: New Mexico State University. Research notes

Zhang, Y., & Song, M. (2013). Deciphering interactions in causal networks without parametric assumptions. *arXiv preprint arXiv:1311.2707*.

Zhu, X., Wang, T., Choy, S. B., & Autchariyapanitkul, K. (2018). Measures of mutually complete dependence for discrete random vectors. In *Predictive econometrics and big data* (pp. 303–317). New York, NY: Springer.

10

Analysis of Asymmetric Dependence for Three-Way Contingency Tables Using the Subcopula Approach

Daeyoung Kim[1] and Zheng Wei[2]

[1]*Department of Mathematics and Statistics, University of Massachusetts Amherst, Amherst, MA, USA*
[2]*Department of Mathematics and Statistics, University of Maine, Orono, ME, USA*

10.1 Introduction

The analysis of asymmetric association has been applied to identify and measure the asymmetric dependence among the variables. The application areas include gene regulatory network inference (Margolin et al., 2006; Nguyen, Tilton, Kemp, & Song, 2017; Sharma, Kumar, Zhong, & Song, 2017; van Someren, Wessels, Backer, & Reinders, 2002), developmental research on attention deficit hyperactivity disorder (Nigg et al., 2008) and aggression in adolescence (von Eye & Wiedermann, 2014), and finances (Aloui, Aissa, & Nguyen, 2013; Balke, Brown, & Yucel, 2002; Hong, Tu, & Zhou, 2007; Wu, Chung, & Chang, 2012; Yang & Hamori, 2014).

In the literature, two types of asymmetric association are considered, directional or non-directional. The directional relationship is that one variable is considered as a dependent (response) variable and the other as an independent variable (predictor). The identification of direction of effects among variables is of main interest in non-experimental research (Wiedermann & von Eye, 2017). For example, in attention-deficit/hyperactivity disorder (ADHD) studies among children, the researchers often want to know if the level lead exposure can be considered as the independent variable for ADHD (Nigg et al., 2008). The nondirectional relationship is the correlational association between the variables with different magnitude. For instance, in biological gene network inference, the importance of directionality among different genes variables can serve as evidence for (recursive/non-recursive) causality (Nguyen et al., 2017; Sharma et al., 2017) and the relation among two genes might affect each other with different degrees.

In the analysis of categorical data, where the main goal is to identify the pattern of association among two discrete variables in a contingency table, a few

Direction Dependence in Statistical Modeling: Methods of Analysis, First Edition.
Edited by Wolfgang Wiedermann, Daeyoung Kim, Engin A. Sungur, and Alexander von Eye.

approaches to the analysis of asymmetric association have been proposed, including Goodman–Kruskal tau index (Goodman & Kruskal, 1954), Theil's uncertainty coefficient (Theil, 1970), directional (non-hierarchical) log-linear models (Wiedermann & von Eye, 2017), and subcopula regression based asymmetric measure (Wei & Kim, 2017).

In a contingency table with two categorical variables X and Y, the Goodman–Kruskal tau index (Goodman & Kruskal, 1954) measures the relative increase in the proportion of correct predictions of the (putative) dependent variable given the (putative) independent variable. The range of the tau index is between 0 (conditional independence) and 1, and it is asymmetric in that the tau index for a case where X causes Y is not necessarily equal to the tau index for a case where Y causes X. One issue with the tau index is that it tends to be smaller as the number of categories in a (putative) dependent variable increases and thus low values of the tau index do not necessarily mean weak asymmetric association between the variables (Agresti, 1990).

The Theil's uncertainty coefficient (Theil, 1970) utilizes the proportional reduction in the variation measure of the (putative) dependent variable. The uncertainty coefficient takes values between 0 (independence between X and Y) and 1, and it is also an asymmetric association measure. The issues of the uncertainty coefficient are that it is also sensitive to the number of categories in a (putative) dependent variable (Agresti, 1990), it is not well defined for contingency tables with zero cell counts, and its asymmetry property is induced by only the variation in the dependent variable.

To formally test and determine the direction of effects in binary categorical variables, Wiedermann and von Eye (2017) proposed a parametric model, namely directional (non-hierarchical) log-linear model. The main idea of this approach is that the univariate probability distribution of the response variable can be explained by the univariate distribution of the independent variable and the interaction between the two variables, and thus omitting the univariate effects of the response variable in the log-linear model does not result in a poorly-fitting model. These authors presented a conceptual link between the proposed directional log-linear model and the method for the determination of direction of effect between continuous variables using linear regression model proposed by Dodge and Rousson (2000, 2001). They also discussed extensions of the proposed directional log-linear model to the case with multiple (putative independent and response) variables.

Wei and Kim (2017) proposed a model-free approach to asymmetric association of a two-way contingency table, subcopula regression based association measure. The proposed measure utilizes the subcopula regression between X and Y to measure the asymmetric predictive powers of the variables. They compared the proposed measure with two existing model-free measures, Goodman–Kruskal tau index (Goodman & Kruskal, 1954) and the Theil's uncertainty coefficient (Theil, 1970), and showed that the proposed measure is not sensitive to the number of categories of the putative response variable and thus the magnitude of the proposed measure can be used as the degree of asymmetric association in the contingency table. Note that Wei and Kim (2017) presented the formulae of the subcopula-based asymmetric association measures for three ordinal variables, without examining their properties and utility.

In this chapter, we investigate the basic properties and the performance of the subcopula-based asymmetric association measure provided by Wei and Kim (2017) for ordinal three-way contingency tables. Specifically, the rest of the chapter is structured as follows. Section 10.2 briefly reviews a subcopula-based measure of asymmetric association for ordinal two-way contingency tables proposed in Wei and Kim (2017). In Section 10.3, we derive the formulae of the subcopula regression-based asymmetric association measures for ordinal three-way contingency tables and study their basic properties. Section 10.5 evaluates the performance and utility of the proposed method using simulation and a real-world data. We end this chapter with some concluding remarks.

10.2 Review on Subcopula Based Asymmetric Association Measure for Ordinal Two-Way Contingency Table

Let X and Y denote two discrete ordinal variables with I and J categories, having the supports $S_0 = \{x_1, \ldots, x_I\}$ and $S_1 = \{y_1, \ldots, y_J\}$, respectively. Consider a two-way $I \times J$ ordinal contingency table with the matrix of the joint probability mass function (p.m.f.) $P = \{p_{ij} = P(X = x_i, Y = y_j)\}$ where $i = 1, \ldots, I, j = 1, \ldots, J$, and $\sum_{i=1}^{I} \sum_{j=1}^{J} p_{ij} = 1$. Furthermore, denote the i-th row marginal p.m.f. for X and the j-th column marginal p.m.f. for Y by $p_{i\cdot} = P(X = x_i) = \sum_{j=1}^{J} p_{ij}$ and $p_{\cdot j} = P(Y = y_j) = \sum_{i=1}^{I} p_{ij}$, respectively. The asymmetric association measure proposed by Wei and Kim (2017) utilized the subcopula and its regression associated with X and Y for the two-way ordinal contingency table. We first recall the subcopula from Nelsen (2006).

Definition 10.1   A *two-dimensional* **subcopula** is a function $C : D_0 \times D_1 \rightarrow [0, 1]$ such that

(a) $D_0 \times D_1$ is the domain of C where D_0 and D_1 are subsets of $[0, 1]$;
(b) For every $(u, v) \in D_0 \times D_1$, $C(u, 1) = u$, $C(1, v) = v$, and $C(u, 0) = C(0, v) = 0$;
(c) $C(u', v') - C(u', v) - C(u, v') + C(u, v) \geq 0$ where $[u, u'] \times [v, v']$ is a rectangle with all vertices in $D_0 \times D_1$.

Given the joint p.m.f. $P = \{p_{ij}\}$ of X and Y in the $I \times J$ ordinal contingency table, denote the joint cumulative distribution of X and Y and their marginal distributions be $H(x, y)$, $F(x)$, and $G(y)$, respectively, and let $U = F(X)$ and $V = G(Y)$. Then U and V have the supports $D_0 = F(S_0) = \{u_0 = 0, u_1, \ldots, u_I = 1\}$ and $D_1 = G(S_1) = \{v_0 = 0, v_1, \ldots, v_J = 1\}$, respectively, where

$$u_i = \sum_{s=1}^{i} p_s. \quad \text{and} \quad v_j = \sum_{t=1}^{j} p_{\cdot t}. \tag{10.1}$$

By the Sklar's theorem (Sklar, 1959), for the joint distribution H for ordinal variables X and Y with marginals U and V, there exists a subcopula C on the supports $D_0 \times D_1$ such that

$$H(x_i, y_j) = \sum_{s=1}^{i} \sum_{t=1}^{j} p_{st} = C(u_i, v_j). \tag{10.2}$$

The marginal p.m.f.s of U and V and the joint p.m.f of C are, respectively,

$$p_0(u_i) = p_i., \quad p_1(v_j) = p_{\cdot j}, \quad c(u_i, v_j) = p_{ij}. \tag{10.3}$$

Furthermore, the conditional p.m.f.s of V given U and of U given V are, respectively,

$$c_{V|U}(v_j \mid u_i) = \frac{c(u_i, v_j)}{p_0(u_i)} = \frac{p_{ij}}{p_i.} \equiv p_{j|i} \quad \text{and} \quad c_{U|V}(u_i \mid v_j) = \frac{c(u_i, v_j)}{p_1(v_j)} = \frac{p_{ij}}{p_{\cdot j}} \equiv p_{i|j}. \tag{10.4}$$

Wei and Kim (2017) defined the subcopula regression functions associated with U and V and utilized them to propose the measurement to quantitatively measure the asymmetric dependence in an $I \times J$ contingency table.

Definition 10.2 (Wei & Kim, 2017) For an $I \times J$ contingency table of two ordinal variables X and Y, let $U = F(X)$ and $V = G(Y)$ be the marginal distributions of X and Y, respectively, and let $C(u, v)$ be the associated subcopula defined on the supports D_0 and D_1 in Eq. (10.1). Then the **subcopula regression functions** of V on U and U on V are defined as follows, respectively:

$$r_{V|U}^C(u_i) \equiv E_C(V \mid U = u_i) = \sum_{v_j \in D_1} v_j c_{V|U}(v_j \mid u_i) = \sum_{j=1}^{J} v_j p_{j|i}, \quad i = 1, \ldots, I.$$

$$r_{U|V}^C(v_j) \equiv E_C(U \mid V = v_j) = \sum_{u_i \in D_0} u_i c_{U|V}(u_i \mid v_j) = \sum_{i=1}^{I} u_i p_{i|j}, \quad j = 1, \ldots, J.$$

The **subcopula regression based asymmetric association measures** of Y on X and of X on Y are defined as follows, respectively:

$$\rho^2_{(X \to Y)} \equiv \frac{\text{Var}(r^C_{V|U}(U))}{\text{Var}(V)} = \frac{E[(r^C_{V|U}(U) - E(V))^2]}{\text{Var}(V)} = \frac{\sum\limits_{i=1}^{I} \left(\sum\limits_{j=1}^{J} v_j p_{j|i} - \sum\limits_{j=1}^{J} v_j p_{\cdot j} \right)^2 p_{i\cdot}}{\sum\limits_{j=1}^{J} \left(v_j - \sum\limits_{j=1}^{J} v_j p_{\cdot j} \right)^2 p_{\cdot j}},$$

(10.5)

$$\rho^2_{(Y \to X)} \equiv \frac{\text{Var}(r^C_{U|V}(V))}{\text{Var}(U)} = \frac{E[(r^C_{U|V}(V) - E(U))^2]}{\text{Var}(U)} = \frac{\sum\limits_{j=1}^{J} \left(\sum\limits_{i=1}^{I} u_i p_{i|j} - \sum\limits_{i=1}^{I} u_i p_{i\cdot} \right)^2 p_{\cdot j}}{\sum\limits_{i=1}^{I} \left(u_i - \sum\limits_{i=1}^{I} u_i p_{i\cdot} \right)^2 p_{i\cdot}}.$$

(10.6)

The association measures in Eqs. (10.5) and (10.6) can be interpreted as the proportion of total variation for the distribution function of a putative response variable that is explained by the subcopula regression based on a putative predictor. Notice that they are defined independent of the supports of X and Y, and they are the functions of the (conditional/marginal) p.m.f.s of X and Y, and the supports of the subcopula, D_0 and D_1.

Wei and Kim (2017) investigated the properties of the association measures in Eqs. (10.5) and (10.6). First, ranging from 0 to 1 (i.e. $0 \leq \rho^2_{(X \to Y)}, \rho^2_{(Y \to X)} \leq 1$), they are asymmetric in that $\rho^2_{(X \to Y)}$ is not equal to $\rho^2_{(Y \to X)}$ in general, except when both X and Y are dichotomous variables with only two categories. Second, $\rho^2_{(X \to Y)} = \rho^2_{(Y \to X)} = 0$ if X and Y are independent. Moreover, zero value of $\rho^2_{(X \to Y)}$ or $\rho^2_{(Y \to X)}$ implies zero correlation between U and V. Note that "one of two measures in Eqs. (10.5) and (10.6) equals to zero" does not necessarily imply that the other one is zero: $\rho^2_{(X \to Y)} = 0$ does not imply $\rho^2_{(Y \to X)} = 0$, and vice versa. Third, $\rho^2_{(X \to Y)} = 1$ if and only if $Y = g(X)$ almost surely for some measurable function g.

To estimate the asymmetric association measures of Eqs. (10.5) and (10.6), Wei and Kim (2017) used the relative frequencies of the observed two-way contingency table. Let $\{n_{ij}\}$, $i = 1, \dots, I$ and $j = 1, \dots, J$, denote the cell counts in an $I \times J$ ordinal contingency table that cross-classifies $n \left(= \sum_{i=1}^{I} \sum_{j=1}^{J} n_{ij} \right)$ subjects/units according to I categories of the row variable X and J categories of the column variable Y. The row sums and columns sums of $\{n_{ij}\}$ are denoted as $n_{i\cdot} = \sum_{j=1}^{J} n_{ij}$ and $n_{\cdot j} = \sum_{i=1}^{I} n_{ij}$, respectively. Then the supports D_0 and D_1 in Eq. (10.1), $p_{ij}, p_{i\cdot}, p_{\cdot j}$,

$p_{i|j}, p_{j|i}$ in Eqs. (10.3) and (10.4) can be estimated by

$$\widehat{D}_0 = \{0, \widehat{u}_1, \ldots, \widehat{u}_I = 1\} \text{ with } \widehat{u}_i = \sum_{s=1}^{i} \widehat{p}_{s\cdot},$$

$$\widehat{D}_1 = \{0, \widehat{v}_1, \ldots, \widehat{v}_J = 1\} \text{ with } \widehat{v}_j = \sum_{t=1}^{j} \widehat{p}_{\cdot t},$$

$$\widehat{p}_{ij} = \frac{n_{ij}}{n}, \quad \widehat{p}_{i\cdot} = \frac{n_{i\cdot}}{n}, \quad \widehat{p}_{\cdot j} = \frac{n_{\cdot j}}{n}, \quad \widehat{p}_{i|j} = \frac{\widehat{p}_{ij}}{\widehat{p}_{\cdot j}}, \quad \widehat{p}_{j|i} = \frac{\widehat{p}_{ij}}{\widehat{p}_{i\cdot}}.$$

Finally, using the estimators obtained above, $\rho^2_{(Y \rightarrow X)}$ and $\rho^2_{(X \rightarrow Y)}$ in Eqs. (10.5) and (10.6) can be estimated by

$$\widehat{\rho}^2_{(X \rightarrow Y)} = \frac{\sum_{i=1}^{I} \left(\sum_{j=1}^{J} \widehat{v}_j \widehat{p}_{j|i} - \sum_{j=1}^{J} \widehat{v}_j \widehat{p}_{\cdot j} \right)^2 \widehat{p}_{i\cdot}}{\sum_{j=1}^{J} \left(\widehat{v}_j - \sum_{j=1}^{J} \widehat{v}_j \widehat{p}_{\cdot j} \right)^2 \widehat{p}_{\cdot j}}$$

and

$$\widehat{\rho}^2_{(Y \rightarrow X)} = \frac{\sum_{j=1}^{J} \left(\sum_{i=1}^{I} \widehat{u}_i \widehat{p}_{i|j} - \sum_{i=1}^{I} \widehat{u}_i \widehat{p}_{i\cdot} \right)^2 \widehat{p}_{\cdot j}}{\sum_{i=1}^{I} \left(\widehat{u}_i - \sum_{i=1}^{I} \widehat{u}_i \widehat{p}_{i\cdot} \right)^2 \widehat{p}_{i\cdot}}. \tag{10.7}$$

Wei and Kim (2017) derived the asymptotic distributions of the estimators $\widehat{\rho}^2_{(X \rightarrow Y)}$ and $\widehat{\rho}^2_{(Y \rightarrow X)}$ in Eq. (10.7) and their difference $\widehat{\rho}^2_{(X \rightarrow Y)} - \widehat{\rho}^2_{(Y \rightarrow X)}$, and showed the construction of the asymptotic confidence intervals.

10.3 Measure of Asymmetric Association for Ordinal Three-Way Contingency Tables via Subcopula Regression

Wei and Kim (2017) presented only the formula of the subcopula-based asymmetric association measures for the ordinal three-way contingency tables. In this section we show the derivation of the subcopula-based asymmetric association measures, investigate their basic theoretical properties and present the corresponding estimators.

10.3.1 Subcopula Regression-Based Asymmetric Association Measures

Suppose that X, Y, and Z are ordinal categorical variables with I, J, and K categories, having the supports $S_0 = \{x_1, \ldots, x_I\}$, $S_1 = \{y_1, \ldots, y_J\}$, and $S_2 = \{z_1, \ldots, z_K\}$, respectively. Consider a three-way ordinal contingency table for X, Y, and Z with the joint p.m.f. $P = \{p_{ijk} = P(X = x_i, Y = y_j, Z = z_k)\}$ where $i = 1, \ldots, I, j = 1, \ldots, J$, $k = 1, \ldots, K$, and $\sum_{i=1}^{I} \sum_{j=1}^{J} \sum_{k=1}^{K} p_{ijk} = 1$. Let $p_{i\cdot\cdot}$, $p_{\cdot j\cdot}$, and $p_{\cdot\cdot k}$ be the i-th row, j-th column, and the k-th tube marginals for X, Y, and Z, respectively, and $p_{ij\cdot}$, $p_{i\cdot k}$, and $p_{\cdot jk}$ be the two-way marginals for (X, Y), (X, Z), and (Y, Z), respectively.

Given the joint p.m.f. $P = \{p_{ijk}\}$ for X, Y, and Z in the $I \times J \times K$ ordinal contingency table, we denote the marginal distributions of X, Y, and Z to be $F_0(x)$, $F_1(y)$, and $F_2(z)$, respectively. Let $U = F_0(X)$, $V = F_1(Y)$, and $W = F_2(Z)$. Then, we define the ranges of U, V, and W to be $D_0 = \{u_0 = 0, u_1, \ldots, u_I = 1\}$, $D_1 = \{v_0 = 0, v_1, \ldots, v_J = 1\}$, and $D_2 = \{w_0 = 0, w_1, \ldots, w_K = 1\}$, respectively, where

$$u_i = \sum_{s=1}^{i} p_{s\cdot\cdot}, \quad v_j = \sum_{t=1}^{j} p_{\cdot t\cdot}, \quad \text{and} \quad w_k = \sum_{r=1}^{k} p_{\cdot\cdot r}. \tag{10.8}$$

For the joint distribution H of ordinal variables X, Y, and Z, there exists a three dimensional subcopula C (Nelsen, 2006) on the domain $D_0 \times D_1 \times D_2$ which satisfies

$$H(x_i, y_j, z_k) = \sum_{s=1}^{i} \sum_{t=1}^{j} \sum_{r=1}^{k} p_{str} = C(u_i, v_j, w_k),$$

with the following three properties,

1) $D_0 \times D_1 \times D_2$ is the domain of C where D_0, D_1, and D_2 are subsets of $[0, 1]$;
2) $C(u, v, w) = 0$ if at least one of u, v, w is zero; for every $u \in D_0, v \in D_1$, and $w \in D_2$, if $C(u,1,1) = u$, $C(1, v, 1) = v$, and $C(1, 1, w) = w$;
3) $C(u', v', w') - C(u', v', w) - C(u', v, w') - C(u, v', w') + C(u', v, w) + C(u, v', w) + C(u, v, w') - C(u, v, w) \geq 0$ where $[u, u'] \times [v, v'] \times [w, w']$ is a rectangle with all vertices in $D_0 \times D_1 \times D_2$.

In addition, the joint p.m.f. of the subcopula C, the conditional p.m.f.s of U given (V, W), V given (U, W), and W given (U, V), are

$$c(u_i, v_j, w_k) = p_{ijk}, \quad c_{U|V,W}(u_i \mid v_j, w_k) = p_{i|jk} = p_{ijk}/p_{\cdot jk},$$
$$c_{V|U,W}(v_j \mid u_i, w_k) = p_{j|ik} = p_{ijk}/p_{i\cdot k}, \quad c_{W|U,V}(w_k \mid u_i, v_j) = p_{k|ij} = p_{ijk}/p_{ij\cdot}.$$

Definition 10.3 below defines the trivariate subcopula regression functions for (U, V, W) associated with the three-way contingency table.

Definition 10.3 The subcopula regression functions of U on (V, W), V on (U, W), and W on (U, V) are defined as follows:

$$r_{U|V,W}^C(v_j, w_k) \equiv E_C(U \mid V = v_j, W = w_k) = \sum_{u_i \in D_0} u_i c_{U|V,W}(u_i \mid v_j, w_k) = \sum_{i=1}^{I} u_i p_{i|jk},$$
$$\tag{10.9}$$

$$r^C_{V|U,W}(u_i, w_k) \equiv E_C(V \mid U = u_i, W = w_k) = \sum_{v_j \in D_1} v_j c_{V|U,W}(v_j \mid u_i, w_k) = \sum_{j=1}^{J} v_j p_{j|ik},$$

$$(10.10)$$

$$r^C_{W|U,V}(u_i, v_j) \equiv E_C(W \mid U = u_i, V = v_j) = \sum_{w_k \in D_2} w_k c_{W|U,V}(w_k \mid u_i, v_j) = \sum_{k=1}^{K} w_k p_{k|ij},$$

$$(10.11)$$

where $i = 1, \ldots, I$, $j = 1, \ldots, J$ and $k = 1, \ldots, K$.

Using the trivariate subcopula regression functions given in Definition 10.3, we can derive the trivariate subcopula regression-based measures to quantify the asymmetric association among the ordinal variables X, Y, and Z in an $I \times J \times K$ contingency table.

Definition 10.4 For the ordinal random variables X, Y, and Z with the trivariate subcopula $C(u, v, w)$ in an $I \times J \times K$ contingency table, let $U = F_0(X)$, $V = F_1(Y)$, and $W = F_2(Z)$. Then, a **measure of subcopula regression-based asymmetric association** of X on Y and Z is,

$$\rho^2_{(Y,Z \to X)} \equiv \frac{\mathrm{Var}(r^C_{U|V,W}(V, W))}{\mathrm{Var}(U)} = \frac{E[(r^C_{U|V,W}(V, W)) - E(U))^2]}{\mathrm{Var}(U)}$$

$$= \frac{\sum_{k=1}^{K} \sum_{j=1}^{J} \left(\sum_{i=1}^{I} u_i p_{i|jk} - \sum_{i=1}^{I} u_i p_{i\cdot\cdot} \right)^2 p_{\cdot jk}}{\sum_{i=1}^{I} \left(u_i - \sum_{i=1}^{I} u_i p_{i\cdot\cdot} \right)^2 p_{i\cdot\cdot}}.$$

Similarly, the **measures of subcopula regression-based asymmetric associations** of Y on X and Z, and of Z on X and Y are

$$\rho^2_{(X,Z \to Y)} = \frac{\sum_{k=1}^{K} \sum_{i=1}^{I} \left(\sum_{j=1}^{J} v_j p_{j|ik} - \sum_{j=1}^{J} v_j p_{\cdot j\cdot} \right)^2 p_{i\cdot k}}{\sum_{j=1}^{J} \left(v_j - \sum_{j=1}^{J} v_j p_{\cdot j\cdot} \right)^2 p_{\cdot j\cdot}},$$

$$\rho^2_{(X,Y \to Z)} = \frac{\sum_{j=1}^{J} \sum_{i=1}^{I} \left(\sum_{k=1}^{K} w_k p_{k|ij} - \sum_{k=1}^{K} w_k p_{\cdot\cdot k} \right)^2 p_{ij\cdot}}{\sum_{k=1}^{K} \left(w_k - \sum_{k=1}^{K} w_k p_{\cdot\cdot k} \right)^2 p_{\cdot\cdot k}}.$$

In Proposition 10.1, we investigate properties of the proposed asymmetric association measures in Definition 10.4.

Proposition 10.1

(a) $0 \leq \rho^2_{(Y,Z \to X)}, \rho^2_{(X,Z \to Y)}, \rho^2_{(X,Y \to Z)} \leq 1$.

(b) If X and (Y, Z) are independent, then $\rho^2_{(Y,Z \to X)} = 0$; Similarly, if Y and (X, Z) are independent, then $\rho^2_{(X,Z \to Y)} = 0$; and if Z and (X, Y) are independent, then $\rho^2_{(X,Y \to Z)} = 0$.

(c) $\rho^2_{(X,Y \to Z)} = 1$ if and only if $Z = g(X, Y)$ almost surely for some measurable function g.

(d) If $W = g(U, V) + \varepsilon$ where ε, being independent of U and V, is a random variable with finite second moments and g is a linear or nonlinear measurable function, then

$$\rho^2_{(X,Y \to Z)} = \frac{\text{Var}(g(U, V))}{\text{Var}(g(U, V)) + \text{Var}(\varepsilon)}.$$

Proof.
See Appendix 10.A.1. □

The Proposition 10.1(a), (b), and (c) indicate that the proposed measures can identify asymmetric nonlinear relations between the ordinal variables in a three-way contingency table. The Proposition 10.1(d) represents that the proposed measures represent the average proportion of variance for the distribution function of the (putative) response variable explained by the subcopula copula regression of the distribution functions for the (putative) predictors.

10.3.2 Estimation

The subcopula regression-based asymmetric association measures given in Definition 10.4 can be estimated utilizing the relative frequencies of the observed three-way contingency table. Let $\{n_{ijk}\}$, $i = 1, \ldots, I, j = 1, \ldots, J$, and $k = 1, \ldots, K$ denote the cell counts in an $I \times J \times K$ ordinal contingency table obtained by classifying $n \left(= \sum_{i=1}^{I} \sum_{j=1}^{J} \sum_{k=1}^{K} n_{ijk} \right)$ units according to the I categories of X, J categories of Y, and K categories of Z. The row sums, column sums, and tube sums (i.e. the marginal frequencies) are denoted as $n_{i..} = \sum_{j=1}^{J} \sum_{k=1}^{K} n_{ijk}, n_{.j.} = \sum_{i=1}^{I} \sum_{k=1}^{K} n_{ijk}$, and

$n_{..k} = \sum_{i=1}^{I} \sum_{j=1}^{J} n_{ijk}$, respectively. We then define the estimators for p_{ijk}, $p_{i..}$, $p_{.j.}$, $p_{..k}$,

and $p_{i|jk} = \frac{p_{ijk}}{p_{.jk}}$, $p_{j|ik} = \frac{p_{ijk}}{p_{i.k}}$ and $p_{k|ij} = \frac{p_{ijk}}{p_{ij.}}$ by

$$\widehat{p}_{ijk} = \frac{n_{ijk}}{n}, \quad \widehat{p}_{i..} = \frac{n_{i..}}{n}, \quad \widehat{p}_{.j.} = \frac{n_{.j.}}{n}, \quad \widehat{p}_{..k} = \frac{n_{..k}}{n},$$

$$\widehat{p}_{i|jk} = \frac{n_{ijk}}{n_{.jk}}, \quad \widehat{p}_{j|ik} = \frac{n_{ijk}}{n_{i.k}}, \quad \widehat{p}_{k|ij} = \frac{n_{ijk}}{n_{ij.}}.$$

The ranges of the marginal distributions of X, Y and Z, D_0, D_1 and D_2 in Eq. (10.8), are estimated by

$$\widehat{u}_i = \sum_{s=1}^{i} \widehat{p}_{s..}, \quad \widehat{v}_j = \sum_{t=1}^{j} \widehat{p}_{.t.}, \quad \text{and} \quad \widehat{w}_k = \sum_{r=1}^{k} \widehat{p}_{..r}.$$

By utilizing the estimators obtained above, we provide the estimators for sub-copula regression-based asymmetric association measures $\rho^2_{(Y,Z \to X)}$, $\rho^2_{(X,Z \to Y)}$, and $\rho^2_{(X,Y \to Z)}$ given in Definition 10.4:

$$\widehat{\rho}^2_{(Y,Z \to X)} = \frac{\sum_{k=1}^{K} \sum_{j=1}^{J} \left(\sum_{i=1}^{I} \widehat{u}_i \widehat{p}_{i|jk} - \sum_{i=1}^{I} \widehat{u}_i \widehat{p}_{i..} \right)^2 \widehat{p}_{.jk}}{\sum_{i=1}^{I} \left(\widehat{u}_i - \sum_{i=1}^{I} \widehat{u}_i \widehat{p}_{i..} \right)^2 \widehat{p}_{i..}},$$

$$\widehat{\rho}^2_{(X,Z \to Y)} = \frac{\sum_{k=1}^{K} \sum_{i=1}^{I} \left(\sum_{j=1}^{J} \widehat{v}_j \widehat{p}_{j|ik} - \sum_{j=1}^{J} \widehat{v}_j \widehat{p}_{.j.} \right)^2 \widehat{p}_{i.k}}{\sum_{j=1}^{J} \left(\widehat{v}_j - \sum_{j=1}^{J} \widehat{v}_j \widehat{p}_{.j.} \right)^2 \widehat{p}_{.j.}},$$

$$\widehat{\rho}^2_{(X,Y \to Z)} = \frac{\sum_{j=1}^{J} \sum_{i=1}^{I} \left(\sum_{k=1}^{K} \widehat{w}_k \widehat{p}_{k|ij} - \sum_{k=1}^{K} \widehat{w}_k \widehat{p}_{..k} \right)^2 \widehat{p}_{ij.}}{\sum_{k=1}^{K} \left(\widehat{w}_k - \sum_{k=1}^{K} \widehat{w}_k \widehat{p}_{..k} \right)^2 \widehat{p}_{..k}}. \tag{10.12}$$

To assess the uncertainty in the estimates of the proposed asymmetric association measures and their pairwise differences, we propose using the bootstrap method (Efron & Tibshirani, 1993). That is, compute the bootstrap (percentile/bias-corrected and accelerated (BCa)) confidence intervals for $\rho^2_{(Y,Z \to X)}$, $\rho^2_{(X,Z \to Y)}$, $\rho^2_{(X,Y \to Z)}$, $\rho^2_{(Y,Z \to X)} - \rho^2_{(X,Z \to Y)}$, $\rho^2_{(Y,Z \to X)} - \rho^2_{(X,Y \to Z)}$, and $\rho^2_{(X,Z \to Y)} - \rho^2_{(X,Y \to Z)}$ using bootstrap samples of size n that are simulated from the saturated log-linear model fitted to the observed three-way contingency table.

10.4   Numerical Examples

In this section, the performance of the proposed subcopula regression association measure is illustrated with simulation studies and a real-world data.

10.4.1   Sensitivity Analysis

Wei and Kim (2017) showed by simulation that the subcopula-based association measure in Eq. (10.7) for two-way ordinal contingency tables is insensitive to the number of categories in the response variable and so the magnitude of their proposed measure can be employed to quantify the degree of association in the two-way contingency tables. In the following paragraphs, we investigate using simulation the sensitivity of the proposed subcopula regression-based measure in Eq. (10.12) to the number of the categories of the response variable in the three-way ordinal contingency tables.

In the simulation, we consider four factors, the type/magnitude of association among three categorical variables (X, Y, Z) with I, J, and K categories, respectively, the sample size, the number of categories in the predictors (X, Z), and the number of categories in the response variables Y. Regarding the type of association, the three-dimensional Gaussian copula for continuous variables is considered and the exchangeable correlation matrix with the correlation parameter θ is employed. For the magnitude of association, we use three values of θ for Gaussian copula, $\theta = (0.2, 0.5, 0.8)$ representing small, moderate and strong associations, respectively. Once the data are generated from the three-dimensional Gaussian copula with the correlation parameter θ, we will convert them into three-way contingency tables. As to the number of categories in the predictors, three values are considered: $(I, K) = (2,2), (4,4), (6,6)$. For the number of categories in the response variable, 11 values are considered: $J = 2, 3, \ldots, 12$.

We simulate data of size $n = 500, 1000, 2000$, and 5000 from the trivariate Gaussian copula with the correlation parameter θ specified above and then create an $I \times J \times K$ contingency table by classifying n data points by the three discrete variables (X, Y, Z) with I, J and K categories. Note that each variable has equal-width categories, $[0, 1/I), [1/I, 2/I), \ldots, [(I-1)/I, 1]$ for X, $[0, 1/J), [1/J, 2/J), \ldots, [(J-1)/J, 1]$ for Y, and $[0, 1/K), [1/K, 2/K), \ldots, [(K-1)/K, 1]$ for Z. For each simulated contingency table, we estimate the subcopula regression-based asymmetric association measure in Eq. (10.12), $\hat{\rho}^2_{(X,Z \to Y)}$. Note that, as the reference value, we compute the true value of the subcopula regression-based asymmetric association measure, denoted as $\tilde{\rho}^2_{(X,Z \to Y)}$, by using the true cell probabilities of an $I \times J \times K$ contingency table under the continuous Gaussian copula with the correlation parameter θ.

Figures 10.1–10.3 show the estimated values of the subcopula regression based asymmetric association measure $\hat{\rho}^2_{(X,Z \to Y)}$ (represented by the square symbols in

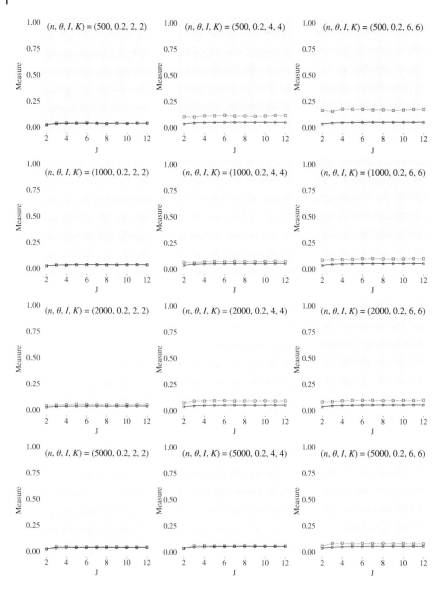

Figure 10.1 The estimated subcopula regression based association measure $\widehat{\rho}^2_{(X,Z \to Y)}$ (squares) with respect to the number of the categories in the response variable J at the number of categories in the predictors $(I, K) = (2, 2), (4, 4), (6, 6)$ (from left column to right column) and the sample size $n = 500, 1000, 2000, 5000$ (from top panel to bottom panel) for the Gaussian copula with the correlation parameter $\theta = 0.2$. Note that the true values of $\rho^2_{(X,Z \to Y)}$ under the continuous Gaussian copula with $\theta = 0.2$ are represented by the circle symbols.

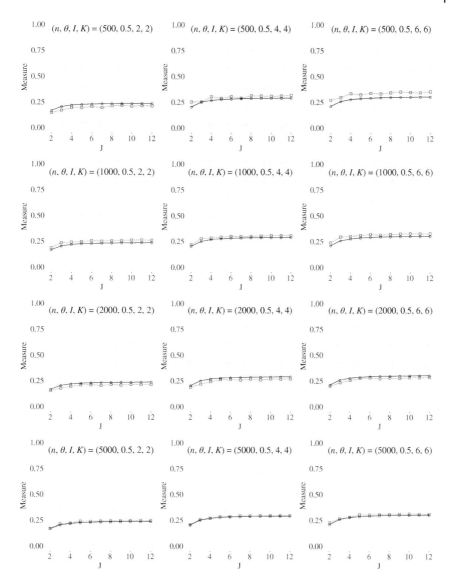

Figure 10.2 The estimated subcopula regression based association measure $\hat{\rho}^2_{(X,Z \to Y)}$ (squares) with respect to the number of the categories in the response variable J at the number of categories in the predictors $(I, K) = (2, 2), (4, 4), (6, 6)$ (from left column to right column) and the sample size $n = 500, 1000, 2000, 5000$ (from top panel to bottom panel) for the Gaussian copula with the correlation parameter $\theta = 0.5$. Note that the true values of $\rho^2_{(X,Z \to Y)}$ under the continuous Gaussian copula with $\theta = 0.5$ are represented by the circle symbols.

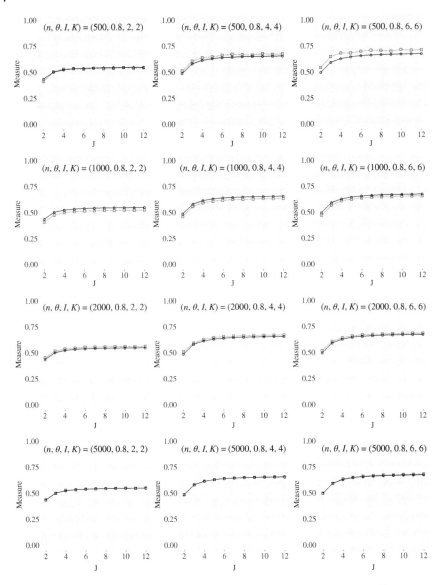

Figure 10.3 The estimated subcopula regression based association measure $\hat{\rho}^2_{(X,Z \to Y)}$ (squares) with respect to the number of the categories in the response variable J at the number of categories in the predictors $(I, K) = (2, 2), (4, 4), (6, 6)$ (from left column to right column) and the sample size $n = 500, 1000, 2000, 5000$ (from top panel to bottom panel) for the Gaussian copula with the correlation parameter $\theta = 0.8$. Note that the true values of $\rho^2_{(X,Z \to Y)}$ under the continuous Gaussian copula with $\theta = 0.8$ are represented by the circle symbols.

each plot) with respect to the number of the categories in the response variable J at the number of categories in the predictors $(I, K) = (2, 2), (4, 4), (6, 6)$ (from left column to right column) and the sample size $n = 500, 1000, 2000, 5000$ (from top panel to bottom panel) for the Gaussian copula with the correlation parameter $\theta = (0.2, 0.5, 0.8)$. Note that the true values of $\rho^2_{(X,Z \to Y)}$ under the continuous Gaussian copula are represented by the circle symbols in each plot.

We first observe that the true and estimated values of the subcopula regression based association measure $\widetilde{\rho}^2_{(X,Z \to Y)}$ and $\widehat{\rho}^2_{(X,Z \to Y)}$ tend to be insensitive to the number of the response categories J, as they are stabilized as the number of the response categories J increases, regardless of the number of categories in the predictors (I, K), the sample size n and the magnitude of association θ. Second, when the magnitude of the association is small $(\theta = 0.2)$, we observe a tendency that the value of $\widehat{\rho}^2_{(X,Z \to Y)}$ is slightly larger than $\widetilde{\rho}^2_{(X,Z \to Y)}$ as the number of categories in the predictors (I, K) increases and/or the sample size is not large. On the other hand, when the magnitude of the association is moderate/strong, the values of $\widehat{\rho}^2_{(X,Z \to Y)}$ and $\widetilde{\rho}^2_{(X,Z \to Y)}$ are very close to each other. Third, for the moderate/strong magnitude of the association, the (true/estimated) values of the subcopula regression based association measure are smaller when the number of the categories in the response variable is small (say, $J = 2, 3$) than when J is larger than or equal to 4. This is because the contingency table with the small number of the categories for the response variable may not completely reflect the (moderate/strong) pattern of the continuous three-dimensional Gaussian copula.

10.4.2 Data Analysis

In this section, we illustrate the subcopula regression-based asymmetric association measure using real data (Valle, 2010) collected from a survey study of job satisfaction of workers employed in the Italian social (health care service related) cooperatives. Several researchers analyzed this job satisfaction data to evaluate workers' job satisfaction in non-profit social cooperatives and study the relationship of job satisfaction with objective and subjective factors of the job quality (Beh & Lombardo, 2014; Valle, 2010; Lombardo & Della Valle, 2011). They found that job satisfaction in social cooperatives is more related with the subjective facets of the job quality (e.g. motivations, fairness, autonomy, complexity, loyalty) than the objective facets (e.g. worker status, job contract, wages/incentives).

Table 10.1 shows the $5 \times 5 \times 4$ contingency table cross-classifying 427 workers according to the three ordinal variables, job motivation, job satisfaction and education. The job motivation (M) has five categories $(M1 = \text{low}, M2 = \text{middle-low}, M3 = \text{middle}, M4 = \text{middle-strong}, M5 = \text{strong})$, job satisfaction (S) has five categories $(S1 = \text{low}, S2 = \text{middle-low}, S3 = \text{middle}, S4 = \text{middle-high}, S5 = \text{high})$, and the level of education (E) has four categories $(E1 = \text{primary}, E2 = \text{secondary}, E3 = \text{degree I [degree received in three years]}, E4 = \text{degree II [degree received in five years]}).$

Table 10.1 Job satisfaction data (Beh & Lombardo, 2014, p. 478).

Satisfaction (*S*)	Motivation (*M*)				
	M1 (low)	M2 (middle-low)	M3 (middle)	M4 (middle-strong)	M5 (strong)
*E*1 (primary)					
S1 (low)	6	0	0	1	1
S2 (middle-low)	11	3	0	2	0
S3 (middle)	2	4	4	4	5
S4 (middle-high)	4	6	14	10	11
S5 (high)	8	0	10	8	18
*E*2 (secondary)					
S1 (low)	1	1	2	0	2
S2 (middle-low)	8	2	2	0	0
S3 (middle)	8	8	10	10	12
S4 (middle-high)	3	5	14	21	19
S5 (high)	5	13	18	19	17
*E*3 (degree I)					
S1 (low)	1	1	2	0	0
S2 (middle-low)	2	1	1	0	1
S3 (middle)	2	2	1	0	0
S4 (middle-high)	2	5	4	2	3
S5 (high)	3	1	7	2	4
*E*4 (degree II)					
S1 (low)	3	0	1	0	0
S2 (middle-low)	2	0	0	0	0
S3 (middle)	2	3	3	4	2
S4 (middle-high)	2	1	3	3	2
S5 (high)	5	1	6	4	1

The goal here is to understand the association structure among the three ordinal variables using the proposed subcopula regression-based asymmetric association measure, particularly, the relationship of the satisfaction of workers and their motivation with their education level. To this end, we assume that job motivation and job satisfaction have a nondirectional (reciprocal) relationship in that they affect each other, possibly with different degree of influence and so we consider the two types of association:

$$(S, E) \to M \quad \text{and} \quad (M, E) \to S,$$

where $(X, Y) \to Z$ means that Z is putatively considered as a response variable and (X, Y) are considered as putative predictors.

Table 10.2 summarizes the results of the subcopula regression-based asymmetric association measure for two types of association and their pairwise difference. Note that in order to compute the bootstrap 95% confidence intervals, we used 10 000 bootstrap samples of size $n = 427$ from the saturated log-linear model fitted to Table 10.1. From Table 10.2, we make the following observations. First, it appears that the degrees of the two association types are numerically different: $\hat{\rho}^2_{((S,E) \to M)} = 0.1864 > \hat{\rho}^2_{((M,E) \to S)} = 0.1182$. Second, each of the two association types is statistically significant because none of the bootstrap (percentile/BCa) confidence intervals for the asymmetric measures include 0. Thus, the putative predictors under each association type are statistically important for the putative response variable. Third, the lower limits of the bootstrap confidence intervals for $\rho^2_{((S,E) \to M)} - \rho^2_{((M,E) \to S)}$ are larger than 0. This result indicates that $(S, E) \to M$ is a more dominant reciprocal association structure than $(M, E) \to S$. From the results above, we see that job satisfaction and job motivation are mutually and asymmetrically associated, given the worker's education level.

Table 10.3 shows the results of the hierarchical analysis for the two association structures $(S, E) \to M$ and $(M, E) \to S$. Note that the values of the subcopula based association measures for $S \to M$, $E \to M$, $M \to S$, and $E \to S$, denoted as $\hat{\rho}^2_{(S \to M)}$, $\hat{\rho}^2_{(E \to M)}$, $\hat{\rho}^2_{(M \to S)}$, and $\hat{\rho}^2_{(E \to S)}$, are computed using Eq. (10.7) for the two-way marginal contingency tables of (S, M), (E, M), and (E, S). For the association structure $(S, E) \to M$, we see that the marginal contribution of job satisfaction

Table 10.2 Analysis of asymmetric association in job satisfaction data.

	$\rho^2_{((S,E) \to M)}$	$\rho^2_{((M,E) \to S)}$	$\rho^2_{((S,E) \to M)} - \rho^2_{((M,E) \to S)}$
Estimate	0.1864	0.1182	0.0682
Bootstrap percentile interval	(0.1559, 0.2855)	(0.0964, 0.2278)	(0.0035, 0.1199)
Bootstrap BCa interval	(0.1105, 0.2169)	(0.0449, 0.1429)	(0.0181, 0.1330)

Table 10.3 Hierarchical analysis for two types of association in job satisfaction data.

Association structure	$\hat{\rho}^2$	Association structure	$\hat{\rho}^2$
$(S, E) \to M$	0.1864	$(M, E) \to S$	0.1182
$S \to M$	0.1336	$M \to S$	0.0900
$E \to M$	0.0253	$E \to S$	0.0023

(S) on job motivation (M) is much larger than the marginal contribution of education level (E) (i.e. $\hat{\rho}^2_{(S \to M)} = 0.1336 > \hat{\rho}^2_{(E \to M)} = 0.0253$). From the association structure $(M, E) \to S$, the marginal contribution of job motivation (M) on job satisfaction (S) is larger than the marginal contribution of education level (E), $\hat{\rho}^2_{(M \to S)} = 0.0900 > \hat{\rho}^2_{(E \to S)} = 0.0023$. This hierarchical analysis indicates that the marginal contribution of education is smaller than both job motivation and job satisfaction in both association structures, and job motivation and job satisfaction are asymmetrically related.

10.5 Conclusion

In this chapter, we have derived a subcopula regression-based measure of the asymmetric association to examine the asymmetric dependence in a three-way contingency table. We have investigated the basic properties of the proposed measure and illustrated its utility using real-world data analysis.

The proposed method is a data-driven approach to explore potential asymmetric (reciprocal/non-reciprocal) dependence in the data and it does not require any parametric model assumption. We believe that the information given by the proposed method may suggest plausible hypotheses in the formal causal modeling approaches. Here, we present a guideline on how the proposed asymmetric measure can be utilized in a proper manner.

- When a priori theoretical considerations on the type of causal mechanism are available, one can apply the proposed measure(s) corresponding to the postulated association structure to empirically estimate the magnitude of the postulated causation. For example, when it is a priori known that X and Z are the predictors and Y is the response variable in the three-way contingency table, compute $\rho^2_{(X,Z) \to Y}$ only, but neither $\rho^2_{(X,Y) \to Z}$ nor $\rho^2_{(Y,Z) \to X}$. Note that one may compare the results from the proposed method with those from the parametric causal models.

- When there is no a priori theory to guide the causal mechanism, one can assume the reciprocal association structure, compute the proposed measures for each of the assumed associations and their pairwise differences. For instance, when the three ordinal variables mutually associated (i.e. no a priori distinction between predictors and response variables), compute the proposed measures for three association structures, $\rho^2_{(Y,Z\to X)}$, $\rho^2_{(X,Z\to Y)}$, $\rho^2_{(X,Y\to Z)}$, and their pairwise differences, $\rho^2_{(Y,Z\to X)} - \rho^2_{(X,Z\to Y)}$, $\rho^2_{(Y,Z\to X)} - \rho^2_{(X,Y\to Z)}$, and $\rho^2_{(X,Z\to Y)} - \rho^2_{(X,Y\to Z)}$. Note that one may use the information obtained from the proposed method for the formal (reciprocal/non-reciprocal) causal modeling approach.

As one of the reviewers pointed out, standard measures of associations are known to be biased by the presence of unobserved confounders that affects both observed variables. A valuable future work would be to investigate the influence of unobserved confounders on the proposed method in terms of bias.

In this chapter, the proposed measure was mainly designed for ordinal three-way contingency tables. A valuable extension of this research would be to extend the proposed method to large high-dimensional contingency tables and investigate its performance, particularly for the large sparse ordinal contingency tables.

10.A Appendix

10.A.1 The Proof of Proposition 10.1

1) By the variance decomposition formula $\text{Var}(W) = E[\text{Var}(W \mid U, V)] + \text{Var}(E[W \mid U, V]) = E[\text{Var}(W \mid U, V)] + \text{Var}(r^C_{W|U,V}(U, V))$, we can rewrite the subcopula-based asymmetric association measure of Z on X, Y as

$$\rho^2_{(X,Y\to Z)} = 1 - \frac{E[\text{Var}(W \mid U, V)]}{\text{Var}(W)} = 1 - \frac{E[\{W - E(W \mid U, V)\}^2]}{\text{Var}(W)}.$$

(10.A.1)

Therefore, $\rho^2_{(X,Y\to Z)}$ is between 0 and 1.

2) If the discrete random variables Z and (X, Y) are independent, we know the corresponding trivariate subcopula C of Z and (X, Y) is also independent. Then the subcopula regression function $r^C_{W|U,V}(u, v) = E[W]$ which is a constant function of u and v. Therefore, in this case, we have $\rho^2_{(X,Y\to Z)} = 0$.

3) By Eq. (10.A.1), $\rho^2_{(X,Y\to Z)} = 1$ if and only if $E[\{W - E(W \mid X, Y)\}^2] = 0$, which is also equivalent to $W = E(W \mid U, V)$ almost surely. Thus, $\rho^2_{(X,Y\to Z)} = 1$ if and only if $W = g(U, V)$ almost surely for some measurable function g. Since $W = F_2(Z)$ and $U = F_0(X)$, $V = F_1(Y)$, are an one-to-one transformation of Z, X, and Y, respectively. $\rho^2_{(X,Y\to Z)} = 1$ if and only if $Z = g(X, Y)$ almost surely for some measurable function g.

4) If $W = g(U, V) + \varepsilon$, then $r^C_{W|U,V}(U, V) = g(U, V) + E[\varepsilon]$, and $\mathrm{Var}(r^C_{W|U,V}(U, V)) = \mathrm{Var}(g(U, V))$. Therefore, we have

$$\rho^2_{(X, Y \to Z)} = \frac{\mathrm{Var}(r^C_{W|U,V}(U, V))}{\mathrm{Var}(W)} = \frac{\mathrm{Var}(g(U, V))}{\mathrm{Var}(g(U, V)) + \mathrm{Var}(\varepsilon)}.$$

References

Agresti, A. (1990). *Analysis of categorical data.* New York, NY: Wiley.

Aloui, R., Aissa, M. S. B., & Nguyen, D. K. (2013). Conditional dependence structure between oil prices and exchange rates: A copula-garch approach. *Journal of International Money and Finance, 32,* 719–738.

Balke, N. S., Brown, S. P. A., & Yucel, M. K. (2002). Oil price shocks and the U.S. Economy: Where does the asymmetry originate? *The Energy Journal, 23,* 27–52.

Beh, E. J., & Lombardo, R. (2014). *Correspondence analysis: Theory, practice and new strategies.* New York, NY: John Wiley & Sons.

Dodge, Y., & Rousson, V. (2000). Direction dependence in a regression line. *Communications in Statistics—Theory and Methods, 29*(9–10), 1957–1972.

Dodge, Y., & Rousson, V. (2001). On asymmetric properties of the correlation coeffcient in the regression setting. *The American Statistician, 55*(1), 51–54.

Efron, R., & Tibshirani, B. (1993). *An introduction to the bootstrap.* London, UK: Chapman & Hall.

Goodman, L. A., & Kruskal, W. H. (1954). Measures of association for cross classifications. *Journal of the American Statistical Association, 49*(268), 732–764.

Hong, Y., Tu, J., & Zhou, G. (2007). Asymmetric correlation of stock returns: Statistical tests and economic evaluation. *Review of Financial Studies, 20,* 1547–1581.

Lombardo, R., & Della Valle, E. (2011). Data mining and exploratory data analysis for the evaluation of job satisfaction. *Business Journal, 3,* 372–382.

Margolin, A. A., Nemenman, I., Basso, K., Wiggins, C., Stolovitzky, G., Favera, R. D., & Califano, A. (2006). Aracne: An algorithm for the reconstruction of gene regulatory networks in a mammalian cellular context. *BMC Bioinformatics, 7*(Suppl 1), S7.

R.B. Nelsen. *An introduction to copulas* (2nd ed.). Springer, New York, NY, 2006.

Nguyen, H. H., Tilton, S. C., Kemp, C. J., & Song, M. (2017). Nonmonotonic pathway gene expression analysis reveals oncogenic role of p27/kip1 at intermediate dose. *Cancer Informatics, 16,* 1176935117740132.

Nigg, J. T., Knottnerus, G. M., Martel, M. M., Nikolas, M., Cavanagh, K., Karmaus, W., & Rappley, M. D. (2008). Low blood lead levels associated with clinically diagnosed attention-deficit/hyperactivity disorder and mediated by weak cognitive control. *Biological Psychiatry, 63*(3), 325–331.

Sharma, R., Kumar, S., Zhong, H., & Song, M. (2017). Simulating noisy, nonparametric, and multivariate discrete patterns. *The R Journal*, *9*(2), 366–377.

Sklar, A. (1959). Fonctions de répartition à *n* dimensions et leurs marges. *Publications de l'Institut de Statistique de L'Université de Paris*, *8*, 229–231.

Theil, H. (1970). On the estimation of relationships involving qualitative variables. *American Journal of Sociology*, *76*(1), 103–154.

E. Della Valle. (2010) *Lánalisi multidimensionale dei dati: La qualitá del lavoro nelle cooperative sociali*. PhD thesis, Economics Faculty, Second University of Naples, Italy. Unpublished thesis.

van Someren, E. P., Wessels, L. F. A., Backer, E., & Reinders, M. J. T. (2002). Genetic network modeling. *Pharmacogenomics*, *3*(4), 507–525.

von Eye, A., & Wiedermann, W. (2014). On direction of dependence in latent variable contexts. *Educational and Psychological Measurement*, *74*(1), 5–30.

Wei, Z., & Kim, D. (2017). Subcopula-based measure of asymmetric association for contingency tables. *Statistics in Medicine*, *36*(24), 3875–3894.

Wiedermann, W., & von Eye, A. (2017). Log-linear models to evaluate direction of effect in binary variables. *Statistical Papers*, *61*, 1–30.

Wu, C., Chung, H., & Chang, Y. (2012). The economic value of co-movement between oil price and exchange rate using copula-based garch models. *Energy Economics*, *34*, 270–282.

Yang, L., & Hamori, S. (2014). Gold prices and exchange rates: A time-varying copula analysis. *Applied Financial Economics*, *24*, 41–50.

Part IV

Applications and Software

11

Distribution-Based Causal Inference

A Review and Practical Guidance for Epidemiologists

Tom Rosenström[1,2] and Regina García-Velázquez[2]

[1] *Department of Mental Disorders, Norwegian Institute of Public Health, Oslo, Norway*
[2] *Department of Psychology and Logopedics, University of Helsinki, Helsinki, Finland*

11.1 Introduction

During training, many empirical researchers have likely heard phrases along the lines of "causality cannot be inferred from cross-sectional data", wherein "inferring causality" refers to distinguishing a cause from its consequence. This statement is actually wrong. It probably reflects the huge role that the correlation coefficient has played in empirical research. One cannot infer causation from a product-moment, or Pearson's, correlation coefficient, without supplementing it with other knowledge. More recent research, however, has derived several other statistics that are able to infer direction of causation in a cross-sectional setting (Wiedermann & von Eye, 2016). So far these have seen relatively little use in epidemiologic research, even though inferring causation (i.e. etiology) is a central topic in epidemiology. This may reflect, in part, a healthy streak of conservatism at the face of novel methods. Eventually, however, too much conservatism may frustrate scientific progress, because many questions of epidemiology do not lend themselves well for experimentation. Neglecting possibilities available for observational data is a luxury we cannot always afford.

This chapter aims to familiarize researchers in epidemiology and related fields with the topic of distribution-based causal inference methods, and to discuss how to build trust in the results from such methods. We review, replicate, and extend some of the few studies that have used the methods in real-world epidemiological issues where the ground truth was not known a priori (Helajärvi et al., 2014; Rosenström et al., 2012). In particular, we concentrate on simple cases of distribution-based causal inference applied to survey data and linear models. These types of data and models permeate much research in epidemiology,

Direction Dependence in Statistical Modeling: Methods of Analysis, First Edition.
Edited by Wolfgang Wiedermann, Daeyoung Kim, Engin A. Sungur, and Alexander von Eye.
© 2021 John Wiley & Sons, Inc. Published 2021 by John Wiley & Sons, Inc.

including our example case of research on causality between sleep problems and other depressive symptoms, introduced below.

11.2 Direction of Dependence in Linear Regression

While correlation does not imply causation, Dodge and Rousson (2000, 2001) were perhaps first to derive "other expressions of the correlation coefficient" that do make it possible to infer which among two *skewed* variables, X and Y, causes the other, or in other words, which variable is the proverbial "cart" and which variable the "horse." Specifically, it is possible to distinguish between two models, one with Y as the dependent variable and the other with X as the dependent, or "causally descendent," variable. This amounts to distinguishing between two systems of equations:

$$\begin{cases} Y = \mu_Y + \beta_x X + \varepsilon_Y \\ X = \mu_X + \varepsilon_X \end{cases} \tag{11.1}$$

and

$$\begin{cases} Y = \mu_Y + \varepsilon_Y \\ X = \mu_X + \beta_y Y + \varepsilon_X \end{cases}, \tag{11.2}$$

where β, μ_Y, and μ_X are constants (i.e. regression coefficient, or slope, and means, or intercepts) and the "residual" variables ε_X and ε_Y are independent of each other and also independent of the predictor. Dodge and Rousson observed that, under Eqs. (11.1) and (11.2), a following relation holds asymptotically for skewed variables:

$$\begin{cases} X \text{ causes } Y \ \text{ if } \ T(X,Y) > 0 \\ Y \text{ causes } X \ \text{ if } T(X,Y) < 0 \end{cases},$$

where the test statistic $T(X, Y) := M(X, Y)_{21} - M(X, Y)_{12}$ is based on sample versions of the difference in squared and centralized third cumulants, defined as $M(X, Y)_{ij} = \{E[(X - \mu_X)^i (Y - \mu_Y)^j]/(\sigma_X^i \sigma_Y^j)\}^2$. Here, E is the expectation operator and σ_X refers to variance of the variable X. As shown by Dodge and Rousson (2000), as well as others with slightly different formulations (Hyvärinen & Smith, 2013), sufficient conditions for causal inference are that one and only one of the two linear models hold (11.1) or (11.2) and that at least one of the variables has a skewed distribution. We do not reproduce the analytic proofs, but instead offer the following intuition and then proceed to more general estimators.

Many frequently studied variables in epidemiology have "right-skewed" population distributions (e.g. alcoholic drinks per week, number of children, and depressive symptoms). Right-skewness means that a variable gets "small" values frequently and "high" values only rarely in comparison to a normal (a.k.a., Gaussian)

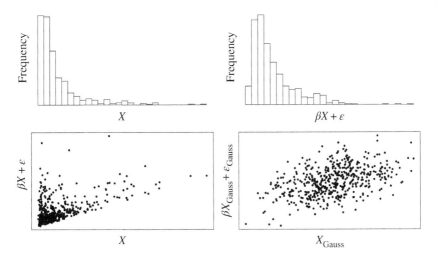

Figure 11.1 Illustration of skewness-based causal signal. Histogram of a Log-normal distributed variable X, and a variable that is a weighted sum of X and another similarly distributed "residual" variable. In the lower-left panel, a scatter plot of X and the weighted-sum variable (the outcome; Y in text) are shown, whereas the lower-right panel shows similarly treated Gaussian variables.

distribution (a case of left-skewed variable can be transformed to right-skewed without loss of information by multiplying by minus one). Then, if a variable Y is a sum of two independent right-skewed variables, βX and ε_Y, it gets high values whenever either one of the two variables gets a high value, which is necessarily more often than for βX and ε_Y on average.[1] Thus, the dependent (causally descendent) variable is less skewed than the independent (causally antecedent) variable and they show a characteristic pattern in bivariate scatter plots (non-symmetry over permutation of axes; Figure 11.1).[2] A similar signal is absent when the data is generated from the same linear model operating on two normally distributed variables (lower-right panel of Figure 11.1). Any weighted sum of normally distributed (Gaussian) variables is also a Gaussian variable, implying no skewness or excess kurtosis, and no information beyond correlations in any of the higher moments that characterize bivariate probability distributions (Hyvärinen, Karhunen, & Oja, 2001; Klenke, 2008).

However, it has later turned out that the measures $M(X, Y)_{21}$ and $M(X, Y)_{12}$ can also be based on kurtosis instead of skewness of the distribution, and even more

---

1 Probability distribution of a sum of two independent random values is a convolution of the two original distributions (Klenke, 2008). "Convolution" operation is a concept of mathematical (functional) analysis, and formalizes a sort of "smearing" of two distributions to a new one.
2 If ε_Y is normally distributed and the causal antecedent X is not, it can be shown analytically that $\gamma_Y = \gamma_X \text{Cor}(X, Y)^3 < \gamma_X$, where γ_X is "skewness coefficient" of X (Dodge & Rousson, 2000).

generally, on *any* type of deviation from normal distribution (Dodge & Yadegari, 2010; Hyvärinen & Smith, 2013; Shimizu, Hoyer, Hyvärinen, & Kerminen, 2006; Wiedermann, 2018). According to the Central Limit Theorem (CLT) of probability theory, sums of *independent* random variables of almost any probability distribution tend towards a normal distribution (Hyvärinen et al., 2001; Klenke, 2008). More precisely, this is Lindeberg's version of CLT, according to the Finnish mathematician Jarl Waldemar Lindeberg (1876–1932), which only requires that the random variables have finite variance and that their sequence satisfies a certain (i.e. "Lindeberg's") regularity condition (e.g. Klenke, 2008). There is necessarily more summation in the causal descendent than in the antecedent in linear model, because the descendent is a weighted sum of the antecedent plus the residual variable. Thus, one can infer the causal antecedent as being the variable that leads to least Gaussian distributions for antecedent and the residual (one of which can even be Gaussian; Hyvärinen & Smith, 2013). Unless, of course, both already are Gaussian, or the linear models does not apply. That is, necessary conditions for pairwise distribution-based causal inference are that (i) at least one of the variables must have non-Gaussian distribution, that (ii) the linear model applies (lest the causal descendent is some other function than the weighted sum required by CLT), and that (iii) the residual variable must be independent of the causal antecedent (again, required to apply CLT to the components βX and ε_Y of the descendent variable Y). The assumption (iii) is already present in (ii), but worth highlighting separately for its important role.

When explicitly stated, the assumption (iii) also suggests an algorithm to evaluate direction of causation between two variables. Assuming the variables are non-Gaussian and one of the above two linear systems of equations holds, (11.1) or (11.2) (i.e. assuming conditions (i) and (ii)), then a result known as Darmois–Skitovich theorem implies that the causally antecedent variable is the variable that is independent of its residual when regressed onto the other variable (Shimizu et al., 2011). That is, we can define the above $M(X, Y)_{12}$ to be $\hat{MI}(X, \varepsilon_Y)$, an estimate of mutual information between X and residual of Y when regressed on X, or $\varepsilon_Y = Y - \frac{\text{Cov}(X,Y)}{\text{Var}(X)} X$, where $\text{Cov}(\cdot, \cdot)$ and $\text{Var}(\cdot)$ are covariance and variance operators, respectively (Shimizu et al., 2011). Analogously, $M(X, Y)_{21}$ is defined to be $MI(Y, \varepsilon_X)$. Then, the statistic $T(X, Y)$ from above will become an estimate of causal direction based on non-Gaussianity and least mutual information between a predictor and its residual. While there are many other estimators for distribution-based causal inference (Hyvärinen & Smith, 2013), here we will concentrate on this "DirectLiNGAM" estimator, which uses a kernel-based estimate of mutual information and has been found useful in previous empirical studies and simulations (Helajärvi et al., 2014; Rosenström et al., 2012; Shimizu et al., 2011). The name refers to a "direct" algorithm for estimating Linear Non-Gaussian Acyclic Models (i.e. LiNGAMs) as opposed to the earlier iterative algorithm (Shimizu et al., 2006, 2011).

Mutual information is a measure for degree of dependence between two random variables. It tells how much information entropy in one variable can be obtained through the other variable. Theoretically, mutual information is defined as an expected difference between true bivariate entropy and entropy of an "independence distribution" (i.e. product of marginal distributions): $E[\log(p(X, Y)) - \log(p(X)p(Y))]$, where $p(\cdot, \cdot)$ and $p(\cdot)$ are the bivariate and marginal probability density functions, respectively, and E is the expectation operation with respect to the bivariate distribution. Thus, mutual information is quantified as departure from bivariate independence. The equation is noteworthy, because often one is interested in what happens in terms of the (population) distribution with respect to which the expectation is taken, rather than what happens for each and every observation per se. In other words, we have no reason to expect that a deviant minority with opposite causal direction would ruin our inferences about dominant population-level direction of causation. This is an important advantage, for example, in our target research problem on causal direction between sleep problems and other depressive symptoms, as there likely are sub-populations that exhibit rather different causal processes in comparison to most cases of depression (e.g. brain trauma patients).

In what follows, we first (in Section 11.3) give a review of previous empirical work and (Section 11.4) introduce a practical research problem in epidemiology, which both represents a novel replication effort and is used as an example data throughout the rest of the chapter. Then, (Section 11.5) we discuss strategies to evaluate the assumptions necessary for distribution-based causal inference, (Section 11.6) analyze the example data and (Sections 11.7 and 11.8) discuss strategies to study robustness and statistical power of distribution-based causal inferences, and (Section 11.9) strategies for, as well as importance of, "triangulation" with multiple methods that are non-overlapping in their assumptions. Finally, in Section 11.10, we conclude the chapter with comments on both present content and other causal-inference methodologies.

11.3 Previous Epidemiologic Applications of Distribution-Based Causal Inference

Psychiatric epidemiology deals with complex disorders whose etiology is not yet understood to large extent. The classic ways of thinking in psychiatric epidemiology do not always fit data and have been challenged by the recent, influential network theory (Borsboom, 2017; Cramer, Waldorp, van der Maas, & Borsboom, 2010; García-Velázquez, Jokela, & Rosenström, 2017). Whereas the traditional diagnostic practice perceives symptoms as passive reflections of an underlying psychiatric root cause, the network theory recognizes the symptoms as causally active entities that can "cause" each other and thereby give rise to syndromes. A classic example in network theory has suggested that sleep problems can gradually give rise to a full-blown depressive syndrome, for example, through inducing fatigue and concentration problems, which then lead to performance issues in daily life, and ultimately to all other depressive symptoms (Borsboom, 2017; Cramer et al., 2010). This is a completely hypothetical example, however, and the direction of causation between sleep and the average of other depressive symptoms (a proxy of the syndrome) remains an open question.

Rosenström et al. (2012) discuss about the multiple difficulties involving the study of causation between human sleep characteristics and depressive disorders. In such topics, it is nearly impossible to design a definitive study. It is not ethically acceptable to experimentally induce depression, because of the involved human suffering and the high risk of suicide in major depressive disorder. Due to possible lagged effects, temporal order of events may not directly inform about the causal order (the horse might as well push the cart as pull it). Furthermore, sleep problems and other depressive symptoms are relatively common phenomena in the population, and it is probably possible to find strong individual cases to argue the causation both ways – as epidemiologists, we were interested in the direction that dominates on average in the population. Rosenström et al. (2012) used the above-introduced distribution-based causality statistic as a research tool fit to assessing population averages. As so often occurs in practice, however, it did not provide a unique answer: sleep problems was estimated to cause depression in most cases, but also opposite findings were obtained. Here, we return to the topic in our running example, using yet another classic real-world dataset, as well as computer simulations.

Another previous application of distribution-based causal inference in epidemiology was a study on direction of causation between average television viewing time and obesity (Helajärvi et al., 2014). Overweight, obesity, insulin resistance, and diabetes have been major public health concerns in the Western world lately. Also the time spent in "sedentary behaviors" has increased in comparison to past times of manual labor. Especially watching television has been recognized as an activity with uncharacteristically low waking-time metabolic rates in evolutionary terms. The relative time spent watching television has been suggested to cause weight gain, but the proposition has been challenged by a reverse causation hypothesis, according to which obese and overweight people may find physical activity less appealing than lean people and therefore spend more time in a substitute activity of watching television. Helajärvi et al. (2014) used distribution-based causal inference to show that high television watching times are more likely to cause changes in weight than the other way around.

That distribution-based causal inference is possible to begin with often comes as a surprise for epidemiology researchers who are well-aware of the fact that an analogous correlation-based causal inference is not possible. Therefore, "toy examples" using real datasets where the direction of causation is obvious have been necessary to demonstrate that the method works. In this category, Shimizu et al. (2011) have shown that the method correctly infers father's education and occupation as causes of his son's education and occupation rather than other way around. Similarly, Rosenström et al. (2012) showed that the method suggests parents' socioeconomic status as a cause of their children's socioeconomic status rather than the other way around. While computer simulations and mathematical analyses are the primary tools to study how well statistical methods function, toy examples with real data are also indispensable in building trust on "black-box" methods that reveal very little about the true mechanism behind the inferred causal effect. The target method of this chapter has so far withstood the test. In the later sections, we discuss about another method that has not always withstood similar tests.

11.4 A Running Example: Re-Visiting the Case of Sleep Problems and Depression

Whereas Rosenström et al. (2012) studied epidemiologic, Finland-based "Young Finns" and USA-based "Wisconsin Longitudinal Study" datasets, here we use data from the Swedish Adoption/Twin Study on Aging (SATSA) that were available to us through the Inter-University Consortium on Political and Social Research (Pedersen, 2015). Specifically, these data include 1439 observations both on a depression score and on average hours slept per night (1326 complete and 1325 valid observations; one person reported no sleep at all). Sleeping hours was a self-reported quantity, whereas the depression score we used was an average of non-sleep-related depressive symptoms assessed by the Center for Epidemiologic Studies Depression (CES-D) scale (Radloff, 1977). The symptom statuses were reported as amount of symptom presence during the past week (0 = "never/almost never," 1 = "Rather seldom/never," 2 = "Quite often," and 3 = "Always/almost always"). Figure 11.2 illustrates the data. Altogether 598 of the subjects were men (41.56%). Average age of participants was 63.42 years at the time of data collection in 1990 (s.d. 13.01 years, range from 32 to 95). SATSA is a twin and adoption study and these participants constitute 134 pairs of monozygotic (i.e. "identical") twins reared apart, 184 monozygotic twin pairs reared together, 345 dizygotic ("fraternal") twin pairs reared apart, and 286 dizygotic twin pairs reared together. This feature of the data will be useful in the causal triangulation section of this chapter.

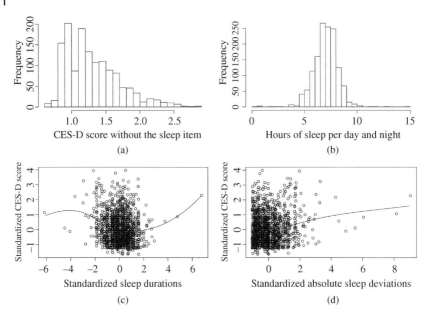

Figure 11.2 Illustrating the SATSA data. Whereas the histograms (a and b) show unstandardized data, data was standardized for bivariate analyses for cross-study comparability, as a wide range of alternative assessment tools exists. Both bipolar (c; direction of deviance matters) and absolute/unipolar (d; both directions equally "bad") sleep deviations were studied. LOESS (local regression) lines in c and d panels show how use of absolute values linearizes the association.

11.5 Evaluating the Assumptions in Practical Work

The assumptions of the DirectLiNGAM approach to distribution-based causal inference (in a bivariate case) are:

i) Linear model: one and only one of the two systems of equations in Eqs. (11.1) and (11.2) hold.
ii) Non-Gaussian continuous variable: the variables have a continuous distribution and at least one of the two independent terms (predictor and residual) has some other distribution than the Normal distribution.
iii) Independence: the predictor (causally antecedent) variable in Eqs. (11.1) and (11.2) is statistically independent of the residual.

Section 11.5.1 discuss strategies for evaluating whether these necessary preconditions of distribution-based causality hold in practice. Applications of the strategies are provided for the example case of sleep problems and depression.

11.5.1 Testing Linearity

It can be very difficult to know whether a given dataset reflects an essentially linear data-generating process. In practice, one is typically willing to accept a linear approximation if both visual inspection and polynomial regression coefficients thus indicate. That is the approach we take here, as well.

To illustrate using our running example, we found a significant regression coefficient for the quadratic effect as well as the linear, when regressing the depression variable of SATSA data on the standardized (z-score transformed to 0 mean and variance of 1) sleep-hours variable and its square ($\beta_{quadratic} = 0.062$, $p < 0.001$), not just for the linear slope ($\beta_{linear} = -0.146$, $p < 0.001$). Standardization of variables is an important part of the polynomial regression method, because polynomials of unstandardized variables can be close to multicollinearity. Usually, it is a good idea to examine some of the higher-order polynomials as well (e.g. cubic transformation of X, or X^3), but typically these explain progressively less variance in noisy epidemiologic data compared to the lower-order polynomials (i.e. 1, X, and X^2).

To understand the nonlinearity we detected, we examined the panel "c" of the figure that illustrates the SATSA data from our running example (Figure 11.2). It shows a scatterplot of the sleep-hours and depression variables, revealing that both much less or much more sleep in comparison to the population average hours appears to be associated with high values of CES-D (i.e. with depression). This is an observation we can readily understand. Typically, researchers consider both insomnia (too little sleep) and hypersomnia (too much sleep) as a symptom of depression, and in fact, diagnostic definitions of depression do not differentiate between insomnia and hypersomnia. Therefore, we considered absolute deviation from the population-average hours slept per night as our new, continuous "sleep problems" variable in the analyses that follow (cf. panel d in Figure 11.2). With this transformation, we both understood what natural phenomenon our transformed variable stands for and were able to remove obvious nonlinearities in the data ($\beta_{quadratic} = 0.012$, $p = 0.231$ for the new sleep deviation variable).

It is generally not advisable to use arbitrary, uninterpretable transformations to linearize data before applying methods for distribution-based causal inference. This is because nonlinear transformations alter substantive meaning of variables, as well as their distributions. One might lose track of what phenomenon is being modeled, and at the same time, manipulate the inferred direction of causation. Thus, some substantive understanding is desirable prior to application of variable transformations in this context. However, the transformation need not be quite as straightforward as in our running example here. For example, Rosenström et al. (2012) discuss more advanced ways to re-interpret nonlinear psychometric data using Item Response Theory models. Similarly, one cannot remove seemingly "outlier" observations to make the data more linear prior to application of distribution-based causal inference, because that alters the distributions in question towards something else than the distributions reflecting the natural data-generating process under investigation (one should of course remove very clear recording errors, etc.; for example, we verified that an individual who appeared to report zero hours of sleep throughout year had no consequences for our analyses). In our running example, the sleep measure derived as absolute deviation from population-mean hours slept per night is a substantively meaningful variable in that it quantifies both hyper- and insomnia, and it fulfills the assumption of linearity with respect to the depression score. Taking absolute values increased skewness, however. Therefore, we also performed sensitivity analyses conducting DirectLiNGAM on separated datasets with hours slept below the mean of 7.21 ($N = 620$) and above it ($N = 706$).

As a cautionary note, even if one can statistically assess whether Y *could* be nonlinear in X or vice versa, undetected complex relationships between the variables typically cannot be fully ruled out by means of empirical analysis. Whether the assumption i (and iii) is reasonable must be assessed also in light of substantive understanding. To illustrate, the first panel of Figure 11.3 shows a sample of apparently stochastic data which can be modeled using a linear model with a statistically significant slope and which shows no quadratic effect, but which has, in fact, been derived from a deterministic nonlinear system.

11.5.2 Testing Non-Normality

Non-normality can be verified by rejecting a hypothesis of normal distribution using, for example, Lilliefors' test, which is an extension of Kolmogorov–Smirnov test (Lilliefors, 1967). The test is readily available in statistical programs, but very sensitive to deviations from normality. As distribution-based causal inference relies on information in the higher, non-Gaussian, moments of statistical distributions, the statistical power of the method depends on the magnitude of the higher moments and is likely to be much lower than the power of the Lilliefors' test. Therefore, the Lilliefors' test and a visual inspection of histograms suit well for establishing the necessary condition (ii) (non-Gaussianity), but they may not be sufficient to ensure good statistical power for causal inference. Statistical power can be assessed by simulation, as further illustrated in Section 11.8. In our running example, Lilliefors' test rejected a null hypothesis of normal distribution for both sleep deviations ($D= 0.163$, $p < 0.001$) and depression ($D= 0.127$, $p < 0.001$). The sample skewness of the standardized sleep deviation and depression variables was 2.58 and 1.03. The estimates of excess kurtosis were 13.39 and 0.78, respectively.

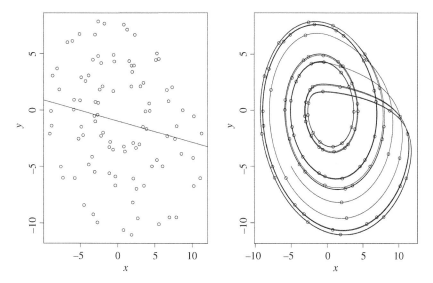

Figure 11.3 Illustrating lurking nonlinearities. The first panel shows a scatterplot of data points that an epidemiologist could legitimately approach via linear regression model. The second panel shows trajectories of the nonlinear Rössler system that was used to generate the data points by taking every 100th iteration from a numeric iteration of the system of differential equations. Rössler's classic parameter values were used ($a = b = 0.2$ and $c = 5.7$; see, e.g. Wikipedia page for Rössler attractor). Human eye and brain are an exceptionally good pattern-detection device, and the reader may see the concentric pattern that hints about the underlying non-randomness even in the left-most panel. However, even a minor degree of measurement noise would destroy the appearance.

11.5.3 Testing Independence

If the LiNGAM model holds, and if X causes Y, then X should be statistically independent of the residual $\varepsilon_Y = Y - \beta X$, where β is the ordinary least squares regression coefficient. By definition, X is uncorrelated with the least squares residual, but it should also be fully independent in the sense that $E[f(X)g(Y - \beta X)] = E[f(X)]E[g(Y - \beta X)]$ holds for all (absolute integrable) functions f and g (Hyvärinen et al., 2001; Klenke, 2008). Studying independence of arbitrary distributions is a difficult task, but several general methods do exist (Einmahl & McKeague, 2003; Gretton & Györfi, 2010; Hoeffding, 1948; Kallenberg & Ledwina, 1999). However, the assumption iii is not strictly necessary in the sense that distribution-based causal inference may work despite confounding (Rosenström et al., 2012) and algorithms designed for estimation in presence of confounding exist (Shimizu & Bollen, 2014). As discussed above, the DirectLiNGAM test statistic compares *expected* values instead of testing strict hypotheses. The best course of action in practice may be to test whether the independence between predictor and estimated residual variable holds fully, and if not, use sensitivity analyses and triangulation (see below) instead of totally abandoning distribution-based causal inference.

In our running example, we observed that there was no *linear* dependence between the depression score and its residual when regressed on sleep deviations ($p = 0.996$), but a clear *statistical* dependence when assessed with tests such as Kallenberg's and Ledwina's V test ($p < 0.001$), Hoeffding's test ($p = 0.005$), and empirical likelihood test ($p < 0.001$), all of which are sensitive to dependencies beyond simple linear relationships (Einmahl & McKeague, 2003; Hoeffding, 1948; Kallenberg & Ledwina, 1999). Similarly, by definition, there was no Pearson's product-moment correlation between sleep deviation and its residual when regressed on depression scores ($p = 0.984$), but there was a clear dependence when assessed using the above nonlinear measures (all $p < 0.001$). That is, whereas the residual and the predictor are *uncorrelated* with each other by definition of Ordinary Least Square regression, they typically are *not necessarily independent* of each other. The residual that is least dependent on the associated predictor may be indicative of causation.

11.6 Distribution-Based Causality Estimates for the Running Example

We report (standardized) DirectLiNGAM and skewness- and kurtosis-based estimators for pairwise causal direction between depression score and sleep deviations, as in our previous work (Rosenström et al., 2012). The latter two estimates may sometimes reveal specific distributional properties most important for the general DirectLiNGAM estimate. The kurtosis-based estimator was previously called "tanh-based" because it is specifically based on hyperbolic-tangent approximation to likelihood ratio. In SATSA data, however, we observed that all the three estimators indicated absolute sleep deviations being a cause of other depressive symptoms rather than the other way around (Table 11.1). That is, whichever non-Gaussian moments of the respective distributions we looked at, they indicated sleep deviations as being a cause of depressive symptoms more likely, or more strongly (i.e. in expected value), than the other way around. The results of the sensitivity analyses we performed supported the same direction of dependence for both hypersomnic and insomnic sleep deviations, with the exception of the DirectLiNGAM estimate in the hypersomnia subsample (Table 11.1). However, we did not further interpret the single deviant result as a key LiNGAM assumption failed in that case: the linear correlation coefficient between depression score and oversleep did not statistically differ from zero

Table 11.1 Estimates of causal direction for depression and sleep variables of the running example.

Method	Depression as cause %	Sleep as cause %	Statistic	Lower CI	Upper CI	Hypersomnia subsample, sleep as cause %	Insomnia subsample, sleep as cause %
DirectLiNGAM	0.20	99.80	−0.0461	−0.0964	−0.0115	40.1	99.65
Skew-based	0.10	99.90	−0.0438	−0.0920	−0.0094	89.8	99.35
Kurtosis-based	0.25	99.75	−0.0034	−0.0066	−0.0009	89.9	83.85

Results are shown for 2000 bootstrap resamples. Percent selected as cause is shown for each estimator, as well as T (depression, sleep) statistic and its 95% bootstrap percentile confidence intervals (CI). The two last columns replicate the analysis in those who sleep more than average ("Hypersomnia subsample") and in those who sleep less than population average hours per night ("Insomnia subsample").

($r = 0.03$, $p = 0.482$). The linear association between insomnia and depression score was statistically significant ($r = 0.27$, $p < 0.001$).

11.7 Conducting Sensitivity Analyses

11.7.1 Convergent Evidence from Multiple Estimators

In the running example, we observed a certain type of indication for robustness, because different estimators, using different types of deviation from Gaussian distributions, converged in their estimates of causal direction (Table 11.1). That is, the estimated causal direction was not sensitive to specific distributional property beyond the necessary requirement of non-Gaussian distribution. Such robustness property does not necessarily hold (e.g. Rosenström et al., 2012), and establishing it can be comforting and evoke trust. It only indicates robustness with respect to distributional characteristics, however, not with respect to model assumptions. Epidemiologic triangulation is a process to establish robustness over model assumptions and modeling approaches, and it will be discussed in Section 11.9.

11.7.2 Simulation-Based Analysis of Robustness to Latent Confounding

As discussed above, the requirement of perfect independence between residual and predictor variable may be unnecessary for causal inference, as well as overly restrictive. When relaxing this assumption, it may be desirable to gather some

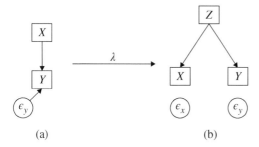

(a)   (b)

Figure 11.4 Illustration of the assumed model in DirectLiNGAM (a; one variable causes the other) and a fully confounded model (b; neither X nor Y is causal despite their correlation). In our simulation protocol, the degree of confounding is manipulated so that Y is a weighted sum of situations (a) and (b) in a gradient from 0 to 100%, where $\lambda = 0$ would correspond to situation (a) and $\lambda = 1$ would correspond to situation (b). The simulation was tailored to inform about the power of DirectLiNGAM to detect the right causal direction specifically in our running example.

insight on possible biases that could result. For that and related purposes, one can conduct brief simulation studies to investigate the extent to which the method is sensitive to the simulated conditions. Unobserved confounding variables are a typical reason for failures of independence between the residual and the predictor variable in regression models. Complete confounding implies that there is no direct effect between X and Y (see Figure 11.4). In other words, experimental manipulations of X have no effect on Y despite their association with each other, unless also the "true" causes of both the variables (variable Z in Figure 11.4) are manipulated. In practice, the possibility of confounding is difficult to definitively test in observational data. However, through simulation we can have a clue of the extent to which unobserved confounders may bias our causal inferences: by generating data which are as similar as possible to the observed data and by manipulating the degree of confounding in it.

For example, we generated a large number of simulated datasets with the same characteristics as our real-world data from the running example, and examined both possible directions of dependence, i.e. X and Y being the cause (i.e. sleep deviation causing other depression symptoms and vice versa). The datasets were manipulated so that they contained different degrees of "unobserved" confounding in X and Y. Then, each one of the datasets was analyzed and the output saved. Finally, we investigated how robust DirectLiNGAM was to different degrees of confounding by obtaining the proportion of success in correctly picking the causally antecedent variable of a given simulation condition.

In what follows, we will walk the reader through the simulation step by step, displaying pseudocode and comments on it. By "pseudocode" we mean an informal code that is not based on a concrete programming language. Instead, it is a general-purpose text that allows one to understand and implement the simulation using whichever programming language that best serves the case.

11.7.2.1 Obtain Data-Based Parameters

We estimated regression models in both directions on the standardized variables, and saved the regression residuals ($\hat{\varepsilon}_Y$ and $\hat{\varepsilon}_X$). The slope coefficient was the same in both models as a consequence of the standardization, regardless of which variable was set as predictor or outcome ($\hat{\beta} = 0.148$) Distributions of the predictor and the residual in the simulation were approximated by a bootstrap distributions of their empirical distribution (Efron & Tibshirani, 1993).

11.7.2.2 Defining Parameters and Simulation Conditions

Once we had the parameters to simulate data akin to our running example, we defined the conditions for our simulation experiment. Our purpose was to check how sensitive DirectLiNGAM is to latent confounders by varying the degree of confounding. The parameter λ quantified the amount of variance due to the latent confounder Z (Figure 11.4). In total, we had four simulation settings coming from two times two conditions: conditions A and B relate to the direction of dependency tested, and conditions 1 and 2 define alternative distributions for the latent confounder Z.

In condition A, the regression model emulated the situation in which the predictor was distributed as the sleep variable in SATSA data, with also other parameters being as in the empiric model (slope and residuals). In condition B, the assumed predictor was distributed as the CESD depression score. Conditions 1 and 2 were included because the distribution of the latent confounder Z is a potential factor affecting the statistical power inferred from our simulation. We addressed this aspect by switching the distribution of Z, so that in the condition 1 it had the same distribution as the antecedent variable, and in the condition 2 it had a different distribution (i.e. bootstrap distribution of the descendent variable).

Because larger sample sizes (N) improve statistical power, we ran the sensitivity analyses using several sample sizes: 200, 500, 1000, 1325 (i.e. the sample size of our running example), and 5000. Thus, we could investigate whether DirectLiNGAM is more or less sensitive to latent confounding depending on sample size. DirectLiNGAM was computed for all combinations of N values and values of λ, totaling 30 parameter combinations for each simulation setting A1, A2, B1, and B2. Finally, one has to set the number of replications (R) the simulation will be run (a single run generates one dataset and the corresponding directLiNGAM estimate). We chose $R = 10\,000$. The larger the R, the more precise information we have on the unavoidable effects of sampling variance.

11.7.2.3 Defining the Simulation Model

The structural model underlying the data-generating process of Y in accordance to LiNGAM assumptions is the linear model (e.g. $Y = \beta_X X + e_y$). Because our purpose was to introduce and investigate latent confounding, we had to generate the confounder Z and use it when generating Y. The data-generating linear model in this simulation is therefore: $Y_{\text{sim}} = \lambda(\beta_Z Z) + (1 - \lambda)(\beta_X X) + e_y$, where λ controls the degree of confounding. The simulated predictor variable X_{sim} is then a weighted sum of "unconfounded" X and a regression on Z, that is, $X_{\text{sim}} = \lambda(\beta_Z Z + e_X) + (1 - \lambda)X$. The distribution of Z cannot be identified from empirical data and was therefore set to a specific distribution under two conditions: in condition 1, Z was bootstrapped from the same distribution as X (e.g. both X and Z were independently bootstrapped from the depression score variable), and in condition 2, Z was bootstrapped from the distribution of the other variable of the pair (e.g. when X was bootstrapped from the depression score, Z was bootstrapped from the sleep variable). Table 11.2 shows the respective roles of the empirical bootstrap distributions (SATSA variables) in the data generating process of the four simulation settings. The residual distributions were also bootstrapped from the empirical distribution. β_Z was set as equal to β_X. The next lines show a brief sketch of the data-generating procedure.

Table 11.2 Role of the variables on generating the data of the four simulation settings.

Simulation setting	Role	SATSA variable
A1	Predictor	Sleep deviation
	Confounder	Sleep deviation
	Residuals	$\hat{e}_{\text{CESD}}$
A2	Predictor	Sleep deviation
	Confounder	CESD score
	Residuals	$\hat{e}_{\text{CESD}}$
B1	Predictor	CESD score
	Confounder	CESD score
	Residuals	$\hat{e}_{\text{sleep}}$
B2	Predictor	CESD score
	Confounder	Sleep deviation
	Residuals	$\hat{e}_{\text{sleep}}$

The variables in the simulation settings were bootstrapped from the standardized SATSA variables and regression residuals $\hat{e}$.

The pseudocode for data simulation of simulation setting A1:

```
R = 10000
lambda = vector(0, .2, .4, .6, .8, 1)
N = vector(200, 500, 1000, sample_size_of (SATSA), 5000)
residuals = residual_cesd
predictor = sleep deviations
confounder = sleep deviations
output = initialize_array(rows = R, cols = length(lambda),
         dim = length(N))
x_PD = parametric_estimate(X)
for each (n in N) do {
  for each (j in lambda) do {
    for each (i in 1 to R) do {
        % generate data
        e = pick_random_with_replacement(residuals), n))
        e2 = pick_random_with_replacement(residuals), n))
        e = e - mean(e)
        e2 = e2 - mean(e2)
        x = pick_random_with_replacement(predictor), n))
        z = pick_random_with_replacement(confounder), n))
        y = ((1-j)*betax*x + e + j*betaz*z)
        x = ((1-j)*x + j*(betaz*z + e2))
        % generate output
        output[i,index_of(j),index_of(n)] = DirectLiNGAM(x, y)
    } endfor
  } endfor
} endfor
```

The "for" statement is used when repeating the same action across all values of a given vector, or from index 1 to another integer. In this case, we repeated the simulation R=10 000 times per value of lambda ($\lambda = 0.0, 0.2, 0.4, 0.6, 0.8, 1$) and per sample size ($N = 200, 500, 1000, 1325, 5000$). This sums up to $6 \times 5 = 30$ experimental conditions, each of which had $R = 10 000$ replications.[3]

11.7.2.4   Run Simulation and Interpret Results

As the output of our simulation experiment, we computed the success rate of DirectLiNGAM in picking up the correct causal direction over all R replications,

---

3 In addition to the pseudo-code shown here, the actual Octave/Matlab code for the simulations can be found from the web page: http://www.iki.fi/tom.rosenstrom

and plotted the success rate with respect to different degrees of confounding (λ) and for different sample sizes (N). Because we have ourselves generated the data, we know the ground truth behind it: (i) that X causes Y rather than vice versa, and (ii) the degree there is a common variable causing them both. Knowing the ground truth makes it possible to estimate how successful the method is *despite* latent confounding. The analysis revealed that the distribution of the confounder Z had a negligible effect on the results, as can be noted comparing simulation settings A1–A2 and B1–B2 (Figure 11.5). Furthermore, switching the causal roles of the original cause and residual distributions in the simulation did not have a mentionable effect on the estimation success (Figure 11.5; a very small bias may be present at the 80% confounding, which could be further investigated in the future).

The success rate in causal estimation remained higher than 90% when introducing up to 40% of latent confounding in samples equal or bigger than $N = 1000$ (Figure 11.5). As expected, when latent confounding was 80–100%, DirectLiNGAM estimates were nearly random, meaning that the method would pick up either X or Y as being the cause with almost the same probability. When the sample size was $N = 5000$, DirectLiNGAM remained robust even up to 60% of latent confounding. In summary, the algorithm may be able to tolerate a considerable amount of latent confounding (violation of assumption iii) without noticeable performance loss in causal inferences.

11.8 Simulation-Based Analysis of Statistical Power

In Section 11.7, we illustrated how to benefit from computer simulation in sensitivity analyses. There, the aim was to simulate controlled experiments that closely resemble the data at hand to investigate consequences of potential partially inaccurate model assumptions (i.e. degree of bias in the final estimate given a known degree of latent confounding). The conclusions one can draw from such sensitivity analyses are context specific by design. However, simulation studies are also helpful in collecting more general knowledge on algorithmic performance under different conditions. Here, we strive to provide the reader with intuition on statistical power of the DirectLiNGAM algorithm in estimation of pairwise directional dependence. Assumption of a non-Gaussian distribution is a *necessary* precondition for the kind of methods discussed here, but it does not automatically provide *sufficient* statistical power. This section tries to provide the reader with a rough intuition on how "big" deviation from a Gaussian distribution is sufficient for a good statistical performance of the causal estimation algorithm.

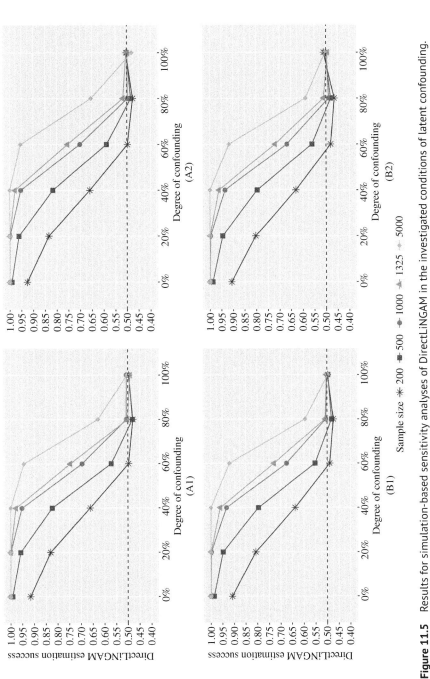

Figure 11.5 Results for simulation-based sensitivity analyses of DirectLiNGAM in the investigated conditions of latent confounding. Different panels correspond to different assigned distributions for the simulated cause, residual, and confounder variables (assigned distributions in Table 11.2).

A "deviation" from Gaussian distribution is commonly quantified using skewness, excess kurtosis, or differential entropy. A Gaussian distribution has a zero skewness and excess kurtosis, and other things being equal, the greater the absolute value of these statistics the less the evaluated distribution resembles a Gaussian distribution. Gaussian distribution is also the distribution of random movement and errors of measurement: on average, observations of a variable with a Gaussian distribution provide the least information imaginable for a continuously distributed variable with a given variance (alternatively, they are the least "surprising" events; Cover & Thomas, 2006). Therefore, Gaussian distribution with variance σ has the maximal entropy (i.e. $\log(2\pi e \sigma^2)/2$), and the lower the entropy the more a distribution "deviates" from a Gaussian one. Thus, also information entropy can quantify deviations from normality for given variance. We arranged a simulation protocol to assess DirectLiNGAM estimation success under deviations of different magnitude, and to answer the question "*how non-Gaussian variables one needs for causal inference.*"

Instead of distributions inferred from data, we used log-normal distributions for the antecedent and residual variables (i.e. for X and e_y), which is a distribution for the exponent of a Gaussian variable (i.e. its logarithm would have a Gaussian distribution). Skewness, excess kurtosis, and entropy of a log-normal variable are simple functions of mean and variance (σ) after log-transformation (i.e. for the generating Gaussian variable). We manipulated these parameters and calculated the estimation success by generating log-normally distributed predictors and residuals with varying scale parameters (eight conditions ranging from $\sigma = 0.05$ to 0.75; β was adjusted to hold $Cor(X,Y)$ at a constant 0.4), with the following sample size conditions: $N = 100, 200, 500, 1000, 5000$. Estimation success was computed as an average over $R = 10\,000$ replications of each condition. The resulting Figure 11.6 provides the reader with intuition on how statistical power of DirectLiNGAM responds to changes in these commonly used quantitative characterizations of statistical distributions. In general terms, the larger the sample is, the smaller the departure from Gaussianity that is sufficient to reach correct detection of causality. Samples of size 200 or less require clear deviations (entropy difference of 0.09, skewness = 1.53, or excess kurtosis = 2.35), while samples of size 500 and larger achieve statistical power above 95% already when showing only small departures from normality (entropy difference of 0.02, skewness = 0.46, or excess kurtosis = 0.37). The largest sample-size condition ($N = 5000$) reached power above 95% with minimal deviations from normality.

While simulations are a good way to answer questions like "*what if* model assumptions are violated or conditions not ideal," triangulation, discussed below, is a technique that may be able to answer questions like "*are* difficult-to-test assumptions violated in the data at hand."

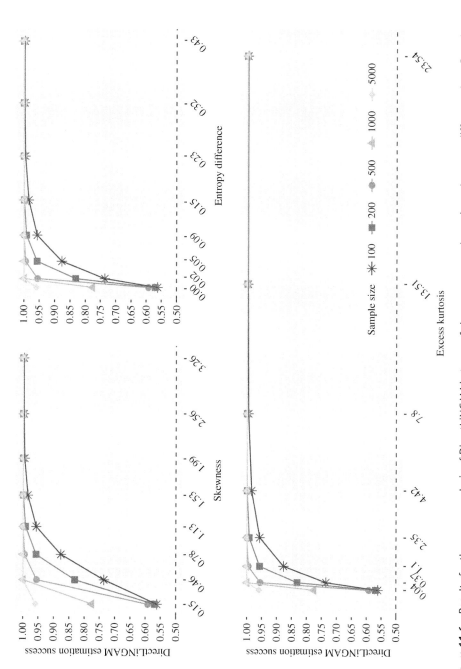

Figure 11.6 Results for the power analysis of DirectLiNGAM in terms of skewness, excess kurtosis, and entropy difference (to a Gaussian variable of equal variance).

11.9 Triangulating Causal Inferences

Epidemiologists frequently deal with issues of life and death, literally. They also have many historical examples on "spurious," or misleading, findings. Thus, both the ramifications of false inferences and the previous experience warrant a cautious attitude toward translation of epidemiologic practice to public health policy. At the same time, doing so is an important part of evidence-based medicine. To cope with these conflicting demands, epidemiologists have introduced the idea of causal triangulation in etiologic epidemiology (Lawlor, Tilling, & Smith, 2017). Triangulation differs from generic attempts to show robustness across several estimators by aiming to show robustness across several estimators that have *different* key assumptions with respect to each other.

Typically, all causal inference techniques involve some assumptions that are difficult to test for, but necessary preconditions for applying the technique. However, it is often possible to find methods that make use of entirely different, or even "opposite" types of information, to derive their inference on causality. For example, we could seek for a cross-sectional observational technique that does not rely on non-Gaussianity of the data when deriving otherwise similar statements on direction of causation between two variables? One such technique would be Direction of Causation (DoC) models studied in behavior genetics (Duffy & Martin, 2010; Heath et al., 1993). DoC models assume that *normally distributed* variables for distinct genetic and environmental influences give rise to the observed (phenotypic) variables and correlations. Typical applications use twin models, where biological knowledge on monozygotic twins' 100% genetic similarity and dizygotic twins' 50% average genetic similarity is used to partition observed variance into additive genetic (A) influences, shared environmental influences common to both twins (C), and non-shared environmental influences unique to each twin (E; see, e.g. Neale & Cardon, 1992).

We consider structural models nested within the path diagram in Figure 11.7, which describes a set of possible causal relationships (arrows) and correlations (arcs) between observed (boxes) and unobserved latent (circles) variables. All the paths are not identified at the same time, but we can test a direct-effect model with no latent correlations (no arcs; "reciprocal causation model") against the full correlational model it nests within (no arrows, but arcs; "no phenotypic causation model"). If the reciprocal causation model is not rejected, we can test if another one of the direct effects could be set to zero, indicating that sleep deviations cause depression, or vice versa.

To intuitively understand how the DoC approach infers causal directions, consider a case where similarity between twins on a causally antecedent trait is explained by genes and the similarity on the causally descendent trait is explained by the shared environment of the twins. Then twins reared apart would show the

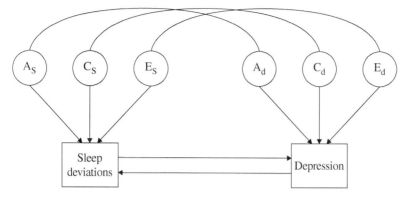

Figure 11.7 Path diagram for Direction of Causation (DoC) models in the running example. By constraining different paths, DoC models study whether family data is best explained by simple correlations between the two observed variables' (boxes) genetic (A) and shared (C) and non-shared (E) environmental influences ("spurious" association), by regression of one variable on the other (causation), or by regression of both variables on each other (reciprocal causation).

same (genetic) cross-trait similarity as twins reared together. With the opposite causation, only twins reared together would show cross-trait correlations because of their shared environment. In practice, one does not necessarily need data on different rearing statuses. For successful causal inference, however, one needs to be able to estimate three or more sources of familial similarity and their composition in the studied phenotypic/trait variables must differ across the variables (Heath et al., 1993).

In our running example, we had data on monozygotic and dizygotic twins reared together and reared apart, which we modeled using DoC models. We direct the reader to behavior genetics literature for more details (Heath et al., 1993; Neale & Cardon, 1992), and simply provide the results here. For the DoC method, we used log-transformations to make the variables closer to being normally distributed. The reciprocal causation model was not rejected in a likelihood-ratio test ($\chi^2 = 0.72$, $df = 1$, $p = 0.396$), allowing us to test unidirectional causal hypotheses. However, both a DoC model with sleep as a cause for depression ($\chi^2 = 3.55$, $df = 1$, $p = 0.059$) and a DoC model with depression as a cause for sleep ($\chi^2 = 3.27$, $df = 1$, $p = 0.071$) were close to being rejected, though not quite statistically significant. In terms of Bayesian Information Criterion (lower values indicate better fit), the unidirectional causal models were practically indistinguishable from each other ($-13\,479.2$ and $-13\,479.4$, respectively), but not from the reciprocal causation model ($-13\,475.9$).

Before we rush to conclude that we have a case of reciprocal causation, however, we must address the limitations of the DoC method in this specific case. First,

Table 11.3 Estimated biometric sources of variance for depression and sleep variables of the running example.

Variable: component	Estimate	Lower CI	Upper CI
Depression score: A	0.216	0.014	0.400
Depression score: C	0.113	−0.071	0.299
Depression score: E	0.671	0.554	0.801
Sleep deviations: A	0.276	0.083	0.446
Sleep deviations: C	−0.001	−0.171	0.171
Sleep deviations: E	0.725	0.597	0.863

"A" refers to additive genetic influences, "C" to shared environmental influences of the twins, "E" to non-shared environmental influences unique to only one member of each twin pair, and "CI" to 95% likelihood-profile confidence intervals.

we did not have the minimum of three biometric sources of variance required for detecting reciprocal causation, because neither of our variables had a statistically significant contribution from shared environmental influences (Table 11.3). Under these conditions, we could not have rejected the hypothesis of reciprocal causation, even if it were false. Second, the inheritance pattern of depression and sleep variables had remarkably similar composition, meaning that we had little power to distinguish between directions of causation (Duffy & Martin, 2010; Heath et al., 1993). Third, although we cannot directly assess the degree of third-variable confounding, a considerable extent is expected in this case and not well-handled by the DoC model (Heath et al., 1993). Fourth, measurement error can bias causal inferences based on DoC models, and therefore efforts to minimize it would be a desirable part of a DoC analysis (Heath et al., 1993). Altogether, promising as it was, the DoC modeling approach provided little causal information in this case. However, it is to be expected in causal triangulation that some approaches turn out more informative than other approaches, and that confidence can be built only gradually.

One could continue the process of causal triangulation using, for example, instrumental variable regression method for causal inference, which is yet another method based on different assumptions than distribution- and DoC-based causal inference (Heath et al., 1993). In instrumental variable regression, one needs an auxiliary variable that is a known cause of a target variable in causal inference and known to affect the other target variable only through the first target variable. For example, genes controlling the circadian clock might serve as an instrument that is causal for sleep deviations, and for depression only through their effect on sleep (Lawlor, Harbord, Sterne, Timpson, & Smith, 2008). Of course, that would be an assumption, and the clock genes might also have other unknown effects on the brain. All in all, every causal-inference method at researcher's disposal is an asset in causal triangulation, as any single method is unlikely to be decisive.

11.10 Conclusion

In this chapter, we briefly reviewed past epidemiologic studies using distribution-based causal inference (DirectLiNGAM in particular), discussed the novel method from the viewpoint of more established epidemiologic research, and replicated previous findings on causality between sleep problems and other depressive symptoms (running example). At the population level, sleep problems were more likely to cause at least mild forms of depressive symptoms than the other way around – a conclusion that may well differ in severely symptomatic clinical samples (Rosenström et al., 2012). In addition, we showed how to conduct simulation-based sensitivity analyses and how to study statistical power of the algorithm in different settings. Finally, we discussed use of DirectLiNGAM as a part of a general process of causal triangulation in etiologic epidemiology. To provide an example of alternative causal inference technique with very different assumptions to DirectLiNGAM, we applied DoC models from behavior genetics. These turned out uninformative in our running example, but nevertheless served to illustrate the general process of triangulation.

DoC models applied herein also illustrated that DirectLiNGAM is, in fact, a rather robust technique for causal inference. Population samples that reflect natural data-generating processes are often available, whereas it can be quite difficult to satisfy the assumptions of DoC or instrumental-variable methods (Duffy & Martin, 2010; Heath et al., 1993; Lawlor et al., 2008). In addition to the above-discussed assumptions, validity of DoC models also depends on the validity of the applied inheritance model; for example, the classic ACE model we used requires a number of conditions to hold, such as non-assortative mating and equal environments for identical twins and other siblings (Neale & Cardon, 1992). Furthermore, the inheritance patterns that DoC methods use often are functions of changing environmental conditions (Heath et al., 1985), which could sometimes lead to surprises in DoC modeling. For example, a colleague once described a lack of trust toward DoC methodology due to having found that recent observations on a variable had "caused" historical observations in the same variable according to his DoC application, thus reversing the "arrow of time" (personal communication). So far DirectLiNGAM has not led to comparable spurious findings. It shows remarkably good robustness properties and statistical power, while making much less stringent assumptions than many alternative methods. However, more research is needed on possible biases of distribution-based causal inference methods in various real-world research problems.

With complex constructs, such as psychiatric disorders, the assumption of no confounding due to third, unobserved variables may not be very realistic. Here and previously, we noted that the DirectLiNGAM approach can be quite robust against confounding. There are also later extensions of the method specifically developed to handle unobserved confounding (Shimizu & Bollen, 2014). Distribution-based causal inference techniques have also been developed for time-series analysis, for some nonlinear models, and for other special cases (Hyvärinen, Zhang, Shimizu, & Hoyer, 2010; Wiedermann & von Eye, 2016). On the methodological side, the field is developing rapidly, whereas within epidemiology, it has yet to demonstrate its value. Distribution-based causal inference methods have essential similarities with a statistical signal-processing technique known as independent component analysis, which has generated much interest and many applications (Hyvärinen et al., 2001; Shimizu et al., 2006). Time will show whether these methods find their place in the standard toolkit of epidemiologists as well.

References

Borsboom, D. (2017). A network theory of mental disorders. *World Psychiatry*, *16*(1), 5–13. doi:10.1002/wps.20375

Cover, T. A., & Thomas, J. A. (2006). *Elements of information theory*. Hoboken, NY: John Wiley & Sons, Inc.

Cramer, A. O. J., Waldorp, L. J., van der Maas, H. L. J., & Borsboom, D. (2010). Comorbidity: A network perspective. *Behavioral and Brain Sciences*, *33*, 137–193. doi:10.1017/S0140525X09991567

Dodge, Y., & Rousson, V. (2000). Direction dependence in a regression line. *Communications in Statistics—Theory and Methods*, *29*(9–10), 1957–1972. doi:10.1080/03610920008832589

Dodge, Y., & Rousson, V. (2001). On asymmetric properties of the correlation coefficient in the regression setting. *The American Statistician*, *55*(1), 51–54. doi:10.1198/000313001300339932

Dodge, Y., & Yadegari, I. (2010). On direction of dependence. *Metrika*, *72*(1), 139–150. doi:10.1007/s00184-009-0273-0

Duffy, D. L., & Martin, N. G. (2010). Inferring the direction of causation in cross-sectional twin data: Theoretical and empirical consideration. *Genetic Epidemiology*, *11*(6), 483–502. doi:10.1002/gepi.1370110606

Efron, B., & Tibshirani, R. J. (1993). *An introduction to bootstrapping*. New York, NY: Chapman & Hall.

Einmahl, J. H. J., & McKeague, I. W. (2003). Empirical likelihood based hypothesis testing. *Bernoulli*, *9*(2), 267–290. Retrieved from http://www.jstor.org/stable/3318940

García-Velázquez, R., Jokela, M., & Rosenström, T. (2017). Symptom severity and disability in psychiatric disorders: The U.S. Collaborative Psychiatric Epidemiology Survey. *Journal of Affective Disorders*, *222*, 204–210. doi:10.1016/j.jad.2017.07.015

Gretton, A., & Györfi, L. (2010). Consistent nonparametric tests of independence. *The Journal of Machine Learning Research*, *11*, 1391–1423.

Heath, A. C., Berg, K., Eaves, L. J., Solaas, M. H., Corey, L. A., Sundet, J., … Nance, W. E. (1985). Education policy and the heritability of educational attainment. *Nature*, *314*(6013), 734–736.

Heath, A. C., Kessler, R. C., Neale, M. C., Hewitt, J. K., Eaves, L. J., & Kendler, K. S. (1993). Testing hypotheses about direction of causation using cross-sectional family data. *Behavior Genetics*, *23*(1), 29–50.

Helajärvi, H., Rosenström, T., Pahkala, K., Kähönen, M., Lehtimäki, T., Heinonen, O. J., … Raitakari, O. T. (2014). Exploring causality between TV viewing and weight change in young and middle-aged adults. The cardiovascular risk in young Finns study. *PLoS ONE*, *9*(7), e101860. doi:10.1371/journal.pone.0101860

Hoeffding, W. (1948). A non-parametric test of independence. *The Annals of Mathematical Statistics*, *19*(4), 546–557. http://www.jstor.org/stable/2236021

Hyvärinen, A., Karhunen, J., & Oja, E. (2001). *Independent component analysis*. New York, NY: Wiley.

Hyvärinen, A., & Smith, S. M. (2013). Pairwise likelihood ratios for estimation of non-gaussian structural equation models. *Journal of Machine Learning Research*, *14*, 111–152. http://jmlr.csail.mit.edu/papers/v14/hyvarinen13a.html

Hyvärinen, A., Zhang, K., Shimizu, S., & Hoyer, P. O. (2010). Estimation of a structural vector autoregression model using non-gaussianity. *Journal of Machine Learning Research*, *11*, 1709–1731. http://dl.acm.org/citation.cfm?id=1756006.1859907

Kallenberg, W. C. M., & Ledwina, T. (1999). Data-driven rank tests for independence. *Journal of the American Statistical Association*, *94*(445), 285–301. doi:10.1080/01621459.1999.10473844

Klenke, A. (2008). *Probability theory: A comprehensive course*. London, UK: Springer-Verlag.

Lawlor, D. A., Harbord, R. M., Sterne, J. A. C., Timpson, N., & Smith, D. G. (2008). Mendelian randomization: Using genes as instruments for making causal inferences in epidemiology. *Statistics in Medicine*, *27*(8), 1133–1163. doi:10.1002/sim.3034

Lawlor, D. A., Tilling, K., & Smith, G. D. (2017). Triangulation in aetiological epidemiology. *International Journal of Epidemiology*, *45*(6), 1866–1886. doi:10.1093/ije/dyw314

Lilliefors, H. W. (1967). On the Kolmogorov–Smirnov test for normality with mean and variance unknown. *Journal of the American Statistical Association, 62*(318), 399–402.

Neale, M. C., & Cardon, L. R. (1992). *Methodology for genetic studies of twins and families.* Dordrecht, NY: Kluwer Academic Publishers.

Pedersen, N. L. (2015). Swedish Adoption/Twin Study on Aging (SATSA), 1984, 1987, 1990, 1993, 2004, 2007, and 2010. Inter-university Consortium for Political. *Social Research (ICPSR) [distributor].* doi:10.3886/ICPSR03843.v2

Radloff, L. S. (1977). The CES-D scale: A self report depression scale for research in the general population. *Applied Psychological Measurement, 1*(3), 385–401.

Rosenström, T., Jokela, M., Puttonen, S., Hintsanen, M., Pulkki-Raback, L., Viikari, J. S., … Keltikangas-Järvinen, L. (2012). Pairwise measures of causal direction in the epidemiology of sleep problems and depression. *PLoS ONE, 7*(11), e50841. doi:10.1371/journal.pone.0050841

Shimizu, S., & Bollen, K. (2014). Bayesian estimation of causal direction in acyclic structural equation models with individual-specific confounder variables and non-Gaussian distributions. *Journal of Machine Learning Research, 15*(1), 2629–2652.

Shimizu, S., Hoyer, P. O., Hyvärinen, A., & Kerminen, A. (2006). A linear non-Gaussian acyclic model for causal discovery. *Journal of Machine Learning Research, 7*, 2003–2030.

Shimizu, S., Inazumi, T., Sogawa, Y., Hyvärinen, A., Kawahara, Y., Washio, T., … Bollen, K. (2011). DirectLiNGAM: A direct method for learning a linear non-Gaussian structural equation model. *Journal of Machine Learning Research, 12*, 1225–1248.

Wiedermann, W. (2018). A note on fourth moment-based direction dependence measures when regression errors are non-normal. *Communications in Statistics: Theory and Methods, 47*(21), 5255–5264. doi:10.1080/03610926.2017.1388403

Wiedermann, W., & von Eye, A. (2016). *Statistics and causality: Methods for applied and empirical research.* Hoboken, NY: Wiley.

12

Determining Causality in Relation to Early Risk Factors for ADHD

The Case of Breastfeeding Duration

Joel T. Nigg[1], Diane D. Stadler[1], Alexander von Eye[2], and Wolfgang Wiedermann[3]

[1] Department of Psychiatry, Oregon Health & Science University, Portland, OR, USA
[2] Department of Psychology, Michigan State University, East Lansing, MI, USA
[3] Department of Educational, School, and Counseling Psychology, College of Education, & Missouri Prevention Science Institute, University of Missouri-Columbia, Columbia, MO, USA

Neurodevelopmental disorders such as ADHD likely emanate from the interplay of genetic and prenatal and post-natal environmental factors, perhaps via epigenetic mechanisms. Determination of early experiential correlates is thus crucial, and many early risk factors have been identified. However, their causal role is often ambiguous and for most risk factors, human randomized trials are infeasible or unethical. Early post-natal experiences may be important for *programming* of later disease risk (Kaplan, Evans, & Monk, 2008; Wells, Chomtho, & Fewtrell, 2007). Etiology of neurodevelopmental conditions such as ADHD likely involves both pre- and post-natal inputs.

Because ADHD is thought to be rooted in altered neural development, early experiences that support neurodevelopment are of particular interest. Perhaps the best-studied early post-natal influence, aside from socialization itself, is initiation and duration of breastfeeding. It can be noted that breastfeeding and psychosocial influences are not entirely distinct. For example, whereas much research on the benefits of breastfeeding focuses on the nutritional and immunological advantages of breast milk, especially the impact of long-chain polyunsaturated fatty acids (e.g. omega 3 fatty acids) (Forsyth et al., 2003; Willatts et al., 2013) and the establishment of a more diverse and complementary microbiome (Foster, 2016), breastfeeding may also promote or be associated with the development of healthy infant–mother attachments highlighted by enhanced maternal sensitivity and responsiveness to the infant (Allen, Lewinsohn, & Seeley, 1998).

The authors (Stadler, Musser, Holton, Shannon, & Nigg, 2016) reported that children with ADHD were breastfed for a shorter duration than comparison

Direction Dependence in Statistical Modeling: Methods of Analysis, First Edition.
Edited by Wolfgang Wiedermann, Daeyoung Kim, Engin A. Sungur, and Alexander von Eye.
© 2021 John Wiley & Sons, Inc. Published 2021 by John Wiley & Sons, Inc.

children, after adjusting for age, sex, parental ADHD, family configuration, and other confounds. Several other studies from different nations using various methods reported a similar correlation (Al Hamed, Taha, Sabra, & Bella, 2008; Field, 2014; Golmirzaei et al., 2013; Julvez et al., 2007; Kadziela-Olech & Piotrowska-Jastrzebska, 2005; Mimouni-Bloch et al., 2013; Sabuncuoglu, Orengul, Bikmazer, & Kaynar, 2014). The association between breastfeeding and reduced likelihood of child ADHD is also supported by two population-wide surveys, one in Germany (Schmitt & Romanos, 2012) and one in the United States (Shamberger, 2012). Two recent meta-analyses suggest the association is robust and substantial in size (Tseng et al., 2019; Zeng et al., 2018).

These reviews all pointed out the need to evaluate causality. What is the causal direction of this correlation? Reasons to think that breastfeeding duration may be causally related to protection against ADHD are several. First, at a general level, ADHD is a neurodevelopmental disorder theorized to emanate from suboptimal early neural development. The early postnatal period is characterized by continued developmental plasticity; nutrition is among the most important supports for such development, such that suboptimal infant nutrition can induce adaptations that may be detrimental to brain development. Breast milk is the best source of nutrition for most infants because of its unique composition (Ballard & Morrow, 2013). By extension, breastfeeding initiation and duration should influence neurodevelopmental risk.

Second, ADHD is associated with higher-order cognitive delays, including subtle reduction in intelligent quotient (IQ). Consistent with the preceding, breastfeeding predicts subsequent cognitive outcomes in children (Al Hamed et al., 2008; Huang, Peters, Vaughn, & Witko, 2014; Quigley et al., 2012), an effect that tends to be dose-dependent in relation to duration of breastfeeding, as noted in a meta-analysis two decades ago (Anderson, Johnstone, & Remley, 1999). Subsequently, this picture has grown stronger. Effects on cognition appear especially important for boys (Oddy, Li, Whitehouse, Zubrick, & Malacova, 2011); ADHD is more common in boys than girls. National and international health organizations recommend that almost all infants be exclusively breastfed for the first six months of life to achieve optimal growth, development, and health (American Academy of Family Physicians, 2007; American College of Obstetricians Gynecologists, 2007; American Public Health Association, 2007; Eidelman & Schanler, 2012; Gartner et al., 2005; World Health Organization & United Nations International Children's Emergency Fund, 2003). When infants are exclusively breastfed for at least six months, cognitive gains are apparent and persistent in childhood (Angelsen, Vik, Jacobsen, & Bakketeig, 2001; Huang et al., 2014; Oddy et al., 2011; Wigg et al., 1998). Moreover, in the United States, while 83% of mothers initiate breastfeeding, and about half (57.6%) maintain some breastfeeding for at least six months, only 24.9% breastfeed exclusively for six months (Center for Disease

Control, 2018), suggesting widespread opportunity for susceptible infants to be exposed to lower amounts or reduced duration of breastfeeding, if in fact reduced breastfeeding is part of the developmental risk profile for ADHD.

Third, duration of breastfeeding has been associated directly with subsequent human brain development, particularly white matter development in infants and toddlers (Herba et al., 2013; Tawia, 2013). More recently, it is notable that breast-feeding duration was associated with increased white matter development in eight year-old children (although again, mainly in boys) in key fiber tracts including superior longitudinal fasciculus, cingulum, body of corpus callosum, and posterior thalamic radiations (Ou et al., 2014). These brain regions are similar to brain regions with altered white matter development in eight year-olds with ADHD that we observed previously (Nagel et al., 2011).

On the other hand, it may also be that child ADHD leads to reduced breastfeeding duration. How could future ADHD reduce duration of breastfeeding in the early months of life? One possibility of course is a genotype-environment correlation. ADHD runs in families and has a strong heritability (Faraone & Larsson, 2019). Mothers who themselves have elevated ADHD symptoms may have more difficulty maintaining breastfeeding. At the same time, parental ADHD is associated with reduced social advantage and less employment success, so these families may be forced into jobs that make breastfeeding and pumping or expressing breast milk impractical while at work.

A second possibility is that early temperament is already harder to manage in children who will have ADHD and this disrupts breastfeeding attempts by the mother. For example, Sullivan et al. (2015) reported that familial ADHD was associated with more infant negative effect at six months of age.

A third possibility is correlated neurodevelopmental problems. For example, ADHD is also associated with subtle neurological soft signs, including alterations in language and speech and motor development. It is possible that oral-motor developmental problems make breastfeeding more difficult for the baby, painful for the mother, or otherwise impede breastfeeding. Of course, it is also possible that these effects are additive, as there is some evidence that parental ADHD complicates the already-difficult task of managing a child with ADHD. If early temperament or neurodevelopmental delays are a precursor to ADHD, then this dynamic may occur in early life as well.

Indeed, the reasons women cease breastfeeding or do not initiate breastfeeding are many – pain on feeding, inability to or perceived inability to produce sufficient milk, infant having difficulties and perception that the infant was not satisfied or lost interest, return to work, effort associated with pumping and work demands, or health problems (Ahluwalia, Morrow, & Hsia, 2005; Li, Fein, Chen, & Grummer-Strawn, 2008; Mangrio, Persson, & Bramhagen, 2018; Newby & Davies, 2016; Odom, Li, Scanlon, Perrine, & Grummer-Strawn, 2013).

Clarifying and confirming this association would help further the specification of early life mechanisms and processes that may influence not only self-regulation, but also ADHD. It is also important to clarify any clinical recommendations that may emanate. In particular, it is possible that inflammation is a key mechanism in ADHD (Dunn, Nigg, & Sullivan, 2019). Breast milk is anti-inflammatory. Thus, breastfeeding or the consumption of breast milk may be an important intervention for at risk children if the association is causal. Yet, in this context, randomized trials are lacking and prospective naturalistic studies are inevitably confounded. Here, the authors report updated large-sample case-control associations and effect sizes for maternal breastfeeding duration and ADHD in well characterized children, and use a novel statistical analysis – direction dependence analysis (DDA; cf. Wiedermann & Li, 2018; Wiedermann & Sebastian, in press; Wiedermann & von Eye, 2015; see also the related Chapter 2 by Wiedermann, Li, and von Eye, in this volume) – to evaluate the causal direction of the correlation between ADHD and breastfeeding duration using one of the largest, well characterized cross-sectional samples to look at this question. Specifically, we use DDA to test which conceptual model better approximates the causal direction of the underlying mechanism: (i) breastfeeding decreases the likelihood of developing ADHD (i.e. *breastfeeding → ADHD*) or (ii) ADHD reduces the duration of breastfeeding (i.e. *ADHD → breastfeeding*).

12.1 Method

12.1.1 Participants

Participants were children aged 7–13 years in an expanded sample from that used in Stadler et al. (2016) and are described in Table 12.1. The new sample provides a more powerful estimate of correlation effect sizes. We provide a recap of recruitment and data collection methods here.

12.1.1.1 Recruitment and Identification

Families were recruited through advertisements and mailings to obtain a broadly representative sample that would not be biased by clinic referred youth. Families volunteering for the study completed a multi-gate screening process to establish eligibility and diagnostic group assignment. A structured clinical diagnostic interview with the primary caregiver (Schedule for Affective Disorders and Schizophrenia for School-Age Children—Epidemiologic Version [KSAD-S-E]) (Puig-Antich & Kiddie, 1996), parent and teacher standardized ratings (Conners, 2008; DuPaul, Power, Anastopoulos, & Reid, 1998), and an IQ screen were completed. The KSADS interviewer and the child test administrator observed

Table 12.1 Sample description.

	Control (*n* = 294)			ADHD (*n* = 496)			
	Mean	SD	SE	Mean	SD	SE	*p*
Age in years	9.25	1.57	0.09	9.50	1.52	0.07	0.029
% Boys	48.6			70.6			0.001
% White, non-Hispanic	82.0			81.0			0.740
IQ score	115.3	12.7	0.74	108.3	13.9	0.63	<0.0001
Father education	6.57	1.72	0.10	5.85	1.88	0.09	<0.0001
Mother education	6.82	1.39	0.08	6.28	1.51	0.07	<0.0001
Mother age child birth	31.2	5.32	0.31	29.88	5.57	0.26	0.003
Father age child birth	32.54	5.67	0.35	32.42	6.33	0.32	0.810
Family income score	4.88	1.72	0.10	4.39	1.94	0.09	0.0007
Occupational score	78.33	19.25	1.14	71.76	24.22	1.11	0.0001
Parent ADHD index	42.12	8.06	0.50	47.39	10.24	0.48	<0.001
SWAN total score	87.71	14.71	0.86	53.35	12.27	0.55	<0.0001
ADHD total RS parent	5.02	6.12	0.36	30.30	10.42	0.47	<0.0001
ADHD total RS teacher	3.09	3.44	0.20	25.65	11.94	0.54	<0.0001

Parent ADHD Index is the ADHD self Conners rating ADHD Index score for the participating parent. Results were unchanged using both parents or spouse ratings.

the child briefly and made detailed notes on the parent and child visit as well. KSADS interviewers were reliability checked against a gold-standard interviewer to achieve inter-interviewer reliability on ADHD and all other child diagnoses (all trained to a standard of $\kappa > 0.70$ on all disorders observed in our sample at greater than 5% base rate). Clinical interviewers were regularly monitored and fidelity checks were carried out by supervisor review of videotapes, and re-calibration training was conducted annually.

A clinical diagnostic team comprising a board-certified psychiatrist and licensed clinical psychologist independently reviewed all case information, including behavioral observations, to arrive at diagnoses using DSM IV (American Psychiatric Association, 2010) and DSM 5 (American Psychiatric Association, 2013) criteria. Their agreement rate was acceptable for all child diagnoses with base rates greater than five percent in the sample (all $\kappa > 0.75$), and agreement for child ADHD was also acceptable (all $\kappa > 0.80$).

12.1.1.2 Parental Psychopathology

The most parsimonious explanation of any ante-natal correlate of ADHD is passive gene-environment correlation (rGE), in that parents with ADHD may both transmit ADHD to the offspring and fail to breastfeed. To provide partial control against this possibility, parent symptoms of ADHD were obtained from both parents by parent self-report on the ADHD Rating Scale (DuPaul et al., 1998) adapted for adults. When self-report was not available, spousal report was obtained. Parental ADHD diagnosis was also obtained on a structured clinical interview (KSADS-E, adapted), but because it was less strongly related to breastfeeding and child ADHD, it was omitted from analysis in lieu of the parent symptom scores to provide a stronger test of the primary hypothesis.

12.1.1.3 Ethical Standards

The study was approved by the Oregon Health & Science University Institutional Review Board and all procedures conformed to the American Psychological Association Ethical Principles of Psychologists and Code of Conduct (American Psychology Association, 2002) and NIH guidelines for protection of human subjects. Parents provided written informed consent and children provided written informed assent, before participating in any study-related activities.

12.1.2 Exclusion Criteria

Children were excluded if parents reported a history of neurological impairments, seizures, traumatic brain-injury, other major medical impairments, or long-acting psychoactive medication (not including stimulants); if the clinical diagnostic team identified current mood disorder, lifetime history of any psychosis, or pervasive developmental disorder, or "sub-threshold" ADHD (defined as five symptoms of either inattention or hyperactivity-impulsivity), or if estimated IQ < 75. In addition, typically developing control children were required to have no prior diagnosis of ADHD, to have four or fewer symptoms of both inattention and hyperactivity-impulsivity, and to not meet criteria for conduct disorder. Their other psychiatric diagnoses were free to vary, to avoid a super-normal comparison group.

12.1.2.1 Assessment of Breastfeeding Duration

Breastfeeding duration was assessed retrospectively by a single item on our developmental history form, in which the mother was asked "To what extent was this baby breastfed?" The response options were 0 months or not breastfed, 1–3, 4–6, 7–12, 13–18, or 19+ months of breastfeeding. The literature suggests that maternal recall of breastfeeding duration is valid even over several years, agreeing with prospective assessment at $\kappa > 0.8$ (Li, Scanlon, & Serdula, 2005; Natland, Andersen, Nilsen, Forsmo, & Jacobsen, 2012). Although a primary predictor of inaccurate recall is having four or more children (Cupul-Uicab, Gladen, Hernandez-Avila, & Longnecker, 2009), in the present sample over 90% of mothers had fewer than four children.

12.1.3   Covariates

We covaried child sex and age at the time of evaluation as a precaution, because both were related to ADHD although neither was related to breastfeeding duration and neither interacted with child diagnostic group in relation to breastfeeding (all $p > 0.30$). We considered parental social disadvantage (SES), family income, and education level as important given strong relations with breastfeeding duration and with ADHD in prior work (Miller et al., 2018). Maternal age was covaried as well, although paternal age (shown later) did not contribute with maternal age in the model and so was not covaried. See detailed data on covariates in results.

12.1.3.1   Parental Education

Each parent reported on their education level on a scale coded as 1 = grade school, 2 = some high school, 3 = high school equivalent, 4 = regular high school degree, 5 = some college but no degree, 6 = associates degree, 7 = bachelor's degree, 8 = Master's degree, law degree, 2–3 years post bachelor's degree, 9 = doctorate, PhD, MD. Parents not participating in the study were rated by their spouse or the participating parent.

12.1.3.2   Primary Residence and Family Income

The current primary residence was defined as that where the child lived at least 50% of the time and total annual income for that household (including child support) was self-reported by parents on the following scale: 1: <\$25k, 2: 25k < \$35k, 3: 35k < 50k, 4: 50k < 75k, 5: 75k < \$100k, 6: 100k < 130k. 7: 130k < 150k, 8: >150k.

12.1.3.3   Parental Occupational Status

Occupations were coded by two reliable raters on the Nam Powers Boyd scale (Miller et al., 2018). The highest occupational status score of the two parents was used here.

12.1.4   Data Reduction and Data Analysis

12.1.4.1   Parental ADHD

Parent ADHD was evaluated with self-ratings on the adult Conners ADHD rating scale. We used the total DSM ADHD score and the ADHD index score as our outcome variables. Results were unchanged using scores on BAARS (Barkley, 2011) childhood recall, or on a modified KSADS interview symptom count.

12.1.4.2   Data Reduction

All data were double entered and cross checked to minimize data entry errors. Data were evaluated for outliers and appropriate distributions for the procedures at hand. Contrary to standard parametric statistical methods, DDA requires non-normally distributed variables. The reason for this is that, under non-normal variable conditions, linear regression models have asymmetry properties that allow one to discern the causal direction of association (if it exists; for more details see, e.g. Wiedermann & Sebastian, in press and the related Chapter 2 by Wiedermann, Li, and von Eye, in this volume). The breastfeeding and ADHD rating scale scores in particular were non-normally distributed; these variables had satisfactory skew (range −0.09 to 0.49), and kurtosis (range 1.85–2.10). To evaluate child ADHD categorically, the authors utilized the diagnostic team consensus diagnosis, which relied on all available information but was blind to breastfeeding history. To evaluate child ADHD symptoms dimensionally, the authors utilized parent and teacher ratings of child ADHD symptoms using the ADHD Rating Scale (DuPaul et al., 1998) total raw score (each ranging 0–27). These were analyzed separately to enable disaggregating of reporter effects and remove source-variance confound. With respect to relevant covariates, Tables 12.1 and 12.2 summarize covariate selection data.

12.1.4.3 Data Analysis

ANCOVA was used to examine group differences while controlling covariates. Follow-up categorical analyses examined whether mothers of children with ADHD were less likely to initiate breastfeeding or to breastfeed for at least six months using logistic regression. The authors did not have data on whether breast

Table 12.2 Breastfeeding duration by ADHD group status.

	Control		ADHD	
	(%)	Cum. (%)	(%)	Cum. (%)
19+ months	20.10	20.10	11.50	11.50
13–18 months	15.60	35.70	13.90	25.40
12 months or more	36		25	
12 months or more, Oregon	52			
7–12 months	33.00	68.70	25.20	50.60
6 months or more	69		51	
6 months or more, Oregon	73			
4–6 months	15.00	83.70	18.30	68.90
1–3 months	11.60	95.30	20.70	89.60
Not breastfed, 0 months	4.80	100.10	10.50	100.10

$X^2(5) = 31.91, p < 0.0001$
"12 months or more" and "6 months or more" summarize the respective group figures to that point for comparison to population norms. The rows ending in "Oregon" provide the corresponding population norms for the state.

milk was expressed and bottle fed or donor milk was used. Regular regression models were computed in MPLUS 7.4, using the Cluster command to adjust for non-independence due to presence of sibling sets in the data, and maximum likelihood to use all available data to estimate parameters in the event of missingness.

Breastfeeding duration and parent- and teacher-rated ADHD scores were used in subsequent DDA. To evaluate the causal direction of the breastfeeding-ADHD association, the authors applied three DDA components, (i) distributional patterns of observed variables, (ii) distributional patterns of model errors, and (iii) independence properties of predictors and errors of the causally competing regression models *breastfeeding* → *ADHD* and *ADHD* → *breastfeeding*. While the first two components are particularly useful to compare two causally competing models, the third component can be used to determine the likelihood of influential confounding (*breastfeeding* ← *confounder* → *ADHD*). Following the decision rules presented by Wiedermann and Li (2018) and Wiedermann and Sebastian (in press), the causally dominant model (i) consists of an outcome variable that is closer to the normal distribution than the predictor, (ii) has an error component that is closer to the normal distribution than the error of the anti-causal model, and (iii) does not violate the assumption of independence of predictors and error while non-independence has to hold for the anti-causal model. Distributional properties of observed variables and model residuals were evaluated using D'Agostino skewness tests and 95% bias corrected accelerated bootstrap confidence intervals (BCa CIs) of skewness and co-skewness differences (based on 2000 re-samples). Independence of predictors and regression residuals was evaluated using the Hilbert–Schmidt Independence Criterion (HSIC) (Gretton et al., 2008). All DDA tests were performed using the R functions dda. vardist, dda.resdist, and dda.indep (for details see http://www.ddaproject.com). The nominal significance level for all significance tests was set to 0.05.

The inclusion of covariates is crucial in any causal analysis. Covariates are commonly included in statistical models to increase the precision of parameter estimates, to increase the statistical power of model significance tests, and to remedy potential confounding (cf. Steiner, Cook, Shadish, & Clark, 2010). When the number of potential covariates is large, covariate selection algorithms (such as best subset selection, backward/forward selection, or stepwise backward/forward selection; for a recent overview see Heinze, Wallisch, & Dunkler, 2018) are commonly applied to find the most parsimonious and well-fitting model. In the present analysis, however, we followed a covariate reduction approach in which all covariates are represented by a smaller number of latent factors. This approach has the advantage that one does not need to select a subset of covariates. Instead, the influence of all covariates are considered in the statistical model through incorporating latent factor scores. Dimension reduction was applied for the covariates (i) mother's age at child birth (father's age at child birth was not considered because of a missing value rate of 18.6%), (ii) mother's education, (iii) fathers' education, (iv) parental occupational status, and (v) primary income. Exploratory factor analysis (EFA)

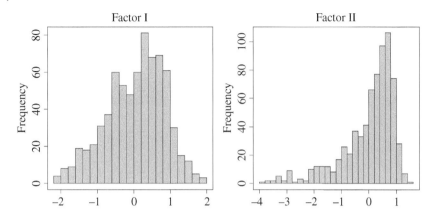

Figure 12.1 Distributions for Thurstone factor scores (mother's age at child birth, mother's education, father's education, and parent's occupational scores are loading on Factor I, prime family income loads on Factor II).

and parallel analysis (Horn, 1965) suggested that two latent factors were necessary to represent the correlational structure among the covariates with a Tucker Lewis Index (TLI) of factor reliability of 0.99 and a root means square error of approximation (RMSEA) of 0.04 suggested an acceptable model fit. The first factor (Factor I) consisted of four indicators (mother's age at child birth, mother's education, fathers' education, and primary income; standardized factor loadings ranged from 0.49 to 0.73). The second factor (Factor II) consisted of parental occupational status as a single indicator (standardized loading of 0.96). Observed distributions for both factors are given in Figure 12.1. Thurstone factor scores for both latent factors significantly deviated from normality (cf. Figure 12.1; Shapiro–Wilk p's < 0.001). Factor I showed a skewness of -0.39 and a kurtosis of 2.67, Factor II showed a skewness of -1.50 and a kurtosis of 5.13. The remaining demographic covariates (child's age, gender, and race) were considered separately. Further, child IQ was not incorporated as a covariate because it is more likely to serve as a causal descendant of ADHD which can bias DDA model selection (a detailed description of covariate criteria is given in Chapter 2 by Wiedermann, Li, and von Eye, in this volume).

12.2 Results

12.2.1 Study Participant Demographic and Clinical Characteristics

Table 12.1 presents demographic and diagnostic characteristics of the group. In brief, consistent with population prevalence data (and as expected, therefore, from our community recruitment strategy), children with ADHD were more likely than typically developing children to be male, to have a lower IQ, to come from lower income families, and to have had parents with more ADHD symptoms and lower educational and occupational scores.

Table 12.2 shows the simple unadjusted breakdown of breastfeeding initiation and duration rates by diagnostic group. As it shows, 4.8% of mothers of children in the control group did not initiate breastfeeding compared to more than twice as many, 10.5%, of mothers of children in the ADHD group reflecting a non-initiation rate similar to the Oregon average of 10.6%. Breastfeeding rates to 6 months of age of the control group, 69%, approximated the Oregon average of 73%, while 36% of mothers continued breastfeeding for 12 months or more (somewhat less than the 52% in Oregon overall). In comparison, 51% of children with ADHD were breast fed for 6 months while 25% were breast fed for 12 months or more.

Table 12.3 outlines our covariate selection process, showing the association of all variables either with breastfeeding (r value) or with ADHD (taken from Table 12.1). It shows the full list of model covariates; note that we excluded paternal age as it was non-informative with maternal age in the model.

Table 12.4 shows the regression models for categorical and dimensional outcomes for parent and teacher ratings. All models agree on a robust association

Table 12.3 Covariate evaluation and selection.

Candidate variable	Breastfeeding r-Value	ADHD p (Table 12.1)	Covaried?
Mother's education	0.302[a]	<0.001	Yes
Father's education	0.293[a]	<0.001	Yes
Mother's age at child birth	0.189[a]	0.003	Yes
Father's age at child birth	0.159[a]	0.810	Yes
Father's age and mother's age together			
Mother's age	0.041[c]		Yes
Father's age	0.241		No
Occupational score	0.157[a]	<0.001	Yes
Family income score	0.154[a]	<0.001	Yes
White/non-hispanic	0.108[b]	0.740	Yes
Sex of child	0.039	<0.001	Yes
Age of child at evaluation	0.002	0.029	Yes
Parent ADHD index	0.002	<0.001	Yes

a) $r \geq 0.14, p < 0.0001$.
b) $r \geq 0.095, p < 0.01$
c) $r \geq 0.072, p < 0.05$

Table 12.4 Regression models (*n* = 829).

	Std estimate	SE	Est/SE	*p* Value (2-tail)
A: ADHD diagnosis as outcome (positive beta: higher score = not ADHD)				
Sex of child	0.261	**0.041**	6.432	0.000
Age of child at evaluation	0.081	0.042	1.910	0.056
White/non-White	0.001	0.044	0.017	0.987
Mother's age at child birth	0.019	0.049	0.396	0.692
Family income score	0.011	0.056	0.190	0.850
Occupational score	0.026	0.056	0.464	0.642
Father's education	0.101	0.052	1.932	0.053
Mother's education	0.095	0.051	1.875	0.061
Parent ADHD score	0.363	0.045	8.139	0.000
Breastfeeding	0.226	0.043	5.240	0.000
B: Parent SWAN Score as outcome (positive beta: less ADHD features, better attention)				
Sex of child	0.203	0.031	6.509	0.000
Age of child at evaluation	0.028	0.033	0.855	0.392
White/non-White	0.006	0.031	0.201	0.841
Mother's age at child birth	0.067	0.039	1.718	0.086
Family income score	−0.046	0.042	−1.074	0.283
Occupational score	0.016	0.040	0.387	0.699
Father's education	0.123	0.040	3.044	0.002
Mother's education	0.074	0.038	1.934	0.053
Parent ADHD score	−0.285	0.032	−8.932	0.000
Breastfeeding	0.098	0.037	2.676	0.007
C: Teacher ADHD Rating Scale Total as outcome (positive beta: high score fewer symptoms, better behavior)				
Mother's age at child birth	0.061	0.041	1.509	0.131
Sex of child	0.285	0.030	9.564	0.000
White/non-White	0.020	0.034	0.604	0.546
Family income score	−0.003	0.044	0.057	0.955
Occupational score	0.022	0.045	0.482	0.630
Father's education	0.087	0.044	1.975	0.048
Mother's education	0.054	0.042	1.295	0.195
Age of child at evaluation	0.085	0.032	2.643	0.008
Parent ADHD score	−0.196	0.031	−6.337	0.000
Breastfeeding	0.115	0.037	3.089	0.002

of reduced duration of breastfeeding with ADHD even with numerous important covariates in the model.

12.2.2 Direction Dependence Analysis

We estimated a series of hierarchical regression models and performed DDA to evaluate the causal mechanism of breastfeeding and child ADHD: Model I accounted for demographic characteristics (gender, age, and race), Model II considered demographic characteristics and Factor I, Model III considered demographic characteristics and Factor II, and Model IV considered demographic characteristics and both factors simultaneously. DDA was performed for both, parent and teacher-rated ADHD scores. Because floor effects can bias DDA outcomes (cf. Wiedermann & Sebastian, in press), all analyses were repeated after discarding subjects without any ADHD symptomology (i.e. only considering ADHD scores > 0, parent- and teacher-rated ADHD scores were analyzed considering 92.3% and 89.0% of the total sample, respectively). Results are summarized in Tables 12.5–12.12.

DDA results for parent-rated ADHD indicate that a causal model of the form $ADHD \rightarrow breastfeeding$ is more likely to approximate the underlying data-generating mechanism than the causally reversed model $breastfeeding \rightarrow ADHD$ when adjusting for basic demographic covariates (Model I in Tables 12.5 and 12.6). This is evidenced by patterns of DDA independence tests (i.e. the HSIC is non-significant in the model $ADHD \rightarrow breastfeeding$ and significant in the causally reversed model), properties of residual distributions (i.e. distributional symmetry holds in the model $ADHD \rightarrow breastfeeding$ and errors are significantly skewed in the causally competing model), and properties of observed variable distributions (i.e. the predictor is more skewed than the outcome variable in the model $ADHD \rightarrow breastfeeding$). However, evidence for the causal model $ADHD \rightarrow breastfeeding$ is weaker when further adjusting for the two latent factor scores as indicated by non-significant HSIC tests for both causally competing models (note that clear cut DDA decisions are only possible for asymmetric HSIC patterns) – distribution-based DDA tests still suggest that $ADHD \rightarrow breastfeeding$ should be preferred over $breastfeeding \rightarrow ADHD$ (cf. Models II–IV in Tables 12.5 and 12.6).

Tables 12.7 and 12.8 summarize DDA results based on participants with parent-rated ADHD scores larger than zero (92.3% of the total sample were included in this subgroup analysis). Because the distribution of ADHD scores is less skewed after temporarily deleting subjects without any ADHD symptomology (i.e. skewness decreased from 0.19 to 0.12), one can expect that DDA patterns are less pronounced. For Models I and III, no causal decisions are possible for the subsample. Here, independence tests are either significant or non-significant in both causally competing models, which does not allow a clear cut decision.

Table 12.5 DDA results for covariate-adjusted models of the form *breastfeeding → parent-rated ADHD* based on the entire sample ($n = 827$).

Variables	Model I Est.	95% CI Lower	Upper	Model II Est.	95% CI Lower	Upper	Model III Est.	95% CI Lower	Upper	Model IV Est.	95% CI Lower	Upper
Breastfeeding	-1.76	-2.44	-1.09	-0.97	-1.76	-0.18	-1.39	-2.15	-0.62	-0.92	-1.71	-0.13
Gender	-5.45	-7.50	-3.40	-5.66	-7.87	-3.46	-5.24	-7.47	-3.00	-5.48	-7.69	-3.27
Age	-0.12	-0.77	0.52	-0.35	-1.05	0.36	-0.18	-0.88	0.53	-0.36	-1.06	0.34
Race: White	-0.91	-3.74	1.92	-0.26	-1.05	0.53	-0.32	-1.11	0.48	-0.21	-1.00	0.58
Factor I	—	—	—	-3.15	-4.58	-1.73	—	—	—	-3.01	-4.44	-1.58
Factor II	—	—	—	—			-1.28	-2.41	-0.15	-1.03	-2.15	0.09
R^2	0.065			0.080			0.062			0.084		
DDA-independence												
HSIC (target)	HSIC = 0.82, p = 0.009			HSIC = 0.39, p = 0.287			HSIC = 0.48, p = 0.164			HSIC = 0.34, p = 0.369		
HSIC (alternative)	HSIC = 0.50, p = 0.130			HSIC = 0.29, p = 0.518			HSIC = 0.31, p = 0.463			HSIC = 0.25, p = 0.626		
DDA-error distr.												
D'Agostino test:	γ = 0.21, p = 0.014			γ = 0.24, p = 0.007			γ = 0.25, p = 0.005			γ = 0.24, p = 0.008		
95% CI skew diff:	Δ = -0.15, [-0.27; -0.02]			Δ = -0.23, [-0.38; -0.13]			Δ = -0.19, [-0.32; -0.06]			Δ = -0.23, [-0.39; -0.12]		
DDA-variable distr.												
D'Agostino test:	γ = 0.22, p = 0.009			γ = 0.24, p = 0.008			γ = 0.27, p = 0.003			γ = 0.24, p = 0.008		
95% CI skew diff:	Δ = -0.13, [-0.26; 0.01]			Δ = -0.23, [-0.39; -0.13]			Δ = -0.19, [-0.33; -0.05]			Δ = -0.23, [-0.37; -0.13]		
95% HOC diff:	Δ = 0.00, [-0.01; 0.01]			Δ = 0.00, [-0.01; 0.01]			Δ = 0.00, [-0.01; 0.01]			Δ = 0.00, [-0.01; 0.01]		
DDA decision:	ADHD → BFD			(Weak) ADHD → BFD			(Weak) ADHD → BFD			(Weak) ADHD → BFD		

Factor I represents mother's age at child birth, mother's education, father's education, and parents occupational scores; Factor II represents primary family income; BFD = breastfeeding duration; HOC = Higher Order Correlation; HSIC = Hilbert–Schmidt Independence Criterion; γ = skewness.

Table 12.6 DDA results for covariate-adjusted models of the form *parent-rated ADHD → breastfeeding* based on the entire sample (n = 827).

Variables	Model I Est.	95% CI Lower	Upper	Model II Est.	95% CI Lower	Upper	Model III Est.	95% CI Lower	Upper	Model IV Est.	95% CI Lower	Upper
Parent-rated ADHD	-0.02	-0.02	-0.01	-0.01	-0.02	0.00	-0.01	-0.02	-0.01	-0.01	-0.01	0.00
Gender	0.02	-0.19	0.23	0.13	-0.08	0.34	0.06	-0.16	0.28	0.11	-0.10	0.32
Age	0.00	-0.07	0.06	0.01	-0.06	0.07	-0.02	-0.09	0.04	0.01	-0.06	0.07
Race: White	0.33	0.04	0.61	0.05	-0.02	0.12	0.07	-0.01	0.14	0.05	-0.03	0.12
Factor I	—	—	—	0.50	0.37	0.63	—	—	—	0.48	0.35	0.61
Factor II	—	—	—	—	—	—	0.14	0.03	0.25	0.10	-0.01	0.20
R^2	0.040			0.101						0.105		
DDA-independence												
HSIC (target)	HSIC = 0.50, p = 0.130			HSIC = 0.29, p = 0.518			HSIC = 0.30, p = 0.462			HSIC = 0.24, p = 0.626		
HSIC (alternative)	HSIC = 0.82, p = 0.009			HSIC = 0.39, p = 0.287			HSIC = 0.47, p = 0.163			HSIC = 0.34, p = 0.386		
DDA-error distr.												
D'Agostino test:	$\gamma = -0.06, p = 0.455$			$\gamma = -0.01, p = 0.837$			$\gamma = -0.06, p = 0.502$			$\gamma = -0.02, p = 0.868$		
95% CI skew diff:	$\Delta = 0.15, [0.02; 0.27]$			$\Delta = 0.23, [0.13; 0.41]$			$\Delta = 0.19, [0.05; 0.34]$			$\Delta = 0.23, [0.13; 0.40]$		
DDA-variable distr.												
D'Agostino test:	$\gamma = -0.09, p = 0.273$			$\gamma = -0.01, p = 0.877$			$\gamma = -0.08, p = 0.365$			$\gamma = -0.01, p = 0.858$		
95% CI skew diff:	$\Delta = 0.13, [-0.01; 0.25]$			$\Delta = 0.23, [0.13; 0.38]$			$\Delta = 0.19, [0.04; 0.34]$			$\Delta = 0.23, [0.12; 0.40]$		
95% HOC diff:	$\Delta = 0.00, [-0.01; 0.01]$			$\Delta = 0.00, [-0.01; 0.01]$			$\Delta = 0.00, [-0.02; 0.01]$			$\Delta = 0.00, [-0.01; 0.01]$		
DDA Decision:	ADHD → BFD			(Weak) ADHD → BFD			(Weak) ADHD → BFD			(Weak) ADHD → BFD		

Factor I represents mother's age at child birth, mother's education, father's education, and parents occupational scores; Factor II represents primary family income; BFD = breastfeeding duration; HOC = Higher Order Correlation; HSIC = Hilbert–Schmidt Independence Criterion; γ = skewness.

Table 12.7 DDA results for covariate-adjusted models of the form *breastfeeding → parent-rated ADHD* based on the subsample with ADHD symptomology (i.e. ADHD scores > 0; $n = 763$).

Variables	Model I Est.	95% CI Lower	Upper	Model II Est.	95% CI Lower	Upper	Model III Est.	95% CI Lower	Upper	Model IV Est.	95% CI Lower	Upper
Breastfeeding	-1.66	-2.34	-0.99	-0.92	-1.71	-0.13	-1.28	-2.05	-0.51	-0.86	-1.65	-0.06
Gender	-4.87	-6.94	-2.80	-4.97	-7.20	-2.73	-4.53	-6.79	-2.27	-4.75	-6.99	-2.51
Age	-0.01	-0.66	0.64	-0.34	-1.05	0.37	-0.19	-0.90	0.52	-0.36	-1.07	0.35
Race: White	0.17	-2.60	2.95	0.10	-3.01	3.21	-0.10	-3.24	3.04	0.30	-2.81	3.42
Factor I	—	—	—	-2.96	-4.39	-1.53	—	—	—	-2.82	-4.25	-1.39
Factor II	—	—	—	—	—	—	-1.33	-2.45	-0.21	-1.12	-2.23	0.00
R^2	0.058			0.070			0.054			0.076		
DDA-independence												
HSIC (target)	HSIC = 0.73, $p = 0.020$			HSIC = 0.40, $p = 0.267$			HSIC = 0.44, $p = 0.211$			HSIC = 0.34, $p = 0.392$		
HSIC (alternative)	HSIC = 0.65, $p = 0.040$			HSIC = 0.36, $p = 0.336$			HSIC = 0.38, $p = 0.306$			HSIC = 0.30, $p = 0.495$		
DDA-error distr.												
D'Agostino test:	$\gamma = 0.15, p = 0.086$			$\gamma = 0.18, p = 0.054$			$\gamma = 0.20, p = 0.041$			$\gamma = 0.18, p = 0.055$		
95% CI skew diff:	$\Delta = -0.09, [-0.24; 0.02]$			$\Delta = -0.17, [-0.33; -0.06]$			$\Delta = -0.13, [-0.28; 0.01]$			$\Delta = -0.16, [-0.32; -0.05]$		
DDA-variable distr.												
D'Agostino test:	$\gamma = 0.15, p = 0.079$			$\gamma = 0.17, p = 0.064$			$\gamma = 0.20, p = 0.034$			$\gamma = 0.17, p = 0.062$		
95% CI skew diff:	$\Delta = -0.06, [-0.19; 0.07]$			$\Delta = -0.16, [-0.32; -0.06]$			$\Delta = -0.12, [-0.27; 0.02]$			$\Delta = -0.15, [-0.30; -0.05]$		
95% HOC diff:	$\Delta = 0.01, [-0.07; 0.11]$			$\Delta = 0.00, [-0.01; 0.01]$			$\Delta = 0.03, [-0.02; 0.16]$			$\Delta = 0.03, [-0.01; 0.12]$		
DDA decision:	Undecided			(Weak) ADHD → BFD			Undecided			(Weak) ADHD → BFD		

Factor I represents mother's age at child birth, mother's education, father's education, and parents occupational scores; Factor II represents primary and parents occupational scores; Factor II represents primary HOC = Higher Order Correlation; HSIC = Hilbert Schmidt Independence Criterion; γ = skewness.

Table 12.6 DDA results for covariate-adjusted models of the form *parent-rated ADHD→ breastfeeding* based on the subsample with ADHD symptomology (i.e. ADHD scores > 0; n = 763).

Variables	Model I Est.	Model I 95% CI Lower	Model I 95% CI Upper	Model II Est.	Model II 95% CI Lower	Model II 95% CI Upper	Model III Est.	Model III 95% CI Lower	Model III 95% CI Upper	Model IV Est.	Model IV 95% CI Lower	Model IV 95% CI Upper
Parent-rated ADHD	-0.02	-0.03	-0.01	-0.01	-0.02	0.00	-0.01	-0.02	-0.01	-0.01	-0.02	0.00
Gender	0.00	-0.22	0.21	0.11	-0.11	0.33	0.04	-0.19	0.27	0.09	-0.13	0.31
Age	0.02	-0.05	0.08	0.01	-0.06	0.08	-0.01	-0.08	0.06	0.01	-0.05	0.08
Race: White	0.34	0.05	0.63	0.23	-0.08	0.53	0.29	-0.02	0.60	0.21	-0.09	0.51
Factor I	—	—	—	0.48	0.34	0.61	—	—	—	0.46	0.33	0.60
Factor II	—	—	—	—	—	—	0.14	0.03	0.25	0.10	-0.01	0.21
R^2	0.038			0.095			0.037			0.099		

DDA-independence

	Model I	Model II	Model III	Model IV
HSIC (target)	HSIC = 0.65, p = 0.040	HSIC = 0.36, p = 0.336	HSIC = 0.38, p = 0.306	HSIC = 0.29, p = 0.494
HSIC (alternative)	HSIC = 0.73, p = 0.020	HSIC = 0.40, p = 0.268	HSIC = 0.44, p = 0.211	HSIC = 0.34, p = 0.392

DDA-error distr.

	Model I	Model II	Model III	Model IV
D'Agostino test:	$\gamma = -0.05$, p = 0.534	$\gamma = -0.01$, p = 0.874	$\gamma = -0.06$, p = 0.538	$\gamma = -0.02$, p = 0.851
95% CI skew diff:	Δ = 0.10, [-0.03; 0.22]	Δ = 0.17, [0.06; 0.35]	Δ = 0.14, [-0.01; 0.28]	Δ = 0.17, [0.06; 0.31]

DDA-variable distr.

	Model I	Model II	Model III	Model IV
D'Agostino test:	$\gamma = -0.09$, p = 0.303	$\gamma = -0.02$, p = 0.878	$\gamma = -0.08$, p = 0.384	$\gamma = -0.02$, p = 0.838
95% CI skew diff:	Δ = 0.06, [-0.06; 0.21]	Δ = 0.16, [0.06; 0.34]	Δ = 0.12, [-0.02; 0.26]	Δ = 0.16, [0.05; 0.31]
95% HOC diff:	Δ = -0.01, [-0.02; 0.01]	Δ = 0.00, [-0.01; 0.01]	Δ = -0.01, [-0.15; 0.03]	Δ = 0.00, [-0.01; 0.01]
DDA decision:	Undecided	(Weak) ADHD → BFD	Undecided	(Weak) ADHD → BFD

Factor I represents mother's age at child birth, mother's education, father's education, and parents occupational scores; Factor II represents primary family income; BFD = breastfeeding duration; HOC = Higher Order Correlation; HSIC = Hilbert–Schmidt Independence Criterion; γ = skewness.

Table 12.9 DDA results for covariate-adjusted models of the form *breastfeeding → teacher-rated ADHD* based on the entire sample ($n = 802$).

Variables	Model I Est.	95% CI Lower	95% CI Upper	Model II Est.	95% CI Lower	95% CI Upper	Model III Est.	95% CI Lower	95% CI Upper	Model IV Est.	95% CI Lower	95% CI Upper
Breastfeeding	-1.62	-2.26	-0.98	-0.90	-1.63	-0.17	-1.26	-1.97	-0.54	-0.86	-1.59	-0.12
Gender	-8.18	-10.13	-6.24	-8.50	-10.56	-6.44	-8.18	-10.26	-6.10	-8.36	-10.42	-6.29
Age	-0.65	-1.26	-0.04	-0.85	-1.51	-0.20	-0.72	-1.38	-0.07	-0.87	-1.52	-0.22
Race: White	-1.97	-4.67	0.74	-0.69	-3.65	2.28	-0.88	-3.87	2.11	-0.56	-3.53	2.40
Factor I	—	—	—	-2.64	-3.96	-1.32	—	—	—	-2.53	-3.86	-1.21
Factor II	—	—	—	—	—	—	-1.04	-2.09	0.01	-0.84	-1.89	0.20
R^2	0.116			0.129			0.115			0.132		
DDA-independence												
HSIC (Target)	HSIC = 1.11, $p < 0.001$			HSIC = 0.57, $p = 0.074$			HSIC = 0.87, $p = 0.005$			HSIC = 0.54, $p = 0.098$		
HSIC (alternative)	HSIC = 0.58, $p = 0.070$			HSIC = 0.41, $p = 0.244$			HSIC = 0.52, $p = 0.115$			HSIC = 0.39, $p = 0.295$		
DDA-error distr.												
D'Agostino test:	$\gamma = 0.39$, $p < 0.001$			$\gamma = 0.40$, $p < 0.001$			$\gamma = 0.40$, $p < 0.001$			$\gamma = 0.40$, $p < 0.001$		
95% CI skew diff:	$\Delta = -0.34$, [-0.49; -0.20]			$\Delta = -0.38$, [-0.56; -0.27]			$\Delta = -0.32$, [-0.48; -0.18]			$\Delta = -0.38$, [-0.54; -0.26]		
DDA-variable distr.												
D'Agostino test:	$\gamma = 0.40$, $p < 0.001$			$\gamma = 0.41$, $p < 0.001$			$\gamma = 0.41$, $p < 0.001$			$\gamma = 0.41$, $p < 0.001$		
95% CI skew diff:	$\Delta = -0.32$, [-0.45; -0.18]			$\Delta = -0.39$, [-0.55; -0.29]			$\Delta = -0.33$, [-0.48; -0.19]			$\Delta = -0.39$, [-0.55; -0.27]		
95% HOC diff:	$\Delta = 0.00$, [-0.01; 0.01]			$\Delta = 0.00$, [-0.02; 0.01]			$\Delta = -0.01$, [-0.03; 0.01]			$\Delta = -0.01$, [-0.08; 0.03]		
DDA decision:	ADHD → BFD			ADHD → BFD			ADHD → BFD			(Weak) ADHD → BFD		

Factor I represents mother's age at child birth, mother's education, father's education, and parents occupational scores; Factor II represents primary ... Hilbert–Schmidt Independence Criterion; γ = skewness)

...results for covariate adjusted models of the form *teacher-rated ADHD → breastfeeding* based on the entire sample (n = 802).

Variables	Model I			Model II			Model III			Model IV		
		95% CI			95% CI			95% CI			95% CI	
	Est.	Lower	Upper	Est.	Lower	Upper	Est.	Lower	Upper	Est.	Lower	Upper
Parent-rated ADHD	-0.02	-0.03	-0.01	-0.01	-0.02	-0.01	-0.01	-0.02	-0.01	-0.01	-0.02	-0.01
Gender	-0.02	-0.23	0.20	0.10	-0.12	0.32	0.03	-0.20	0.25	0.09	-0.13	0.31
Age	-0.02	-0.08	0.05	-0.01	-0.08	0.06	-0.04	-0.11	0.03	-0.01	-0.08	0.06
Race: White	0.33	0.04	0.62	0.22	-0.08	0.52	0.28	-0.03	0.59	0.20	-0.10	0.50
Factor I	—	—	—	0.50	0.37	0.63	—	—	—	0.48	0.35	0.62
Factor II	—	—	—	—	—	—	0.14	0.03	0.25	0.10	-0.01	0.20
R^2	0.040			0.103			0.038			0.107		
DDA-independence												
HSIC (target)	HSIC = 0.58, p = 0.070			HSIC = 0.41, p = 0.244			HSIC = 0.52, p = 0.115			HSIC = 0.39, p = 0.295		
HSIC (alternative)	HSIC = 1.11, p < 0.001			HSIC = 0.57, p = 0.074			HSIC = 0.88, p = 0.005			HSIC = 0.54, p = 0.098		
DDA-error distr.												
D'Agostino test:	γ = -0.05, p = 0.391			γ = -0.02, p = 0.832			γ = -0.07, p = 0.431			γ = -0.02, p = 0.810		
95% CI skew diff:	Δ = 0.34, [0.21; 0.49]			Δ = 0.39, [0.27; 0.56]			Δ = 0.32, [0.18; 0.47]			Δ = 0.38, [0.25; 0.54]		
DDA-variable distr.												
D'Agostino test:	γ = -0.07, p = 0.367			γ = -0.02, p = 0.871			γ = -0.08, p = 0.374			γ = -0.02, p = 0.832		
95% CI skew diff:	Δ = 0.32, [0.20; 0.46]			Δ = 0.40, [0.28; 0.58]			Δ = 0.33, [0.19; 0.47]			Δ = 0.39, [0.28; 0.55]		
95% HOC diff:	Δ = 0.09, [0.01; 0.20]			Δ = 0.01, [-0.05; 0.07]			Δ = 0.00, [-0.01; 0.02]			Δ = 0.01, [-0.04; 0.09]		
DDA decision:	ADHD → BFD			ADHD → BFD			ADHD → BFD			(Weak) ADHD → BFD		

Factor I represents mother's age at child birth, mother's education, father's education, and parents occupational scores; Factor II represents primary family income; BFD = breastfeeding duration; HOC = Higher Order Correlation; HSIC = Hilbert–Schmidt Independence Criterion; γ = skewness.

Table 12.11 DDA results for covariate-adjusted models of the form *breastfeeding → teacher-rated ADHD* based on the subsample with ADHD symptomology (i.e. ADHD scores > 0; $n = 714$).

	Model I			Model II			Model III			Model IV		
		95% CI			95% CI			95% CI			95% CI	
Variables	Est.	Lower	Upper	Est.	Lower	Upper	Est.	Lower	Upper	Est.	Lower	Upper
Breastfeeding	−1.48	−2.13	−0.83	−0.76	−1.52	−0.01	−1.16	−1.89	−0.43	−0.74	−1.49	0.02
Gender	−6.91	−8.96	−4.86	−7.14	−9.33	−4.96	−6.88	−9.09	−4.67	−7.08	−9.27	−4.89
Age	−0.66	−1.30	−0.02	−0.90	−1.59	−0.21	−0.75	−1.44	−0.06	−0.91	−1.60	−0.22
Race: White	−1.82	−4.59	0.96	−0.68	−3.76	2.39	−0.89	−4.00	2.23	−0.57	−3.65	2.52
Factor I	—	—	—	−2.80	−4.17	−1.42	—	—	—	−2.72	−4.11	−1.33
Factor II	—	—	—	—	—	—	−0.76	−1.83	0.31	−0.54	−1.60	0.53
R^2	0.094			0.109			0.088			0.110		
DDA-independence												
HSIC (target)	HSIC = 0.87, p = 0.005			HSIC = 0.53, p = 0.105			HSIC = 0.73, p = 0.021			HSIC = 0.51, p = 0.125		
HSIC (alternative)	HSIC = 0.52, p = 0.113			HSIC = 0.44, p = 0.200			HSIC = 0.50, p = 0.129			HSIC = 0.43, p = 0.221		
DDA-error distr.												
D'Agostino test:	γ = 0.30, p = 0.002			γ = 0.31, p = 0.002			γ = 0.30, p = 0.004			γ = 0.30, p = 0.002		
95% CI skew diff:	Δ = −0.27, [−0.42; −0.15]			Δ = −0.30, [−0.47; −0.18]			Δ = −0.23, [−0.38; −0.09]			Δ = −0.30, [−0.48; −0.20]		
DDA-variable distr.												
D'Agostino test:	γ = 0.30, p = 0.001			γ = 0.32, p = 0.002			γ = 0.31, p = 0.002			γ = 0.31, p = 0.002		
95% CI skew diff:	Δ = −0.26, [−0.41; −0.14]			Δ = −0.30, [−0.45; −0.16]			Δ = −0.25, [−0.41; −0.12]			Δ = −0.30, [−0.47; −0.19]		
95% HOC diff:	Δ = −0.01, [−0.01; 0.01]			Δ = −0.01, [−0.11; 0.02]			Δ = −0.04, [−0.16; 0.01]			Δ = −0.01, [−0.11; 0.02]		
DDA decision:	ADHD → BFD			(Weak) ADHD → BFD			ADHD → BFD			(Weak) ADHD → BFD		

Factor I represents mother's age at child birth, mother's education, father's education, and parents occupational scores; Factor II represents primary family income; BFD = breastfeeding duration; HOC = Higher Order Correlation; HSIC = Hilbert–Schmidt Independence Criterion; γ = skewness.

Table 11.12 DDA results for covariate-adjusted models of the form *teacher-rated ADHD → breastfeeding* based on the subsample with ADHD symptomology (i.e. ADHD scores > 0; n = 714).

	Model I			Model II			Model III			Model IV		
		95% CI			95% CI			95% CI			95% CI	
Variables	Est.	Lower	Upper	Est.	Lower	Upper	Est.	Lower	Upper	Est.	Lower	Upper
Parent-rated ADHD	-0.02	-0.03	-0.01	-0.01	-0.02	-0.01	-0.01	-0.02	-0.01	-0.01	-0.02	0.00
Gender	-0.03	-0.27	0.20	0.11	-0.13	0.34	0.04	-0.21	0.28	0.10	-0.14	0.34
Age	0.00	-0.07	0.08	0.01	-0.07	0.08	-0.03	-0.10	0.05	0.01	-0.06	0.08
Race: White	0.36	0.05	0.67	0.24	-0.08	0.57	0.30	-0.03	0.64	0.23	-0.10	0.55
Factor I	—	—	—	0.51	0.37	0.65	—	—	—	0.50	0.36	0.64
Factor II	—	—	—	—	—	—	0.13	0.01	0.24	0.08	-0.03	0.19
R^2	0.037			0.101			0.034			0.104		
DDA-independence												
HSIC (target)	HSIC = 0.52, p = 0.113			HSIC = 0.44, p = 0.200			HSIC = 0.50, p = 0.130			HSIC = 0.43, p = 0.221		
HSIC (alternative)	HSIC = 0.87, p = 0.005			HSIC = 0.53, p = 0.105			HSIC = 0.73, p = 0.021			HSIC = 0.51, p = 0.125		
DDA-error distr.												
D'Agostino test:	γ = -0.02, p = 0.808			γ = 0.01, p = 0.916			γ = -0.06, p = 0.556			γ = 0.01, p = 0.958		
95% CI skew diff:	Δ = 0.27, [0.16; 0.43]			Δ = 0.29, [0.19; 0.47]			Δ = 0.23, [0.09; 0.39]			Δ = 0.30, [0.20; 0.49]		
DDA-variable distr.												
D'Agostino test:	γ = -0.03, p = 0.301			γ = 0.02, p = 0.842			γ = -0.06, p = 0.557			γ = 0.01, p = 0.893		
95% CI skew diff:	Δ = 0.26, [0.14; 0.40]			Δ = 0.30, [0.17; 0.47]			Δ = 0.25, [0.10; 0.38]			Δ = 0.30, [0.19; 0.49]		
95% HOC diff:	Δ = 0.10, [0.03; 0.24]			Δ = 0.01, [-0.02; 0.11]			Δ = 0.01, [-0.01; 0.03]			Δ = 0.00, [-0.01; 0.01]		
DDA decision:	ADHD → BFD			(Weak) ADHD → BFD			ADHD → BFD			(Weak) ADHD → BFD		

Factor I represents mother's age at child birth, mother's education, father's education, and parents occupational scores; Factor II represents primary family income; BFD = breastfeeding duration; HOC = Higher Order Correlation; HSIC = Hilbert Schmidt Independence Criterion; γ = skewness.

Further, 95% bootstrap CIs of skewness differences are non-significant for observed variable distributions and model residuals. Weak evidence for the causal model *ADHD → breastfeeding* is obtained for distribution-based DDA measures for Models II and IV. Although HSIC tests are, again, non-significant in both causal models, 95% bootstrap CIs of skewness differences for observed variables and model residuals suggest that the effect is more likely to be transmitted from ADHD to breastfeeding.

To account for potential reporter effects, direction dependence modeling was repeated using teacher-based ADHD ratings. Tables 12.9 and 12.10 summarize DDA results for the total sample ($n = 802$). Here, the mechanism *ADHD → breastfeeding* is selected as the causally dominant model independent of covariate adjustment. In other words, for all estimated models (Models I–IV) we observe that the focal predictor and model residuals tend to be (more) independent when using teacher-rated ADHD as the predictor and breastfeeding duration as the outcome. Further, outcome scores and model residuals tend to be closer to normality under *ADHD → breastfeeding* than under the causally competing model.

Finally, to evaluate the robustness of the teacher-based results, we, again, repeated DDA after temporarily deleting cases without any ADHD symptomology (subgroup analysis is based on 89.0% of the total sample; cf. Tables 12.11 and 12.12). Again, removing these cases decreased the magnitude of skewness of teacher-rated ADHD from 0.50 to 0.40. Here, no changes in DDA decisions were observed for Models I and III, that is, all three DDA components suggest that child ADHD is more likely to influence breastfeeding duration and not vice versa. For Models II and IV (i.e. after incorporating Factor I), HSIC tests were again non-significant in both models and, therefore, suggesting only weak evidence for the causal model *ADHD → breastfeeding*.

12.3 Discussion

Breastfeeding is one of the most widely studied and important influences not only on infant health but on early cortical development and thus, likely on self-regulation and neurodevelopmental conditions. While it may be a proxy for a number of other variables – including socio-economic status or parental psychopathology, and while its mechanisms may be biological (protective effects of breast milk; e.g. via long chain polyunsaturated fatty acids; for a review see (Koletzko et al., 2008)) or psychological (maternal-child contact), establishing its association is necessary to flesh out these diverse models of early emergence of ADHD risk.

Here, we confirm in one of the largest, well characterized samples of children with ADHD to date that this association is robust. However, we then add causal analysis using new methods of evaluating association within the linear model framework in correlational data.

Those results, interestingly, point primarily toward the argument that child ADHD is the driver of reduced breastfeeding, and not the other way around. This result has been confirmed for two different ADHD measures (using both, parent and teacher-based ADHD ratings) and proved quite robust under various scenarios of covariate adjustment. Although some covariate adjustment strategies led to weaker causal evidence for the model *ADHD → breastfeeding*, none of the in total 32 estimated models pointed at the causally reversed mechanism *breastfeeding → ADHD*. Interestingly, results of HSIC tests were ambiguous after adjusting for educational and financial characteristics. This effect was more pronounced for parent-rated ADHD scores. One possible explanation may be that the factor scores (Factor I and II) possess characteristics similar to a collider (Elwert & Winship, 2014), i.e. a construct that is a common effect of ADHD and breastfeeding (or causal descendants thereof).

12.3.1 Limitations

Our primary limitation was reliance on retrospective recall of breastfeeding practice through use of a single questionnaire item to assess initiation and duration of breastfeeding; the present study also did not examine the duration of *exclusive* breastfeeding. However, prior literature suggests that maternal recall of breastfeeding initiation and duration is reasonably accurate, provided relatively small family sizes, as is the case in our sample (Li et al., 2005). Further, the present results converge with prior findings using other kinds of methods and samples and controlling for exclusivity of breastfeeding, but lacking our strong ADHD-related analysis. The authors also did not differentially record different kinds of breastfeeding such as partial breastfeeding (combined with some formula bottle feeding), feeding pumped (expressed) breast milk, and the use of breast milk obtained from breast milk banks, in relation to ADHD outcome. Nor did the authors explore reasons why breastfeeding was discontinued.

12.3.2 Question of Causality

It is important to reflect on whether the breastfeeding association is causal or is explained by effects such as passive gene-environment correlation (rGE). Although our covariance analysis strongly suggests that parental ADHD or "genetic loading" do not account for breastfeeding difficulties in children who will go on to ADHD, we cannot fully rule out unmeasured parental effects related to ADHD. Another possibility, however, is that child effects related to associated early developmental difficulties (difficult temperament, oral-motor delay) are in play. Our results were not explained by social disadvantage or parental education, even though these were associated with breastfeeding, suggesting that effects are not due merely to lack of opportunity to breastfeed.

Complex reasons are associated with the decision to initiate and continue breastfeeding. Direct, causally informative designs, such as sibling comparisons, are still needed to examine this potential effect, now that it has been supported using a retrospective approach. Causally informative designs of this nature have supported a causal association of low birth weight with ADHD (Thapar et al., 2009) but called into question the association of maternal smoking in pregnancy with ADHD (D'Onofrio et al., 2008; Thapar et al., 2009). It appears here that breastfeeding may also not be as directly causal in preventing ADHD as the strong correlational data have suggested.

At the same time, the emerging picture regarding ADHD and breastfeeding from our work here and the literature that preceded us raises other questions. If breastfeeding is causally associated with ADHD, its effect is likely to be part of a liability (or else susceptibility) by experience interplay (or more narrowly, genotype by environment interplay).

Examination of breastfeeding in gene × environment interaction studies will be of keen interest. Additional focus on early infant developmental risk factors for ADHD appears to be an important future direction for the field.

If difficulties in breastfeeding are a marker of child liability for ADHD, then it will be valuable to look for potential behavioral or biological markers of risk in the population of women who find breastfeeding difficult and to refer them to lactation consultants for alternative breastfeeding strategies and support. It may be that child oral motor delays, frequent/excessive feedings, early disinterest in breastfeeding or difficult temperament, sleep problems, or other markers of breastfeeding challenge may signal future ADHD risk that can be amenable to early intervention.

Acknowledgments

The authors report no conflicts of interest related to this work. This work was supported by NIMH R37MH59105 (Nigg).

References

Ahluwalia, I. B., Morrow, B., & Hsia, J. (2005). Why do women stop breastfeeding? Findings from the pregnancy risk assessment and monitoring system. *Pediatrics*, *116*(6), 1408–1412. doi:10.1542/peds.2005-0013

Al Hamed, J. H., Taha, A. Z., Sabra, A. A., & Bella, H. (2008). Attention deficit hyperactivity disorder (ADHD) among male primary school children in Dammam, Saudi Arabia: Prevalence and associated factors. *Journal Egypt Public Health Association*, *83*(3–4), 165–182.

Allen, N. B., Lewinsohn, P. M., & Seeley, J. R. (1998). Prenatal and perinatal influences on risk for psychopathology in childhood and adolescence. *Journal Development and Psychopathology*, *10*(3), 513–529.

American Academy of Family Physicians. (2007). Breastfeeding, family physicians supporting (position paper). Retrieved from http://www.aafp.org/about/policies/all/breastfeeding-support.html

American College of Obstetricians Gynecologists. (2007). Committee opinion no. 361: Breastfeeding: maternal and infant aspects. *Obstetrics & Gynecology Journal*, *109*(2 Pt 1), 479–480.

American Psychiatric Association. (2010). *Diagnostic and statistical manual of mental disorders. 4th Edition*. Author.

American Psychiatric Association. (2013). *Diagnostic and statistical manual of mental disorders. 5th Edition*. Author.

American Psychology Association. (2002). Ethical principles of psychologists and code of conduct. *American Psychology Association*, *57*(12), 1060–1073.

American Public Health Association. (2007). Policy statement database: A call to action on breastfeeding: A fundamental public health issue. *Policy number 200714. Policy Date: November 6, 2007* Retrieved from www.apha.org/advocacy/policy/policysearch/default.htm?id=1360

Anderson, J. W., Johnstone, B. M., & Remley, D. T. (1999). Breast-feeding and cognitive development: A meta-analysis. *American Journal of Clinical Nutrition*, *70*(4), 525–535.

Angelsen, N. K., Vik, T., Jacobsen, G., & Bakketeig, L. S. (2001). Breast feeding and cognitive development at age 1 and 5 years. *Archives of Disease in Childhood*, *85*(3), 183–188.

Ballard, O., & Morrow, A. L. (2013). Human milk composition: Nutrients and bioactive factors. *Pediatric Clinics of North America*, *60*(1), 49–74. doi:10.1016/j.pcl.2012.10.002

Barkley, R. A. (2011). *Barkley adult ADHD rating scale-IV (BAARS-IV)*. New York, NY: Guilford Press.

Center for Disease Control. (2018). Breastfeeding report card. Retrieved from http://www.cdc.gov/breastfeeding/data/reportcard.htm

Conners, K. C. (2008). *Conners* (3rd ed.). Toronto, ON: Multi-Health Systems.

Cupul-Uicab, L. A., Gladen, B. C., Hernandez-Avila, M., & Longnecker, M. P. (2009). Reliability of reported breastfeeding duration among reproductive-aged women from Mexico. *Maternal & Child Nutrition*, *5*(2), 125–137.

D'Onofrio, B. M., Van Hulle, C. A., Waldman, I. D., Rodgers, J. L., Harden, K. P., Rathouz, P. J., & Lahey, B. B. (2008). Smoking during pregnancy and offspring externalizing problems: An exploration of genetic and environmental confounds. *Development and Psychopathology, 20*(1), 139–164. doi:10.1017/S0954579408000072

Dunn, G. A., Nigg, J. T., & Sullivan, E. L. (2019). Neuroinflammation as a risk factor for attention deficit hyperactivity disorder. *Pharmacology Biochemistry and Behavior, 182*, 22–34. doi:10.1016/j.pbb.2019.05.005

DuPaul, G. J., Power, T. J., Anastopoulos, A. D., & Reid, R. (1998). *ADHD rating scale-IV: Checklists, normsn, and clinical cinterpretationi*. New York, NY: Guilford Press.

Eidelman, A. I., & Schanler, R. J. (2012). Breastfeeding and the use of human milk. *Pediatrics, 129*(3), e827–e841. doi:10.1542/peds.2011-3552

Elwert, F., & Winship, C. (2014). Endogenous selection bias: The problem of conditioning on a collider variable. *Annual Review of Sociology, 40*, 31–53. doi:10.1146/annurev-soc-071913-043455

Faraone, S. V., & Larsson, H. (2019). Genetics of attention deficit hyperactivity disorder. *Molecular Psychiatry, 24*(4), 562–575. doi:10.1038/s41380-018-0070-0

Field, S. S. (2014). Interaction of genes and nutritional factors in the etiology of autism and attention deficit/hyperactivity disorders: A case control study. *Journal of Medical Hypotheses, 82*(6), 654–661.

Forsyth, J. S., Willatts, P., Agostoni, C., Bissenden, J., Casaer, P., & Boehm, G. (2003). Long chain polyunsaturated fatty acid supplementation in infant formula and blood pressure in later childhood: Follow up of a randomised controlled trial. *British Medical Journal, 326*(7396), 953. doi:10.1136/bmj.326.7396.953

Foster, J. A. (2016). Gut microbiome and behavior: Focus on neuroimmune interactions. *International Review of Neurobiology, 131*, 49–65. doi:10.1016/bs.irn.2016.07.005

Gartner, L. M., Morton, J., Lawrence, R. A., Naylor, A. J., O'Hare, D., Schanler, R. J., & Eidelman, A. I. (2005). Breastfeeding and the use of human milk. *Pediatrics, 115*(2), 496–506.

Golmirzaei, J., Namazi, S., Amiri, S., Zare, S., Rastikerdar, N., Hesam, A. A., … Asadi, S. (2013). Evaluation of attention-deficit hyperactivity disorder risk factors. *International Journal of Clinical Pediatrics, 953103*(10), 953103.

Gretton, A., Fukumizu, K., Teo, C. H., Song, L., Schölkopf, B., & Smola, A. J. (2008). A kernel statistical test of independence. *Advances in Neural Information Processing Systems, 20*, 585–592.

Heinze, G., Wallisch, C., & Dunkler, D. (2018). Variable selection—A review and recommendations for the practicing statistician. *Biometrical Journal, 60*, 431–449. doi:10.1002/bimj.201700067

Herba, C. M., Roza, S., Govaert, P., Hofman, A., Jaddoe, V., Verhulst, F. C., & Tiemeier, H. (2013). Breastfeeding and early brain development: The Generation R study. *Maternal & Child Nutrition, 9*(3), 332–349. doi:10.1111/mcn.12015

Horn, J. L. (1965). A rationale and test for the number of factors in factor analysis. *Psychometrika, 30,* 179–185. doi:10.1007/bf02289447

Huang, J., Peters, K. E., Vaughn, M. G., & Witko, C. (2014). Breastfeeding and trajectories of children's cognitive development. *Journal Developmental Science, 17*(3), 452–461. doi:10.1111/desc.12136

Julvez, J., Ribas-Fito, N., Forns, M., Garcia-Esteban, R., Torrent, M., & Sunyer, J. (2007). Attention behaviour and hyperactivity at age 4 and duration of breast-feeding. *Acta Paediatrica, 96*(6), 842–847. doi:10.1111/j.1651-2227.2007.00273.x

Kadziela-Olech, H., & Piotrowska-Jastrzebska, J. (2005). The duration of breastfeeding and attention deficit hyperactivity disorder. *Rocz Akad Med Bialymst Journal, 50,* 302–306.

Kaplan, L. A., Evans, L., & Monk, C. (2008). Effects of mothers' prenatal psychiatric status and postnatal caregiving on infant biobehavioral regulation: Can prenatal programming be modified? *Early Human Development, 84*(4), 249–256.

Koletzko, B., Lien, E., Agostoni, C., Bohles, H., Campoy, C., Cetin, I., … World Association of Perinatal Medicine Dietary Guidelines Working Group. (2008). The roles of long-chain polyunsaturated fatty acids in pregnancy, lactation and infancy: Review of current knowledge and consensus recommendations. *Journal of Perinatal Medicine, 36*(1), 5–14. doi:10.1515/JPM.2008.001

Li, R., Fein, S. B., Chen, J., & Grummer-Strawn, L. M. (2008). Why mothers stop breastfeeding: Mothers' self-reported reasons for stopping during the first year. *Pediatrics, 122*(Suppl 2), S69–S76. doi:10.1542/peds.2008-1315i

Li, R., Scanlon, K. S., & Serdula, M. K. (2005). The validity and reliability of maternal recall of breastfeeding practice. *Nutrition Reviews, 63*(4), 103–110.

Mangrio, E., Persson, K., & Bramhagen, A. C. (2018). Sociodemographic, physical, mental and social factors in the cessation of breastfeeding before 6 months: A systematic review. *Scandinavian Journal of Caring Sciences, 32*(2), 451–465. doi:10.1111/scs.12489

Miller, L. L., Gustafsson, H. C., Tipsord, J., Song, M., Nousen, E., Dieckmann, N., & Nigg, J. T. (2018). Is the association of ADHD with socio-economic disadvantage explained by child comorbid externalizing problems or parent ADHD? *Journal of Abnormal Child Psychology, 46*(5), 951–963. doi:10.1007/s10802-017-0356-8

Mimouni-Bloch, A., Kachevanskaya, A., Mimouni, F. B., Shuper, A., Raveh, E., & Linder, N. (2013). Breastfeeding may protect from developing attention-deficit/hyperactivity disorder. *Breastfeed Medicine Journal, 8*(4), 363–367.

Nagel, B. J., Bathula, D., Herting, M., Schmitt, C., Kroenke, C. D., Fair, D., & Nigg, J. T. (2011). Altered white matter microstructure in children with attention-deficit/hyperactivity disorder. *Journal of the American Academy of Child and Adolescent Psychiatry, 50*(3), 283–292. doi:10.1016/j.jaac.2010.12.003

Natland, S. T., Andersen, L. F., Nilsen, T. I., Forsmo, S., & Jacobsen, G. W. (2012). Maternal recall of breastfeeding duration twenty years after delivery. *BMC Medical Research Methodology, 12*(179), 179.

Newby, R. M., & Davies, P. S. (2016). Why do women stop breast-feeding? Results from a contemporary prospective study in a cohort of Australian women. *European Journal of Clinical Nutrition, 70*(12), 1428–1432. doi:10.1038/ejcn.2016.157

Oddy, W. H., Li, J., Whitehouse, A. J., Zubrick, S. R., & Malacova, E. (2011). Breastfeeding duration and academic achievement at 10 years. *Pediatrics, 127*(1), e137–e145. doi:10.1542/peds.2009-3489

Odom, E. C., Li, R., Scanlon, K. S., Perrine, C. G., & Grummer-Strawn, L. (2013). Reasons for earlier than desired cessation of breastfeeding. *Pediatrics, 131*(3), e726–e732. doi:10.1542/peds.2012-1295

Ou, X., Andres, A., Cleves, M. A., Pivik, R. T., Snow, J. H., Ding, Z., & Badger, T. M. (2014). Sex-specific association between infant diet and white matter integrity in 8-y-old children. *Pediatric Research, 76*(6), 535–543. doi:10.1038/pr.2014.129

Puig-Antich, J. R., & Kiddie, N. (1996). *Schedule for affective disorders and schizophrenia*. Pittsburgh, PA: Western Psychiatric Institute.

Quigley, M. A., Hockley, C., Carson, C., Kelly, Y., Renfrew, M. J., & Sacker, A. (2012). Breastfeeding is associated with improved child cognitive development: A population-based cohort study. *The Journal of Pediatrics, 160*(1), 25–32. doi:10.1016/j.jpeds.2011.06.035

Sabuncuoglu, O., Orengul, C., Bikmazer, A., & Kaynar, S. Y. (2014). Breastfeeding and parafunctional oral habits in children with and without attention-deficit/hyperactivity disorder. *Breastfeed Medicine Journal, 9*, 244–250.

Schmitt, J., & Romanos, M. (2012). Prenatal and perinatal risk factors for attention-deficit/hyperactivity disorder. *Archives of Pediatrics and Adolescent Medicine, 166*(11), 1074–1075.

Shamberger, R. (2012). Attention-deficit disorder associated with breast-feeding: A brief report. *Journal of the American College of Nutrition, 31*(4), 239–242.

Stadler, D. D., Musser, E. D., Holton, K. F., Shannon, J., & Nigg, J. T. (2016). Recalled initiation and duration of maternal breastfeeding among children with and without ADHD in a well characterized case–control sample. *Journal of Abnormal Child Psychology, 44*(2), 347–355. doi:10.1007/s10802-015-9987-9

Steiner, P. M., Cook, T. D., Shadish, W. R., & Clark, M. H. (2010). The importance of covariate selection in controlling for selection bias in observational studies. *Psychological Methods, 15*(3), 250–267. doi:10.1037/a0018719

Sullivan, E. L., Holton, K. F., Nousen, E. K., Barling, A. N., Sullivan, C. A., Propper, C. B., & Nigg, J. T. (2015). Early identification of ADHD risk via infant temperament and emotion regulation: A pilot study. *Journal of Child Psychology and Psychiatry, and Allied Disciplines, 56*(9), 949–957. doi:10.1111/jcpp.12426

Tawia, S. (2013). Breastfeeding, brain structure and function, cognitive development and educational attainment. *Breastfeeding Review, 21*(3), 15–20.

Thapar, A., Rice, F., Hay, D., Boivin, J., Langley, K., van den Bree, M., … Harold, G. (2009). Prenatal smoking might not cause attention-deficit/hyperactivity disorder: Evidence from a novel design. *Biological Psychiatry, 66*(8), 722–727. doi:10.1016/j.biopsych.2009.05.032

Tseng, P. T., Yen, C. F., Chen, Y. W., Stubbs, B., Carvalho, A. F., Whiteley, P., … Lin, P. Y. (2019). Maternal breastfeeding and attention-deficit/hyperactivity disorder in children: A meta-analysis. *European Child and Adolescent Psychiatry, 28*(1), 19–30. doi:10.1007/s00787-018-1182-4

Wells, J. C., Chomtho, S., & Fewtrell, M. S. (2007). Programming of body composition by early growth and nutrition. *Proceedings of the Nutrition Society, 66*(3), 423–434.

Wiedermann, W., & Li, X. (2018). Direction dependence analysis: A framework to test the direction of effects in linear models with an implementation in SPSS. *Behavior Research Methods, 50*, 1581–1601. doi:10.3758/s13428-018-1031-x

Wiedermann, W., & Sebastian, J. (in press). Direction dependence analysis in the presence of confounders: Applications to linear mediation models. *Multivariate Behavioral Research*. doi:10.1080/00273171.2018.1528542

Wiedermann, W., & von Eye, A. (2015). Direction-dependence analysis: A confirmatory approach for testing directional theories. *International Journal of Behavioral Development, 39*, 570–580. doi:10.1177/0165025415582056

Wigg, N. R., Tong, S., McMichael, A. J., Baghurst, P. A., Vimpani, G., & Roberts, R. (1998). Does breastfeeding at six months predict cognitive development? *The Australian and New Zealand Journal of Public Health, 22*(2), 232–236.

Willatts, P., Forsyth, S., Agostoni, C., Casaer, P., Riva, E., & Boehm, G. (2013). Effects of long-chain PUFA supplementation in infant formula on cognitive function in later childhood. *The American Journal of Clinical Nutrition, 98*(2), 536s–542s. doi:10.3945/ajcn.112.038612

World Health Organization, & United Nations International Children's Emergency Fund. (2003). *Global strategy for infant and young child feeding*. Geneva, Switzerland: WHO.

Zeng, Y., Tang, Y., Tang, J., Shi, J., Zhang, L., Zhu, T., … Mu, D. (2018). Association between the different duration of breastfeeding and attention deficit/hyperactivity disorder in children: A systematic review and meta-analysis. *Nutritional Neuroscience*, 1–13. doi:10.1080/1028415x.2018.1560905

13

Direction of Effect Between Intimate Partner Violence and Mood Lability

A Granger Causality Model

G. Anne Bogat, Alytia A. Levendosky, Jade E. Kobayashi, and Alexander von Eye

Department of Psychology, Michigan State University, East Lansing, MI, USA

13.1 Introduction

Intimate partner violence (IPV) includes physical, sexual, and psychological violence directed from one romantic partner toward another. IPV is a major public health problem affecting large numbers of individuals in the United States. About 1 in 3 women and 1 in 4 men will experience some type of IPV over their lifetimes (Black et al., 2011). For most individuals, the frequency of IPV is highest during young adulthood and then declines (Johnson, Giordano, Manning, & Longmore, 2015; Shortt et al., 2012). For example, in one study, a significant number of women (69%) reported that their first experience of IPV occurred before age 25 (Catalano, 2012). And research generally shows that for young adults, IPV occurs in about 50% of their relationships (Renner & Whitney, 2012). Notably, there are similar rates of perpetration and victimization for young adult men and women (e.g. Langhinrichsen-Rohling, Misra, Selwyn, & Rohling, 2012).

Injury and health problems that result from IPV are significant (e.g. Brownridge, 2006; Kaura & Lohman, 2007). IPV has been associated with mental health problems for both men and women, including mood disorders such as depression and anxiety (Beydoun, Beydoun, Kaufman, Lo, & Zonderman, 2012; Coker et al., 2002). Studies also show greater mood lability (emotional dysregulation) among individuals in relationships characterized by IPV (Gratz, Paulson, Jakupcak, & Tull, 2009; McNulty & Hellmuth, 2008; Shorey, Brasfield, Febres, & Stuart, 2011, 2015). Depression/depressive symptoms and mood lability are related. Depressed individuals often demonstrate features of mood lability that include "tearfulness, irritability, brooding, [and] obsessive rumination" (American Psychiatric Association, 2013, p. 164).

Direction Dependence in Statistical Modeling: Methods of Analysis, First Edition.
Edited by Wolfgang Wiedermann, Daeyoung Kim, Engin A. Sungur, and Alexander von Eye.
© 2021 John Wiley & Sons, Inc. Published 2021 by John Wiley & Sons, Inc.

There are two types of studies that examine mood or depression and their relationship to IPV. The first category of research examines broad relationships between depression and IPV. Types or rates of IPV are assessed retrospectively over a specified length of time (e.g. past month, past six months) and depressive symptoms, usually in the last two weeks (the time frame for many measures of depression; e.g. the Beck Depression Inventory (Beck, Ward, Mendelson, Mock, & Erbaugh, 1961)). Thus, because of the manner in which depression and depressive symptoms are assessed (per DSM-5 diagnostic criteria), understanding the association between IPV and depression necessitates examining this relationship over a period of at least two weeks. Another aspect of this category of research is that it generally focuses on women, and it assumes that IPV victimization causes depression, not that depression causes IPV victimization. This is in keeping with the broader trauma literature that documents the effects of significant traumatic events on an individual's mental health (in this case, IPV is the traumatic event). And, in fact, several recent longitudinal studies suggest that this is the direction of the relationship (e.g. Ouellet-Morin et al., 2015).

The second category of research examines *specific incidents* of violence and what factors provoke the *perpetration*. Researchers measure a number of factors that they believe might be antecedents of abuse, including mood lability (i.e. emotion dysregulation). As noted above, mood lability is one marker of depression. In this literature, the assumption is made that mood lability increases the chance that an individual will perpetrate IPV. But to our knowledge, no studies have examined whether mood lability increases the chance of victimization or whether victimization increases the chance of mood lability (the latter being analogous to how the depression and IPV literature is conceptualized). Understanding the direction of the relationship has very specific implications for the types of interventions researchers develop to combat IPV.

The present research explored the direction of causation between mood lability and IPV victimization, using Granger causal analysis. Granger causal analysis requires multiple measurements of the same variables over time to determine cause and effect. Our data are an in-depth, micro-longitudinal examination of daily relationship violence among college students for 28 days. Using a web-based survey method (described below), we obtained daily information from 104 couples (208 individuals) about their experiences of IPV and their mood lability.

13.1.1 Definitions and Frequency of IPV

IPV consists of physical, psychological (emotional), and sexual actions from one partner to another. Physical violence involves any unwanted or threat of unwanted physical contact between two individuals. It can also involve the destruction of the property of one individual by another. Psychological violence encompasses a wide range of behaviors that are typically either coercive or controlling. Sexual violence includes any action in which an individual is forced or coerced to engage in an unwanted sexual act.

Early research on IPV focused solely on women as victims and men as perpetrators (Langhinrichsen-Rohling, 2010). In part, this was because feminist theory (power and control of men as it relates to perpetration of IPV) guided much of the early research. Researchers have questioned the validity and utility of this model on several grounds, including the relatively low effectiveness of interventions for perpetrators based on this model (e.g. Eckhardt et al., 2013). In fact, the vast majority of recent research finds that IPV is perpetrated by both men and women, and it is often bidirectional (e.g. Archer, 2000; Langhinrichsen-Rohling et al., 2012; Shorey et al., 2011). Studies examining whether men and women perpetrate different *types* of IPV at different rates are mixed (e.g. Archer, 2000; Tjaden & Thoennes, 2000); however, most research agrees that men perpetrate more sexual IPV (e.g. Hines & Saudino, 2003).

Most research on IPV has examined physical violence (see meta-analyses by Archer, 2000, 2002). Physical violence is serious, can reach the standard of criminal behavior, and can result in significant injury and trauma responses. Injury and health problems that result from IPV are significant (e.g. Brownridge, 2006; Kaura & Lohman, 2007). Between 15 and 53% of women (of all ages) report sustaining a physical injury from the abuse (e.g. Amar & Gennaro, 2005; Brownridge, 2006; Hines & Saudino, 2003); the average number of injuries ranges from 2.3 to 7.8 (Brownridge, 2006; Hines & Saudino, 2003). For example, in one study, 32% of women who experienced IPV reported a physical injury (Amar & Gennaro, 2005), ranging from minor (89%; e.g. sore muscles) to significant (13%; e.g. welts, black eyes). Rates of injury for male victims of IPV in opposite-gender relationships are consistently lower than female victims, due to lower comparative rates of IPV victimization overall. In community samples, 1–2% of men report injuries ranging from broken bones and teeth to injuries to sensory organs (e.g. eyes, nose; Cascardi, Langhinrichsen, & Vivian, 1992). Other studies have suggested that injuries due to IPV are underreported by men; men sustain up to one-third of injuries and one-third of deaths caused by IPV (Catalano, 2012; Straus, 2005). Data from a national hotline indicated that 20% of male victims reported severe violence that resulted in injury (e.g. scalded with boiling water, hurt with a knife; Hines, Brown, & Dunning, 2007). Rates of injury are more comparable for men and women when violence is bidirectional; one study found similar rates of emergency department visits for men and women who reported being both victims and perpetrators of IPV (Kothari et al., 2015).

However, physical IPV typically co-occurs with other types of IPV (e.g. Archer, 2000; Coker, Smith, McKeown, & King, 2000; Sullivan, McPartland, Armeli, Jaquier, & Tennen, 2012; Tjaden & Thoennes, 2000). The co-occurrence of additional types of IPV has been linked to more problematic mental and physical health outcomes for victims (Edwards, Black, Dhingra, McKnight-Eily, & Perry, 2009; Hedtke et al., 2008), but, on any given day, the co-occurrence differs for individuals (Sullivan et al., 2012). Thus, it is important to measure all types of abuse in order to capture an individual's experience of victimization over time.

Rates of IPV are highest in young adulthood and then, for most individuals, decline over time (Johnson et al., 2015; Shortt et al., 2012). For example, one study indicated that almost half of young adult men and women report some type and level of IPV in their romantic relationships (Renner & Whitney, 2012). However, the rates of IPV differ according to the type of IPV perpetrated. Although the rates of IPV are higher in young adult romantic couples, it is also the case, compared with older adults, that the aggression that occurs tends to be milder (Holtzworth Munroe & Stuart, 1994; Woodin, Caldeira, & O'Leary, 2013). Psychological IPV is more common than physical IPV across all age groups (Coker et al., 2000), but it occurs at an even higher rate in young adults (e.g. Scott & Straus, 2007). For example, Shorey, Cornelius, and Bell (2008) note that numerous studies find that, among dating couples, psychological aggression is most common, followed by physical violence, and then sexual violence. Importantly, rates of victimization and perpetration between men and women within this age group tend to be quite similar (e.g. Langhinrichsen-Rohling et al., 2012).

Studies examining college students, specifically, find that IPV occurs with great frequency (Follingstad, Bradley, Laughlin, & Burke, 1999; Gover, Kaukinen, & Fox, 2008; Langhinrichsen-Rohling, 2010; Straus, 2008). Between 20 and 36% of college women and men experience or perpetrate physical IPV (Arias, Samios, & O'Leary, 1987; White & Koss, 1991); again, psychological IPV is even more prevalent (e.g. Eshelman & Levendosky, 2012; Fossos, Neighbors, Kaysen, & Hove, 2007; Neufield, McNamera, & Ertl, 1999). However, even though rates of IPV among college student men and women are similar, research finds that women rate these behaviors as more abusive than do men (Dardis, Edwards, Kelley, & Gidycz, 2017), perhaps, in part, because women suffer more serious consequences (see meta-analyses by Archer (2000, 2002)).

Importantly, rates of IPV perpetration and victimization differ according to the timeframe examined. Typical measurement of IPV is cross-sectional and occurs by asking individuals to report how often particular events occurred in a prior time period – usually over the past month or six months (Bogat, Levendosky, DeJonghe, Davidson, & von Eye, 2004; Bogat, Levendosky, Theran, von Eye, & Davidson, 2003). Relatively little prospective research has examined frequency of IPV over time. However, studies that use prospective methods find evidence for the continuation of violence committed by the same partner (e.g. Aldarondo, 1996; Capaldi, Shortt, & Crosby, 2003; Feld & Straus, 1989; Mihalic, Elliott, & Menard, 1994), the escalation of same partner violence over time (e.g. Capaldi et al., 2003;

Walker, 1984), as well as violence that escalates early in the relationship and then remains stable (e.g. Follingstad, Laughlin, Polek, Rutledge, & Hause, 1991), especially when the violence is severe (Feld & Straus, 1989; Wofford, Elliott, & Menard, 1994). The variability in IPV frequency is striking when it is assessed more often, with shorter times between assessments (e.g. Gondolf & Beeman, 2003); thus daily diary studies are important methods to capture the occurrence of IPV. In the present research, assessing IPV over the course of 28 days allowed for a prospective assessment of the frequency of IPV and factors associated with it.

13.1.2 Depression, Mood and IPV

As noted earlier, there are two types of research examining mood and IPV. The first examines depressive symptoms or diagnoses and their association with IPV. These studies focus on overall mental health problems, such as depression or depressive symptoms, that result from IPV. The implication being that IPV causes depression. The second category of studies examines specific incidents of IPV and whether these are *instigated* by mood. That is, research asks the question about whether lability of mood leads to IPV perpetration. However, regarding the second category, we argue that it is important to examine whether specific incidents of IPV *victimization* result in changes to mood or whether mood results in incidents of IPV *victimization*.

13.1.2.1 Depression and IPV

Most research on the association of depression and IPV has focused on women and their experience of IPV victimization. The prevalence rates for depressive symptoms in women experiencing IPV are quite high, ranging from 30% for community samples (Beydoun et al., 2012), 55% for clinical samples (Beydoun et al., 2012), and 78% for college samples (Eshelman & Levendosky, 2012). For example, in a meta-analysis by Beydoun et al. (2012), the rates of depressive symptoms in women experiencing IPV were two times higher than women who were not being abused, and the rates of major depressive disorder were three times higher.

Depression and depressive symptoms are also present among male victims of IPV (Cascardi et al., 1992; Hines & Saudino, 2003; Simonelli & Ingram, 1998; Stets, Straus, & Smith, 1990). Authors often provide the caveat that men may underreport depressive symptoms (Chuick et al., 2009) and that depression may exhibit differently for men than for women (e.g. alcohol abuse; anger – Cochran & Rabinowitz, 2000). Coker et al. (2002) emphasized that for men, the type of abuse has different associations with depression. These authors noted that psychological abuse, compared to other types of abuse, has the highest association with depression for men.

Early research on depression and IPV took either an explicit or implicit position that IPV caused depression. There were some researchers who questioned this, suggesting there were reasons to think depression led to IPV. First, research showed that some women in violent relationships were depressed prior to entering the relationship (e.g. Moffitt & Caspi, 1998). Second, women with mental health problems, such as depression, might be more likely to enter into violent relationships as a result of factors such as low self-esteem (e.g. Cascardi & O'Leary, 1992). However, recent research, as we have shown, indicates IPV victimization is common for both men and women. Evidence for whether IPV victimization is associated with depression for both men and women is somewhat mixed. Graham, Bernards, Flynn, Tremblay, and Wells (2012) found that men were more likely to be depressed when they perpetrated IPV whereas women more likely to be depressed when they were victimized. However, another study found that men and women who experience IPV had the same rates of depression (Follingstad, Wright, Lloyd, & Sebastian, 1991).

There has been research that addresses the question of directionality directly. An early, large-scale epidemiological study that obtained retrospective accounts of mental health problems found that women had no overall pattern indicating that *premarital* depression predicted subsequent IPV (Kessler, Molnar, Feurer, & Appelbaum, 2001). Thus, these researchers concluded that IPV victimization leads to women's depression. A systematic review of longitudinal studies was inconclusive as to whether IPV victimization caused depression (Devries et al., 2013). However, a seven-year prospective longitudinal study, that accounted for important covariates (e.g. history of child maltreatment), found that "women who experienced partner violence were more likely to report new-onset depression compared to those who had not, both at the end of the partner violence reporting period … and 2 years later" (Ouellet-Morin et al., 2015; p. 319).

13.1.2.2 Mood and IPV

The research on mood and IPV is quite different from that on depression and IPV – it focuses on specific episodes of IPV and assesses both men and women. In this research, mood is considered a potential "trigger" for IPV, and it is equally likely to serve as an instigator for both male and female *perpetration*.

For example, Finkel (e.g. Finkel, 2007, 2008, 2014; Finkel & Slotter, 2009; Finkel et al., 2012) argues that emotional conflict in romantic relationships is normal because of the deep investment that each member has in the other. However, in some individuals, this emotion turns to violence and in others it does not. Yet, even so, on any given day, most individuals do not necessarily engage in violence. The question then becomes – in which situations do couples' interactions lead to violence? According to Finkel (2007), there are factors that inhibit violence and factors that instigate it. Instigating factors provide cues to the partner that provoke or trigger aggression. Lability of mood or affective lability can be one of those cues. Thus, although depression is related to mood lability, the research on mood and IPV focuses on how it leads to *perpetration*, not victimization.

In this chapter, we argue that understanding specific incidents of IPV victimization and their relationship to mood is also important. It is surprising that research has not examined this. By way of definitions – affective lability is frequent change in mood and is related to various forms of psychopathology such as depression (e.g. Berking, Wirtz, Svaldi, & Hofmann, 2014; Joormann & Gotlib, 2010). Research on mood and IPV focuses on emotional dysregulation. Mood lability and emotional dysregulation are synonyms. Emotional *regulation* is conceptualized as the manner in which emotions are managed, whether that is an automatic or an effortful, controlled process (Gross & Thompson, 2007). In order to manage mood, individuals have to be aware of their emotions, control their negative emotions, and then find the appropriate emotional strategies to respond to a situation (Gratz & Roemer, 2004).

We propose that mood lability is likely to influence victimization. Interestingly, in a study of couples, women's emotional dysregulation had a stronger effect on men than men's dysregulation had on women (Lee, Rodriguez, Edwards, & Neal, 2019). In other words, women's dysregulation was more likely to lead to their victimization than vice versa. Perhaps the perpetrator interprets the victim's mood lability as dissatisfaction with the relationship and thus becomes emotionally dysregulated and physically lashes out. In fact, some researchers have even suggested that IPV perpetration is a (problematic) strategy that some individuals use to regulate their emotions (e.g. Bushman, Baumeister, & Phillips, 2001; Jakupcak, Lisak, & Roemer, 2002).

In contrast, when one is victimized it might make one emotionally labile. This could occur for several reasons. For instance, research indicates that a significant number of women experiencing IPV also have insecure attachment styles (Hellemans, Loeys, Dewitte, De Smet, & Buysse, 2015; Kuijpers, Van Der Knaap, & Winkel, 2012; Oka, Sandberg, Bradford, & Brown, 2014) – unfortunately, we know less about men who are abused and their attachment styles. When IPV occurs, victims might perceive the abuse as proof that their internal working models (e.g. I am not a good person) are valid, resulting in negative mood states. A second reason for IPV leading to emotional lability is that the victim might have posttraumatic stress disorder or a significant number of posttraumatic stress symptoms. Again, research is mainly with women, but it shows that this is a common response to severe IPV (e.g. Black et al., 2011; Martinez-Torteya, Bogat, von Eye, Levendosky, & Davidson, 2009). Symptoms associated with PTSD involve negative alterations in mood. If the PTSD is the result of childhood maltreatment or prior IPV, the current experience of IPV may trigger mood lability again.

Unfortunately, the research on IPV and emotional dysregulation focuses mainly on perpetration. Research finds a positive correlation between IPV perpetration and emotional dysregulation (Gratz et al., 2009; McNulty & Hellmuth, 2008; Shorey et al., 2011, 2015; Stuart et al., 2006). One study implicated both individuals in the couple – when newlyweds were able to regulate their emotions, there was less physical IPV perpetration (McNulty & Hellmuth, 2008). Implicit assumptions are often made about the causal direction between the two variables (e.g. "motivations for IPV perpetration may partly stem from emotional dysregulation" (Bliton et al., 2016, p. 375)).

Recently, in an attempt to better understand specific incidents of IPV, researchers have begun to conduct daily diary studies (e.g. Crane, Testa, Derrick, & Leonard, 2014; Derrick, Testa, & Leonard, 2014). Again, this research has focused mainly on proximal factors that lead to *perpetration* of episodes of IPV. However, there are some exceptions. Derrick et al. (2014) found that verbal aggression predicted next day negative mood among victims. There is also one study that suggests victimization leads to mood dysregulation. Derrick and Testa (2017) found that within three hours of experiencing verbal IPV, victims were twice as likely to consume alcohol as when they had not experienced IPV. Alcohol use might be considered an action employed by victims to moderate emotional dysregulation. Thus, there is some indirect evidence that when examining daily reports of intimate relationships, victimization leads to mood dysregulation.

13.1.3 Summary

The current research addressed the question of causation between the victimization of IPV and mood lability. A Granger causality analysis tested causation in both directions. In addition, we examined whether fear of the partner leaving the relationship was a covariate in this relationship.

A daily, web-survey was administered for 28 days to college age participants. There were several benefits to this approach. (i) Most current research cannot elucidate the causative relationship between IPV and a given situational determinant, such as an individual's mood. It can only determine, on average, that IPV is associated with a risk factor or an outcome (see critiques by e.g. Bogat, Levendosky, & von Eye, 2005; Nurius & Macy, 2010). (ii) Current research relies mainly on retrospective reporting of IPV experiences (e.g. past six months, past year) leading to errors in recollection the longer the time from the episode to the report (e.g. Hamberger & Guse, 2002). A more accurate picture of IPV and the factors related to it can be elucidated if IPV is assessed in near "real-time." Daily diary data, analyzed using Granger causation analysis, can begin to answer whether mood lability leads to IPV or whether IPV leads to mood lability. That is, are the findings similar to the causation found between depression and IPV, or do specific incidents of IPV have a different causative path?

13.2 Methods

13.2.1 Participants

The participants were 208 college undergraduates (104 couples) all of whom experienced IPV within their relationship. There were eight criteria for participation for each individual/couple. We asked seven of the screening questions on the initial web survey to establish eligibility. These included: heterosexual relationship, marital status (must be unmarried as rates of IPV differ among married and unmarried individuals), age (both individuals must be between 18 and 24 years old), length of romantic relationship (must be ≥ 4 weeks), couple contact (must be physically present at least 2× per week), IPV in current relationship, and must own and use a smart phone with a data plan. The final criterion, both members of the couple must participate freely, without coercion, was assessed in an in-person interview (see Section 13.2.3).

The mean age of participants was 19.6 years old (SD = 1.3). Participants were primarily undergraduates (87%), participants who were not university students but who were partnered with students (11%), and graduate students (2%). The participants' race/ethnicity was Caucasian (78%), African American (7%), Asian American (7%), Latina/Latino (6%), and bi- or multi-racial (2%).

13.2.2 Measures

13.2.2.1 Daily Diary Questions

Participants were emailed every morning and asked to report on what happened and how they felt the day before. Participants that did not complete the questions by that evening were sent an email reminder with another link to the survey.

Regarding **IPV**, men and women were asked five questions regarding whether there was Emotional Violence (i.e. name calling, yelling, or use of put downs), Controlling Behavior (i.e. monitoring and isolating from family or friends), Physical Violence (i.e. pushing, shoving, or grabbing a partner), Sexual Coercion, (i.e. insisting on having sex even if the partner does not want to), and Severe Violence (i.e. acts that involved a weapon or led to injury). During the previous 24 hours (YES/NO), questions were summed to determine whether IPV occurred on a given day or not.

To determine **mood**, each respondent was asked to respond to this statement: "In the past 24 hours, my mood was up and down." They rated this question on a 7-point scale from "very untrue of me" to "very true of me."

To determine fear of leaving (this variable was used as a time-varying covariate), each respondent was asked to respond to this statement: "In the past 24 hours, I felt like my romantic partner might leave me." They rated this question on a 7-point scale ranging from "very untrue of me" to "very true of me."

13.2.3 Procedures

Couples were recruited two ways. First, all students who were sophomores, juniors, and seniors enrolled at a Midwestern university were contacted over the course of two consecutive academic year semesters via email. The Registrar's Office sent the study's email solicitation to about 1500 students at a time. For those students interested in participating, a hotlink within the email was provided that, when clicked, sent them to a web survey where screening questions were administered to determine whether couples met the first six criteria for participation, described above. Second, students enrolled in psychology classes and eligible for extra credit could also participate. These students completed the screening online through a system that provides information about all studies available for extra credit in the psychology department in a given semester. Again, the screening questions were administered and eligible students were identified.

If the student met the screening criteria, a research assistant (RA) emailed with the student, explaining the research study and answering any questions. The RA asked whether the student thought his or her partner would also like to participate and whether the student felt safe/comfortable asking the partner to participate. If the student did not feel safe or comfortable, the student was not allowed to participate. Participants came to project offices and answered in-person questionnaires. They earned $15 for completing these questions (no data from these questionnaires are reported in this chapter). The RA spoke to each member of the couple separately and privately. If either partner had concerns about participating, both members of the couple were thanked for their time and given the $15. If enrolled, participants earned $0.50 for each day they completed a daily survey. Participants could earn a $5 bonus for completing all surveys within a seven-day period. Finally, at the end of the 28-days, participants came to project offices for follow-up questions (again, these data were not included in this chapter), for which they could earn another $15.

13.3 Results

13.3.1 Data Consolidation

The dataset consisted of 104 couples (208 individuals) $\times$ 28 days = 5824 observations. Attrition across the 28-day period was 8%, and missing data across all participants ranged from 3.4 to 12.5% on any given day of data collection. Inspection of person-by-person scatterplots did not indicate any significant outliers or apparent time trends across the variables of interest. Due to the amount of missing data, observations across the 28 days were aggregated into four consecutive seven-day waves. In general, IPV (victimization and perpetration) is a low-occurrence behavior; this variable was summed across each seven-day period in order to increase variability. Daily mood lability and fear of partner leaving were treated as averages across each wave.

For the models that we describe in subsequent sections, we used all available data. No case was excluded because of missing data. Instead of estimating and imputing missing data points, we estimated the model based on the full information maximum likelihood (FIML) method that is implemented in LISREL.

13.3.2 Descriptive Statistics

In our sample, daily experience of IPV was a low-occurrence event; across the entire sample, average victimization by all forms of IPV was 0.07 (SD = 0.29), and average perpetration of all forms of IPV was 0.11 (SD = 0.38). On any given day, up to 13% of the total sample experienced emotional violence, which was the most commonly reported form of daily IPV, followed by controlling behavior (up to 7.7%), physical violence (up to 2.9%), sexual coercion (2.9%), and severe violence (1%). Rates were similar for IPV perpetration.

Men and women reported similar rates of violence across all forms of IPV, suggesting that much of the violence endorsed in the current sample is bidirectional. On average, women reported higher perpetration and lower victimization of physical and psychological IPV than their male partners. Men reported higher perpetration and lower victimization of more severe forms of IPV: sexual coercion and physical violence, although these forms of violence were low overall.

Mood lability was relatively low across the entire sample ($\overline{X} = 2.68$, SD = 1.95) and did not differ significantly between men ($\overline{X} = 2.52$, SD = 1.89) and women ($\overline{X} = 2.84$, SD = 1.99). Fear of partner leaving was also low $\overline{X}$ (= 1.66, SD = 1.38) and was similar across genders ($\overline{X} = 1.66$, SD = 1.40 for men; $\overline{X} = 1.65$, SD = 1.38 for women).

13.3.3 Model Development

The idea of a Granger-causal analysis is as follows: (i) there are at least two series of data. The series of data are fit individually. The best result is that they fit. However, if fit is less than perfect, there may be room for improvement of the models. (ii) One series is added to the model, and cross-lagged (and contemporaneous) paths from one series to the other (but not vice versa) are included in the model. If this results in improved fit (e.g. better RMSEA), the added series is considered a cause for the other. (iii) The cross-lagged and contemporaneous paths are reversed. If this works as well, the two series cause each other. If not, the model from Step (ii) is kept.

In many instances, researchers include covariates in a causality model. Usually, there are two groups of reasons for including covariates. These reasons are statistical and substantive. The statistical reasons mostly concern the exclusion of possible confounders and the magnitude of error in the estimation of parameters. When, by way of including covariates in a model, the error of the estimated parameters can be reduced, it can be expected that the estimates suggest significant parameters more often than without the covariates.

Substantive reasons concern the explanation of a causal relation. In empirical behavioral science research, virtually all relations, including causal ones, are multivariate. That is, more than one factor acts as cause, and causes themselves are dependent upon other causes. Therefore, meaningful covariates serve to specify a more complete model. This model will be more complex than one without covariates. However, the explanation of the process under study will also be deeper and more complete.

For these analyses, we included a question from the daily survey that asked how much the individual worried that their partner might leave them. We believed that concern about a partner leaving the relationship could result in becoming a victim of IPV or being a victim of IPV could increase concerns about the partner leaving. Concerns about a partner leaving a relationship could have a valid basis, but it might also be enhanced in individuals with insecure attachments (Brennan & Shaver, 1995). Research indicates that individuals with insecure attachment styles who worry that their partner might leave them generally have low relationship satisfaction (Hazan & Shaver, 1987; Levy & Davis, 1988). Thus, IPV victimization might reinforce an individual's concern that the partner will leave. The victim might construe the violence as an indication that the partner is not committed to the relationship or that the victim is not worthy of the relationship. In either case, these thoughts might lead to concerns about the partner leaving. On the other hand, individuals might ruminate about their concerns that the partner will leave. Rumination can amplify negative feelings about a relationship (Cloven & Roloff, 1991) as well as anxiety about the security of the relationship (Carson & Cupach, 2000). Inevitably, then, the concern will be a prominent component of the relationship. Some or all aspects of this concern might lead to arguments and result in verbal or physical outbursts from the partner.

In the following section, we describe the development of and present results for the model that we, then, interpret in a Granger-causality sense. The model is developed in three steps.

Step 1: Separately for each of the three series that we consider, intimate partner violence (IPV), lability of mood (MOOD), and fear of whether the person thinks their partner will leave them (LEAVE), we estimate autoregression parameters. These are the parameters that describe the regression of a variable observed at $t > 1$ onto itself, where t indicates the waves of data collection. The goals of this first step are, first, to determine whether variables are time-stable in the sense that positive autoregressions exist (other characteristics of time stability such as time-constant means or stationarity are not considered in the present context). This goal is considered reached when all first-order autoregression parameters are significant. Technically, higher order effects can be considered, but, here, we focus on first order effects. Under the first goal, we do not require that the models exhibit good fit. All we require is that all autoregressions exist.

Step 2: The second goal, pursued in the second step of model development, is to determine whether a model in which parameters are estimated that relate two series to each other sufficiently describes the data. Granger causality is said to exist when adding a second series to a model results in a fitting model, and when cross-series relations exist. In addition, one series is said to Granger-cause the second when the cross-series paths originate in only one series. Taking a developmental perspective, however, we also consider models in which the direction of causal effects change over time. Without discussing the temporal nature of causality, we also consider cross-sectional cross-series paths. This is done because some of the effects of interest can be short-term. The second goal is considered reached when cross-series relations exist and when a model, overall, describes the data well. In the present context, we relate the two series of IPV and MOOD to each other.

Step 3: The third goal is to determine whether a covariate series results in a well-fitting model and the covariate is related to one series or both. Here, we consider LEAVE as the covariate.

13.3.4 Granger Causality Analyses

Granger-causality (Granger, 1969) involves evaluating prediction errors and states that a variable X "Granger-causes" a variable Y if the prediction error variance of Y_t given *a universal set of information* up to time point t is smaller than the prediction error variance of Y_t given t without the information of X. The crucial element which justifies causality statements is that the set of information is defined as all information in the universe which is relevant for the prediction of Y_t. Please note that, in real-world applications, "all information in the universe" obviously reduces to "the observed variables" which opens the door for confoundedness and spurious effects.

Step 1: Table 13.1 displays the autoregression parameters for the three series that we examine. It shows that each of the autoregression parameters is significant. Each parameter exceeds the threshold of 1.96. Table 13.1 also shows that none of the models describes the data sufficiently well. Each of the RMSEA estimates is larger than the threshold for acceptable fit, 0.08. We conclude that the first goal of analysis has been reached. All three variables, IPV, LEAVE, and MOOD are time-stable in the sense that autoregressions exists.

Table 13.1 Autoregression parameter estimates for IPV, LEAVE, and MOOD (standardized solutions given).

	IPV		LEAVE		MOOD	
	Parameter	*t*	Parameter	*t*	Parameter	*t*
$b_{1\to2}$	0.57	8.47	0.70	12.26	0.68	11.66
$b_{2\to3}$	0.44	5.91	0.74	13.19	0.72	12.54
$b_{3\to4}$	0.50	6.49	0.69	10.47	0.79	14.35
RMSEA	0.285		0.208		0.208	

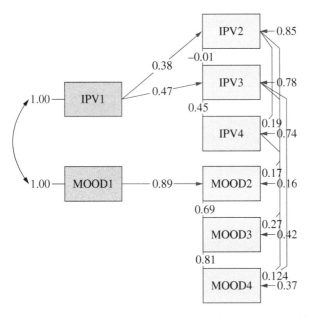

Figure 13.1 Granger causality model in which IPV causes MOOD lability (standardized solution given).

Step 2: In Step 2, we proceeded from the assumption that IPV, over time, is causally related to MOOD. Therefore, we first estimated a model in which the causal effects originate in IPV. Figure 13.1 displays the final model. Note that the dependence structure of dyads is ignored in this analysis.

The model in Figure 13.1 describes the data very well. The RMSEA is 0.047, and the test of close fit allows retaining the null hypothesis ($p = 0.50$). What follows, are details of this model.

1. With only one exception, each of the proposed paths is significant. The exception is the path from IPV2 to IPV3. This path was significant when IPV was modeled in isolation and only first-order autoregression co-efficients were estimated. Here, it is no longer significant. This is compensated by the path from IPV1 to IPV3. This path had not been considered in the above model. In addition to IPV1 → IPV3, we also included IPV2 → IPV4 in the model. These two paths suggest that IPV exhibits a complex pattern of relations over time, a pattern that exists above and beyond first-order autocorrelation/regression. This is not the case for MOOD.
2. The connection between IPV and MOOD originates exclusively in IPV, thus supporting the hypothesis that IPV is the cause and MOOD is the effect.
3. Specifically, the following causal paths are supported:
    a. IPV2 → MOOD2, IPV3 → MOOD3, and IPV4 → MOOD4. These relations suggest that, with the exception of the first time of observation, there is a contemporaneous effect of partner violence on MOOD.
    b. Only IPV3 has a longer-term effect. It also affects MOOD at the fourth observation point in time, that is, in the final week of the data collection.
4. In sum, we conclude that the causal flow originates in violence and has clear effects on MOOD lability. These effects are mostly contemporaneous but longer-term effects do also exist. Reversing the paths in direction can destroy the model in the sense that it will no longer fit, and reverse-direction paths may not be significant. For example, reversing the cross-series paths in Figure 13.1 results in a model with an RMSEA of 0.86, and the path from MOOD 4 to IPV4 is non-significant ($t = 1.43$, *ns*). In future research, this conclusion could be tested using DDA-type methods as suggested by von Eye and Wiedermann (2013).
5. *Step 3*: In Step 3, we proceed from the model that was retained in the second step, and ask whether LEAVE predicts the series of IPV or MOOD; that is, does LEAVE play the role of a covariate or is LEAVE predicted from either series. Figure 13.2 displays the final model.

The model shown in Figure 13.2 describes the data well. Its RMSEA is 0.067, and the *p*-value for test of close fit of 0.099 supports the hypothesis of close fit. More specifically, this model has the following characteristics:

1. Each of the proposed relations (paths and correlation) is significant.
2. The part of the model that is carried over from the model in Figure 13.1 is unchanged. That is, IPV Granger-causes MOOD.
3. LEAVE and MOOD are causally unrelated when only unidirectional relations are considered. Neither predicts the respective other.
4. However, the relation between the series of LEAVE and MOOD is characterized by a bidirectional relation at observation point 4. The signs of both paths

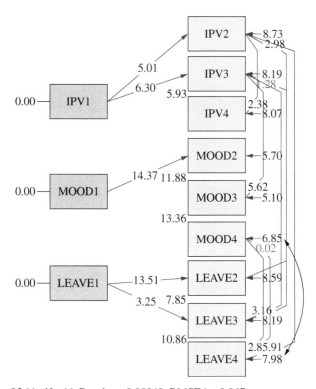

Chi–Square = 85.11, df = 46, P–value = 0.00040, RMSEA = 0.067

Figure 13.2 Granger causality model in which IPV causes MOOD lability, under consideration of LEAVE (standardized solution given).

are positive. This suggests that the fear of the partner leaving and agitated mood reinforce each other contemporaneously. From a developmental causality perspective, this suggests that, at least on occasion (here, at observation point 4), the causal relation between IPV and MOOD is influenced by fear of partner leaving.

5. As in the model in Figure 13.1, reversing paths mostly results in non-significant path co-efficients and ill-fitting models. As in the case above, DDA-type methods might be considered to test this conclusion in more detail.

In sum, we conclude that the hypothesis can be retained that IPV Granger-causes MOOD. This relation is unidirectional and mostly short-term. That is, partner violence results in a contemporaneous lability of MOOD. At Time 4, this is reinforced by LEAVE in the sense that the fear of a partner leaving results in agitated mood and vice versa.

13.4 Discussion

The findings of this study support the theoretical proposition that IPV victimization *causes* mood lability, rather than the reverse. The methods of our current study, including the repeated observations of mood and IPV over a 28-day period and the use of Granger causality analysis allowed for a rigorous testing of this proposition.

Theoretically, the causal effects of IPV on mood lability may be explained by attachment theory. Many individuals who experience IPV have insecure attachment (e.g. Hellemans et al., 2015; Kuijpers et al., 2012; Oka et al., 2014) and an incident of IPV re-affirms their negative internal working model of the self, leading to negative affect. The findings might also be explained by PTSD or posttraumatic symptoms. Those who experience IPV often experience PTSD or its symptoms (e.g. Black et al., 2011; Martinez-Torteya et al., 2009), which include negative alterations in mood. Thus, an incident of IPV may trigger a negative affect response.

Notably, the effects of IPV victimization are mostly contemporaneous (e.g. IPV at Time 2 affects mood lability at Time 2), suggesting that longer-term mood is not typically affected by IPV incidents. Because mood lability/negative affect is a core feature of depression, our results suggest that perhaps IPV does not cause depression directly. However, another explanation may be that the findings reflect the way that mood lability was assessed. Our daily question asked the participant's strength of agreement with the following statement – "In the past 24 hours, my mood was up and down." This question defines mood lability as fluctuations within one day. However, mood lability also occurs across days. A question that enables the respondent to compare their mood from one day to the next (e.g. "Compared to yesterday, my mood is better/worse today.") across numerous days would better assess long-term mood lability or stability. It might be "long-term" mood that relates to depression. Future research, examining daily mood and IPV over a longer period of time and/or developing different ways of measuring mood lability, may shed more light on this issue.

The findings that IPV victimization affect mood lability are consistent with several prior studies finding that IPV perpetration was associated with mood lability in survey research (Gratz et al., 2009; McNulty & Hellmuth, 2008; Shorey et al., 2011, 2015; Stuart et al., 2006) and in daily diary studies (Derrick et al., 2014). The high level of both victimization and perpetration among men and women in young adult couples suggests a complex relationship between mood and both perpetration and victimization. Perhaps there is a cascading effect such that IPV victimization leads to increased mood lability in one partner, which leads to an increased likelihood of this partner then perpetrating IPV, which leads to increased mood lability in the partner experiencing IPV, and so forth. Future research should examine the possible transactional relationship of mood as it relates to the cycle of victimization and perpetration of IPV. Prior research, including this study, has focused only on either victimization or perpetration.

In order to test a more complete model, we also examined the relationship of an individual's fear of their partner leaving them as a covariate in the relationship between IPV victimization and mood lability. We found that at one time point, fear of the partner's leaving and mood lability both affected each other such that increased fear led to more negative mood and vice versa. Attachment theory is also a useful explanation for these findings. Individuals with insecure attachment can feel threatened and upset by indications that their partner may be leaving them. Thus, individuals may experience IPV as a sign of the partner's dissatisfaction with the relationship and, thus, their potential departure. This could serve to increase negative mood, and negative mood can, in turn, increase the sense of threat that the partner is leaving. However, because this relationship was found at only one of the four time points, strong conclusions cannot be made about this relationship. However, this is a promising area for future research.

The primary strength of this study was its prospective, daily diary design. Although research on IPV is beginning to use daily diary methods, much of the extant literature is retrospective, asking individuals, typically women, to report on IPV they experienced over the past six months or year. Retrospective designs obviously suffer from problems with recall. The daily assessment of IPV, mood lability, and fear of the partner leaving allowed for a prospective and, thus, more granular understanding of the cause and effect relationship of these variables. However, because *daily* IPV is an infrequent occurrence in most couples, even those who are violent, we consolidated the data into seven-day waves in order to increase variability in the sample. Future research using Granger causal analysis may require larger sample sizes to examine the relationship of these variables in a more fine-grained manner. In addition, Granger causality models rest on unconfoundedness assumptions. In other words, omitted variables can bias path co-efficients which, in turn, may bias causal conclusions. There are likely other variables that influence the relationship between IPV and mood (e.g. alcohol) that should be examined in future research.

Notably, this study included both men and women. Although most of the extant research on IPV victimization only examines women's experiences, research over the last decade has begun to focus on the importance of understanding men's victimization. This growing body of research has focused mainly on adolescents and young adults, and daily diary studies with these groups are beginning to appear in the literature. The inclusion of both men and women is both a strength and a weakness of the current study. Because recent literature finds that IPV victimization in heterosexual couples is as common among young men as it is among young women (Langhinrichsen-Rohling et al., 2012; Renner & Whitney, 2012), it is important for research to include both sexes. However, the extant literature also suggests that there may be different relationships between mood and IPV victimization in men and women. For example, Graham et al. (2012) found that depression in men predicted perpetration of IPV while IPV victimization predicted women's depression, suggesting that mechanisms affecting these actions and moods may operate differently in men and women. This would suggest that analyzing separately by gender would be advisable; however, we were unable to do so because of the size of our sample. Finally, this study collected data on couples, but analyzed them as individuals. Thus, data independence assumptions were violated. We cannot know whether this data clustering affects interpretation, and, if it does, in which way.

In summary, our study demonstrated the utility of Granger causal analysis to resolve an important question in the field of IPV research. Our findings supported a causal model in which IPV victimization increases mood lability, mostly concurrently. We interpret these findings, including those about the fear of partner leaving, in the context of attachment theory such that IPV serves as a relationship threat which increases negative affect and fears about losing the partner, each of which positively reinforces each other. Questions still remain about the causal direction of the relationship between IPV and depression. Granger analysis using longitudinal data, examining IPV and depression over many time points with many subjects, could provide important information. In addition, future research should examine the complex transactional relationships between mood lability and IPV perpetration and victimization, especially in a young adult population, where mutual violence occurrence is so common. This, and the results of the current study, could possibly inform the development of better interventions or therapies for those couples who are struggling with mood and violence.

References

Aldarondo, E. (1996). Cessation and persistence of wife assault: A longitudinal analysis. *American Journal of Orthopsychiatry, 66*, 141–151.

Amar, A. F., & Gennaro, S. (2005). Dating violence in college women: Associated physical injury, healthcare usage, and mental health symptoms. *Nurs Res, 54*, 235–242.

American Psychiatric Association. (2013). *Diagnostic and statistical manual of mental disorders: DSM5* (fifth Edition ed.). Washington, DC: American Psychiatric Publishing.

Archer, J. (2000). Sex difference in aggression between heterosexual partners: A meta-analytic review. *Psychological Bulletin, 126*, 651–680.

Archer, J. (2002). Sex differences in physically aggressive acts between heterosexual partners: A meta-analytic review. *Aggression and Violent Behavior, 7*, 313–351.

Arias, I., Samios, M., & O'Leary, K. D. (1987). Prevalence and correlates of physical aggression during courtship. *Journal of Interpersonal Violence, 2*(1), 82–90.

Beck, A. D., Ward, C. H., Mendelson, M., Mock, J., & Erbaugh, J. (1961). An inventory for measuring depression. *Arch Gen Psychiatry, 4*, 53–63.

Berking, M., Wirtz, C. M., Svaldi, J., & Hofmann, S. G. (2014). Emotion regulation predicts symptoms of depression over five years. *Behaviour Research and Therapy, 57*, 13–20.

Beydoun, H. A., Beydoun, M. A., Kaufman, J. S., Lo, B., & Zonderman, A. B. (2012). Intimate partner violence against adult women and its association with major depressive disorder, depressive symptoms and postpartum depression: A systematic review and meta-analysis. *Social Science & Medicine, 75*, 959–975.

Black, M. C., Basile, K. C., Smith, S. G., Walters, M. L., Merrick, M. T., Chen, J., & Stevens, M. R. (2011). *National intimate partner and sexual violence survey 2010 summary report.* Retrieved from https://www.cdc.gov/violenceprevention/pdf/nisvs_report2010-a.pdf.

Bliton, C. F., Wolford-Clevenger, C., Zapor, H., Elmquist, J., Brem, M. J., Shorey, R. C., & Stuart, G. L. (2016). Emotion dysregulation, gender, and intimate partner violence perpetration: An exploratory study in college students. *Journal of Family Violence, 31*, 371–377.

Bogat, G. A., Levendosky, A. A., DeJonghe, E. S., Davidson, W. S., & von Eye, A. (2004). Pathways of suffering: The temporal effects of domestic violence on women's mental health. *Maltrattamento e abuso all'infanzia, 6*(2), 97–112.

Bogat, G. A., Levendosky, A. A., Theran, S. A., von Eye, A., & Davidson, W. S. (2003). Predicting the psychosocial effects of interpersonal partner violence (IPV): How much does a woman's history of IPV matter? *Journal of Interpersonal Violence, 18*, 121–142.

Bogat, G. A., Levendosky, A. A., & von Eye, A. (2005). The future of research on intimate partner violence (IPV): Person-oriented and variable-oriented perspectives. *American Journal of Community Psychology, 36*(1/2), 49–70.

Brennan, K. A., & Shaver, P. R. (1995). Dimensions of adult attachment, affect regulation, and romantic relationship functioning. *Personality and Social Psychology Bulletin, 21*(3), 267–283. doi:10.1177/0146167295213008

Brownridge, D. A. (2006). Intergenerational transmission and dating violence victimization: Evidence from a sample of female university students in Manitoba. *Canadian Journal of Community Mental Health, 25*, 75–93.

Bushman, B. J., Baumeister, R. F., & Phillips, C. M. (2001). Do people aggress to improve their mood? Catharsis beliefs, affect regulation opportunity, and aggressive responding. *Journal of Personality and Social Psychology, 81*, 17–32.

Capaldi, D. M., Shortt, J. W., & Crosby, L. (2003). Physical and psychological aggression in at-risk young couples: Stability and change in young adulthood. *Merrill-Palmer Quarterly, 49*, 1–27.

Carson, C. L., & Cupach, W. R. (2000). Fueling the flames of the green-eyed monster: The role of ruminative thought in reaction to romantic jealousy. *Western Journal of Communication, 64*(3), 308–329. 10.1080/10570310009374678

Cascardi, M., Langhinrichsen, J., & Vivian, D. (1992). Marital aggression. Impact, injury, and health correlates for husbands and wives. *Arch Intern Med, 152*(6), 1178–1184.

Cascardi, M., & O'Leary, K. D. (1992). Depressive symptomatology, self-esteem, and self-blame in battered women. *Journal of Family Violence, 7*(4), 249–259.

Catalano, S. M. (2012). *Intimate partner violence, 1993-2010*. Washington, DC. Retrieved from https://www.bjs.gov/content/pub/pdf/ipv9310.pdf

Chuick, C. D., Greenfeld, J. M., Greenberg, S. T., Shepard, S. J., Cochran, S. V., & Haley, J. T. P. O. M. M. (2009). A qualitative investigation of depression in men. *Psychology of Men & Masculinity, 10*, 302–313.

Cloven, D. H., & Roloff, M. E. (1991). Sense-making activities and interpersonal conflict: Communicative cures for the mulling blues. *Western Journal of Speech Communication, 55*, 134–158.

Cochran, S. V., & Rabinowitz, F. E. (2000). *Men and depression: Clinical and empirical perspectives*. San Diego, CA: Academic Press.

Coker, A. L., Davis, K. E., Arias, I., Desai, S., Sanderson, M., Brandt, H. M., & Smith, P. H. (2002). Physical and mental health effects of intimate partner violence for men and women. *Am J Prev Med, 23*, 260–268.

Coker, A. L., Smith, P. H., McKeown, R. E., & King, M. J. (2000). Frequency and correlates of intimate partner violence by type: Physical, sexual, and psychological battering. *American Journal of Public Health, 90*, 553–559.

Crane, C. A., Testa, M., Derrick, J. L., & Leonard, K. E. (2014). Daily associations among self-control, heavy episodic drinking, and relationship functioning: An examination of actor and partner effects. *Aggressive Behavior, 40*, 440–450.

Dardis, C. M., Edwards, K. M., Kelley, E. L., & Gidycz, C. A. (2017). Perceptions of dating violence and associated correlates: A study of college young adults. *Journal of Interpersonal Violence, 32*(1), 3245–3271.

Derrick, J. L., & Testa, M. (2017). Temporal effects of perpetrating or receiving intimate partner aggression on alcohol consumption: A daily diary study of community couples. *J Stud Alcohol Drugs, 78*(2), 213–221. doi:10.15288/jsad.2017.78.213

Derrick, J. L., Testa, M., & Leonard, K. E. (2014). Daily reports of intimate partner verbal aggression by self and partner: Short-term consequences and implications for measurement. *Psychology of Violence, 4*(4), 416–431.

Devries, K. M., Mak, J. Y., Bacchus, L. J., Child, J. C., Falder, G., Petzold, M., … Watts, C. H. (2013). Intimate partner violence and incident depressive symptoms and suicide attempts: A systematic review of longitudinal studies. *PLoS Med, 10*, e1001439.

Eckhardt, C., Murphy, C., Whitaker, D., Sprunger, J., Dykstra, R., & Woodard, K. (2013). The effectiveness of intervention programs for perpetrators and victims of intimate partner violence: Findings from the Partner Abuse State of Knowledge Project. *Partner Abuse*, *4*, 175–195.

Edwards, V., Black, M., Dhingra, S., McKnight-Eily, L., & Perry, G. (2009). Physical and sexual intimate partner violence and reported serious psychological distress in the 2007 BRFSS. *International Journal of Public Health*, *54*, 37–42.

Eshelman, L., & Levendosky, A. A. (2012). Dating violence: Mental health consequences based on type of abuse. *Violence and Victims*, *27*(2), 215–228.

Feld, S. L., & Straus, M. A. (1989). Escalation and desistance of wife assault in marriage. *Criminology*, *27*, 141–161.

Finkel, E. J. (2007). Impelling and inhibiting forces in the perpetration of intimate partner violence. *Review of General Psychology*, *11*, 193–207.

Finkel, E. J. (2008). Intimate partner violence perpetration: Insights from the science of self-regulation. In J. Forgas & J. Fitness (Eds.), *Social relationships: Cognitive, affective, and motivational processes* (pp. 271–288). New York, NY: Psychology Press.

Finkel, E. J. (2014). The I3 model: Metatheory, theory, and evidence. *Advances in Experimental Social Psychology*, *49*, 1–104.

Finkel, E. J., DeWall, C. N., Slotter, E. B., McNulty, J. K., Pond, R. S., & Atkins, D. C. (2012). Using I3 theory to clarify when dispositional aggressiveness predicts intimate partner violence perpetration. *Journal of Personality and Social Psychology*, *102*(3), 533–549.

Finkel, E. J., & Slotter, E. B. (2009). An I^3 theory analysis of human sex differences in aggression. *Brain and Behavior Sciences*, *32*, 31.

Follingstad, D. R., Bradley, R. G., Laughlin, J. E., & Burke, L. (1999). Risk factors and correlates of dating violence: The relevance of examining frequency and severity levels in a college sample. *Violence and Victims*, *14*, 365–380.

Follingstad, D. R., Laughlin, J. E., Polek, D. S., Rutledge, L. L., & Hause, E. S. (1991). Identification of patterns of wife abuse. *Journal of Interpersonal Violence*, *6*(2), 187–204.

Follingstad, D. R., Wright, S., Lloyd, S., & Sebastian, J. A. (1991). Sex differences in motivations and effects in dating violence. *Family Relations*, *40*(1), 51–57.

Fossos, N., Neighbors, C., Kaysen, D., & Hove, M. C. (2007). Intimate partner violence perpetpration and problem drinking among college students: The roles of expectancies and subjective evaluations of alcohol aggression. *J Stud Alcohol Drugs*, *68*(5), 706–713.

Gondolf, E. W., & Beeman, A. K. (2003). Women's accounts of domestic violence versus tactics-based outcome categories. *Violence Against Women*, *9*(3), 278–301.

Gover, A., Kaukinen, C., & Fox, K. (2008). The relationship between violence in the family of origin and dating violence among college students. *Journal of Interpersonal Violence, 23*, 1667–1693.

Graham, K., Bernards, S., Flynn, A., Tremblay, P. F., & Wells, S. (2012). Does the relationship between depression and intimate partner aggression vary by gender, victim-perpetrator role, and aggression severity? *Violence and Victims, 27*(5), 730–743.

Granger, C. W. J. (1969). Investigating causal relations by econometric models and cross-spectral methods. *Econometrica, 37*, 424–438.

Gratz, K. L., Paulson, A., Jakupcak, M., & Tull, M. T. (2009). Exploring the relationship between childhood maltreatment and intimate partner abuse: Gender differences in the mediating role of emotion dysregulation. *Violence and Victims, 24*(1), 41–54.

Gratz, K. L., & Roemer, L. (2004). Multidimensional assessment of emotion regulation and dysregulation: Development, factors structure, and initial validation of the Difficulties in Emotion Regulation Scale. *Journal of Psychopathology and Behavioral Assessment, 26*, 41–54.

Gross, J. J., & Thompson, R. A. (2007). Emotion regulation: Conceptual foundations. In J. J. Gross (Ed.), *Handbook of emotion regulation* (pp. 3–24). New York, NY: Guilford Press.

Hamberger, L. K., & Guse, C. E. (2002). Men's and women's use of intimate partner violence in clinical samples. *Violence Against Women, 8*, 1301–1331.

Hazan, C., & Shaver, P. (1987). Romantic love conceptualized as an attachment process. *Journal of Personality and Social Psychology, 52*(3), 511–524.

Hedtke, K. A., Ruggiero, K. J., Fitzgerald, M. M., Zinzow, H. M., Saunders, B. E., Resnick, H. S., & Kilpatrick, D. G. (2008). A longitudinal investigation of interpersonal violence in relation to mental health and substance use. *Journal of Consulting and Clinical Psychology, 76*, 633–647.

Hellemans, S., Loeys, T., Dewitte, M., De Smet, O., & Buysse, A. (2015). Prevalence of intimate partner violence victimization and victims' relational and sexual well-being. *Journal of Family Violence, 30*(6), 685–698. doi:10.1007/s10896-015-9712-z

Hines, D. A., Brown, J., & Dunning, E. (2007). Characteristics of callers to the domestic abuse helpline for men. *Journal of Family Violence, 22*(2), 63–72. doi:10.1007/s10896-007-9091-1

Hines, D. A., & Saudino, K. J. (2003). Gender differences in psychological, physical, and sexual aggression among college students using the Revised Conflict Tactics Scales. *Violence and Victims, 18*, 197–217.

Holtzworth Munroe, A., & Stuart, G. L. (1994). Typologies of male batterers: Three subtypes and the differences among them. *Psychological Bulletin, 116*(3), 476–497.

Jakupcak, M., Lisak, D., & Roemer, L. (2002). The role masculine ideology and masculine gender role stress in man's perpetration of relationship violence. *Psychology of Men & Masculinity, 3*(2), 97–106.

Johnson, W. L., Giordano, P. C., Manning, W. D., & Longmore, M. A. (2015). The age–IPV curve: Changes in the perpetration of intimate partner violence during adolescence and young adulthood. *Journal of Youth and Adolescence, 44*(3), 708–726.

Joormann, J., & Gotlib, I. H. (2010). Emotion regulation in depression: Relation to cognitive inhibition. *Cognition and Emotion, 24*(2), 281–298.

Kaura, S. A., & Lohman, B. J. (2007). Dating violence victimization, relationship satisfaction, mental health problems, and acceptability of violence: A comparison of men and women. *Journal of Family Violence, 22*, 367–381.

Kessler, R. C., Molnar, B. E., Feurer, I. D., & Appelbaum, M. (2001). Patterns and mental health predictors of domestic violence in the United States: Results from the National Comorbidity Survey. *International Journal of Law and Psychiatry, 24*, 487–508.

Kothari, C. L., Rohs, T., Davidson, S., Kothari, R. U., Klein, C., Koestner, A., … Kutzko, K. (2015). Emergency department visits and injury hospitalizations for female and male victims and perpetrators of intimate partner violence. *Advances in Emergency Medicine.* doi:10.1155/2015/502703

Kuijpers, K. F., Van Der Knaap, L. M., & Winkel, F. W. (2012). Risk of revictimization of intimate partner violence: The role of attachment, anger and violent behavior of the victim. *Journal of Family Violence, 27*(1), 33–44. doi:10.1007/s10896-011-9399-8

Langhinrichsen-Rohling, J. (2010). Controversies involving gender and intimate partner violence in the United States. *Sex Roles, 62*, 179–193.

Langhinrichsen-Rohling, J., Misra, T., Selwyn, C., & Rohling, M. (2012). A systematic review of rates of bidirectional versus unidirectional intimate partner violence. *Partner Abuse, 3*, 199–230.

Lee, K. D. M., Rodriguez, L. M., Edwards, K. M., & Neal, A. M. (2019). Emotional dysregulation and intimate partner violence: A dyadic perspective. *Psychology of Violence.* doi:10.1037/vio0000248

Levy, M., & Davis, K. (1988). Love styles and attachment styles compared: Their relations to each other and to various relationship characteristics. *Journal of Social and Personal Relationships, 5*, 439–471.

Martinez-Torteya, C., Bogat, G. A., von Eye, A., Levendosky, A. A., & Davidson, W. S. (2009). Women's appraisals of intimate partner violence stressfulness and their relationship to depressive and posttraumatic stress disorder symptoms. *Violence Vict, 24*(6), 707–722.

McNulty, J. K., & Hellmuth, J. C. (2008). Emotion regulation and intimate partner violence in newlyweds. *Journal of Family Psychology, 22*(5), 794–797.

Mihalic, S. W., Elliott, D. A., & Menard, S. (1994). Continuities in marital violence. *Journal of Family Violence, 9*, 195–225.

Moffitt, T. E., & Caspi, A. (1998). Annotation: Implications of violence between intimate partners for child psychologists and psychiatrists. *Journal of Child Psychology & Psychiatry, 39*, 137–144.

Neufield, J., McNamera, J. R., & Ertl, M. (1999). Incidence and prevalence of dating partner abuse and its relationship to dating practices. *Journal of Interpersonal Violence, 14*, 125–137.

Nurius, P. S., & Macy, R. J. (2010). Person-oriented methods in partner violence research: Distinct biopsychosocial profiles among battered women. *Journal of Interpersonal Violence, 25*(6), 1064–1093. doi:10.1177/0886260509340541

Oka, M., Sandberg, J. G., Bradford, A. B., & Brown, A. (2014). Insecure attachment behavior and partner violence: Incorporating couple perceptions of insecure attachment and relational aggression. *Journal of Marital and Family Therapy, 40*(4), 412–429. doi:10.1111/jmft.12079

Ouellet-Morin, I., Fisher, H. L., York-Smith, M., Fincham-Campbell, S., Moffitt, T. E., & Arseneault, L. (2015). Intimate partner violence and new-onset depression: A longitudinal study of women's childhood and adult histories of abuse. *Depression and Anxiety, 32*, 316–324.

Renner, L. M., & Whitney, S. D. (2012). Risk factors for unidirectional and bidirectional intimate partner violence among young adults. *Child Abuse & Neglect, 36*(1), 40–52.

Scott, K., & Straus, M. (2007). Denial, minimization, partner blaming, and intimate aggression in dating partners. *Journal of Interpersonal Violence, 22*(7), 851–871.

Shorey, R. C., Brasfield, H., Febres, J., & Stuart, G. L. (2011). An examination of the association between difficulties with emotion regulation and dating violence perpetration. *Journal of Aggression, Maltreatment, & Trauma, 20*(8), 870–885.

Shorey, R. C., Brasfield, H., Febres, J., & Stuart, G. L. (2015). An examination of the association between difficulties with emotion regulation and dating violence perpetration. *Journal of Aggression, Maltreatment, & Trauma, 20*(8), 870–885.

Shorey, R. C., Cornelius, T. L., & Bell, K. M. (2008). A critical review of theoretical frameworks for dating violence: Comparing the dating and marital fields. *Aggression and Violent Behavior, 13*(3), 185–194.

Shortt, J. W., Capaldi, D. M., Kim, H. K., Kerr, D. C., Owen, L. D., & Feingold, A. (2012). Stability of intimate partner violence by men across 12 years in young adulthood: Effects of relationship transitions. *Prevention Science, 13*(4), 360–369.

Simonelli, C. J., & Ingram, K. M. (1998). Psychological distress among men experiencing physical and emotional abuse in heterosexual dating relationships. *Journal of Interpersonal Violence, 13*, 667–681.

Stets, J. E., Straus, M. A., & Smith, C. (1990). Gender differences in reporting marital violence and its medical and psychological consequences. In M. A. Straus & J.

Gelles Richard (Eds.), *Physical violence in American families* (pp. 151–165). New Brunswick, NJ: Transaction Publishers.

Straus, M. A. (2005). Women's violence toward men is a serious social problem. In D. R. Loseke, R. J. Gelles, & M. M. Cavanaugh (Eds.), *Current controversies on family violence* (2nd ed., pp. 55–77). Newbury Park, CA: Sage.

Straus, M. A. (2008). Dominance and symmetry in partner violence by male and female university students in 32 nations. *Children and Youth Services Review, 30,* 252–275.

Stuart, G. L., Moore, T. M., Gordon, K. C., Hellmuth, J. C., Ramsey, S. E., & Kahler, C. W. (2006). Reasons for intimate partner violence perpetration among arrested women. *Violence Against Women, 12*(7), 609–621.

Sullivan, T. P., McPartland, T. S., Armeli, S., Jaquier, V., & Tennen, H. (2012). Is it the exception or the rule? Daily co-occurence of physical, sexual, and psychological partner violence in a 90-day study of substance using, community women. *Psychology of Violence, 2*(2), 154–164.

Tjaden, P., & Thoennes, N. (2000). *Extent, nature, and consequences of intimate partner violence: Findings from the National Violence Against Women Survey* (NCJ 181867). Retrieved from https://www.ncjrs.gov/pdffiles1/nij/181867.pdf.

von Eye, A., & Wiedermann, W. (2013). On direction of dependence in latent variable contexts. *Educational and Psychological Measurement, 74,* 5–30. doi:10.1177/0013164413505863

Walker, L. (1984). *The battered woman syndrome.* New York, NY: Springer.

White, J. W., & Koss, M. P. (1991). Courtship violence: Incidence in a national sample of higher education students. *Violence and Victims, 6*(4), 247–256.

Wofford, S., Elliott, D., & Menard, S. (1994). Continuities in marital violence. *Journal of Family Violence, 9*(3), 195–225.

Woodin, E. M., Caldeira, V., & O'Leary, K. D. (2013). Dating aggression in emerging adulthood: Interactions between relationship processes and individual vulnerabilities. *Journal of Social & Clinical Psychology, 32*(6), 619–650.

14

On the Causal Relation of Academic Achievement and Intrinsic Motivation

An Application of Direction Dependence Analysis Using SPSS Custom Dialogs

Xintong Li[1] and Wolfgang Wiedermann[2]

[1] *Assessment Resource Center, University of Missouri, Columbia, MO, USA*
[2] *Department of Educational, School, and Counseling Psychology & Missouri Prevention Science Institute, University of Missouri, Columbia, MO, USA*

Establishing cause–effect relations between variables is one of the focal interests of researchers in educational, behavioral, and social sciences. Although conventionally, causal effects are best analyzed using experimental designs, observational data have received renewed attention due to the enormous progress that has been made in the development of statistical methods for the analysis of quasi-experimental designs – including, but not restricted to, propensity score methods (Rosenbaum & Rubin, 1983), instrumental variable approaches (Imbens & Angrist, 1994), or regression discontinuity designs (Thistlewaite & Campbell, 1960). While these methods were developed to enable researchers to estimate the magnitude of the causal effects (provided that the corresponding statistical assumptions are fulfilled), a related line of causal inference research focuses on the development of methods to learn and critically evaluate the causal direction of effects (for an overview see the related Chapter 2 by Wiedermann, Li, and von Eye and Chapter 5 by Shimizu and Blöbaum, in this volume).

Correlational analysis, regression modeling, and structural equation models are routinely used to estimate the magnitude of causal effects. However, these methods shed no light on the causal direction of effects of observed variables (Dodge & Rousson, 2000; Wiedermann & Hagmann, 2016; Wiedermann & von Eye, 2015b). The reason for this is that these methods are based on covariances, whose inherent symmetry properties (i.e. the covariance does not depend on the causal order of variables) prevent researchers from distinguishing directionally competing causal models. For example, in the simple linear regression case, the estimated slope parameters of two causally competing models will be identical when the predictor (x) and the outcome (y) are standardized (von Eye & DeShon, 2012; Wiedermann & von Eye, 2015b). Therefore, it is impossible to confirm a hypothesized causal

Direction Dependence in Statistical Modeling: Methods of Analysis, First Edition.
Edited by Wolfgang Wiedermann, Daeyoung Kim, Engin A. Sungur, and Alexander von Eye.

direction of effect by simply exchanging the status of the hypothesized dependent and independent variables (i.e. $x \to y$ versus $y \to x$).

To make use of observational data for causal analysis, a framework summarizing several so-called asymmetry properties of the linear regression model called Direction Dependence Analysis (DDA) was recently proposed, which utilizes these asymmetry properties to determine and critically evaluate the data generating mechanism (Wiedermann & Li, 2018). DDA has largely been proposed as a confirmatory method, which aims to validate a postulated explanatory model against plausible alternative models with reversed flow of causality (Wiedermann & von Eye, 2015a).

Currently, statistical software implementations for DDA are scarce and mainly consist of R scripts (Wiedermann & Li, 2019) and SPSS macros (Wiedermann & Li, 2018). SPSS (IBM Corp., 2017) is one of the most popular statistical software programs in the educational, behavioral, and social sciences because of its ease-of-use (Muenchen, 2017). The present chapter introduces SPSS Custom Dialogues (SCDs) for DDA, which can be installed as an add-on to the SPSS statistics software. DDA SCDs are based on DDA SPSS macros discussed in Wiedermann and Li (2018), and translate Graphic User Interface (GUI) inputs into SPSS macro commands. This makes DDA accessible via the regular SPSS dropdown menu. Compatibility of DDA SCDs was tested on IBM SPSS 21 through 25, though for version 24 and 25, the SCDs can only be installed using the compatibility mode. DDA SCDs as well as detailed installation instructions are available from https://ddaproject.com.

We start with a brief summary of the direction dependence framework. For a detailed discussion of the DDA approach, see the related Chapter 2 by Wiedermann, Li, and von Eye and Chapter 1 by Dodge and Rousson, in this volume. We then introduce DDA SCDs using a worked data example focusing on the causal relation of intrinsic motivation and academic (mathematics) achievement (Deci, 1975; Deci & Ryan, 1980). Data are publicly available from the High School Longitudinal Study of 2009 (HSLS: 09; Ingels et al., 2011).

14.1 Direction of Dependence in Linear Regression

DDA is a framework that consists of an inventory of tests that aim to distinguish competing linear models. The models have the same focal variables (x and y) but reversed causal flow (cf. Figure 14.1), i.e. x causes y (i.e. $x \to y$) or y causes x ($y \to x$). To simplify presentation, we assume that $x \to y$ corresponds to the "true" underlying causal mechanism. DDA for continuous variables assumes that the "true" predictor is a non-normally distributed, exogenous variable. That is, the distribution of the predictor is assumed to deviate from the normal (Gaussian) distribution and the cause of the predictor lies outside the model of interest.

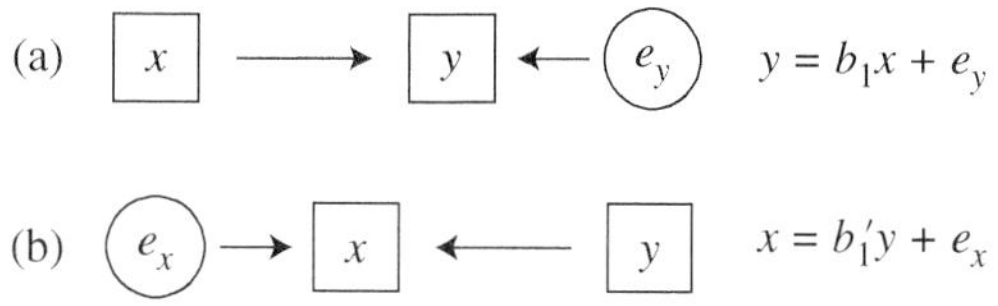

(a) $\boxed{x} \longrightarrow \boxed{y} \longleftarrow \left(\! e_y \!\right) \qquad y = b_1 x + e_y$

(b) $\left(\! e_x \!\right) \longrightarrow \boxed{x} \longleftarrow \boxed{y} \qquad x = b_1' y + e_x$

Figure 14.1 Simplified competing models with standardized focal variables *x* and *y*. Rectangles represent observed variables and circles represent latent residual terms. The constant b_1 represents the causal effect of *x* on *y*, and e_y is the error term of the "true" model (a). The constant b_1' represents the causal effect associated with the predictor in the mis-specified model (b).

Various approaches to evaluate asymmetry properties of causally competing models are available in the direction dependence framework. These include approaches that focus on distributional characteristics (skewness and kurtosis) of observed variables (Dodge & Rousson, 2000, 2001) and residuals (Wiedermann, 2015; Wiedermann, Hagmann, Kossmeier, & von Eye, 2013), higher order correlations (HOCs) of observed variables (Wiedermann, 2017; Wiedermann, Li, & von Eye, 2019), and asymmetry in the independence of predictors and residuals (Shimizu et al., 2011; Wiedermann, Artner, & von Eye, 2017; Wiedermann & Li, 2018). In the following sections, we briefly discuss these asymmetry properties and present proper methods for statistical inference.

14.1.1 Distributional Properties of *x* and *y*

In the linear regression model with additive normal errors, third and fourth powers of the Pearson correlation coefficient can be expressed as the ratio of the skewnesses (γ_x and γ_y) and kurtosis values (κ_x and κ_y) of the two variables *x* and *y*, i.e. $\rho_{xy}^3 = \frac{\gamma_y}{\gamma_x}$ and $\rho_{xy}^4 = \frac{\kappa_y - 3}{\kappa_x - 3}$. These facets of the Pearson correlation reflect asymmetry properties that emerge in the simple linear regression model when variables are non-normally distributed and hold as long as $\gamma_x \neq 0$ and $\kappa_x - 3 \neq 0$ (Dodge & Rousson, 2000, 2001; Dodge & Yadegari, 2010). Because the absolute value of the Pearson correlation cannot exceed 1 ($|\rho_{xy}| \leq 1$), the absolute value of the skewness and excess kurtosis (kurtosis – 3) of the "true" outcome *y* (see Figure 14.1a) will always be smaller than that of the predictor *x*. In other words, when the "true" errors is normally distributed, *y* will always be closer to the normal distribution than *x*. This asymmetry property provides the key to examine the causal direction of effects based on distributional features of *x* and *y*, and can be evaluated using, for example, omnibus normality tests, skewness tests (e.g. D'Agostino, 1971), and kurtosis tests (e.g. Anscombe & Glynn, 1983). For a discussion of proper decision guidelines see von Eye and DeShon (2008) and von Eye and DeShon (2012). Pornprasertmanit and Little (2012) suggested nonparametric bootstrap confidence intervals (CIs) for the difference of absolute values of higher moments, i.e. $\Delta(\gamma) = |\gamma_x| - |\gamma_y|$ and $\Delta(\kappa) = |\kappa_x| - |\kappa_y|$.

Asymmetric properties of x and y can also be examined using higher-order correlation (HOC) tests (Wiedermann, 2017; Wiedermann et al., 2019). Compared to skewness- and kurtosis-based DDA measures, HOC-based approaches have the advantage that they do not rely on distributional assumptions about the "true" errors. In general, HOCs are formally defined as $cor(x,y)_{ij} = \frac{\sum (x-\bar{x})^i(y-\bar{y})^j/n}{\sigma_x^i \sigma_y^j}$, where $\sum (x-\bar{x})^i(y-\bar{y})^j/n$ refers to the ij-th higher-order covariance of x and y and σ_x^i and σ_y^j reflect the ij-th powers of the standard deviations of x and y. In the "true" model $x \to y$, using i and $j \in \{1, 2\}$, one obtains $\rho_{xy} = \frac{cor(x,y)_{12}}{cor(x,y)_{21}}$, with $cor(x,y)_{12}$ and $cor(x,y)_{21}$ being the co-skewnesses of x and y. Thus, if $x \to y$ describes the "true" data-generating mechanism, the difference in squared co-skewnesses, $\Delta HOC_s = cor(x,y)_{21}^2 - cor(x,y)_{12}^2$ will always be greater than zero whenever $\rho_{xy} \neq 0$ and $\gamma_x \neq 0$. Similarly, using i and $j \in \{1, 3\}$, the co-kurtosis based difference, $\Delta HOC_c = sgn(\kappa_x - 3)\{\rho_{xy}^2(\kappa_x - 3)(1 - \rho_{xy}^2)[\rho_{xy}^2(\kappa_x - 3) + \kappa_x + 3]\}$ will always be greater than zero, with $sgn(\kappa_x - 3)$ being the signum function returning the sign of $(\kappa_x - 3)$ (Wiedermann, 2017). Since both, $\Delta HOC_s > 0$ and $\Delta HOC_c > 0$, will hold for the "true" model $x \to y$, bootstrap CIs of ΔHOC_s and ΔHOC_c can be used for statistical inference. Note that, for co-kurtosis-based inference, one needs to assume that $(\kappa_x - 3)$ and $(\kappa_y - 3)$ have the same sign (for details see Wiedermann, 2017).

14.1.2 Distributional Properties of e_x and e_y

Distributional properties of error terms in the two competing linear models, e_y and e_x, can also bring in asymmetry features that enable researchers to examine the flow of causality between two focal variables (Wiedermann, 2015; Wiedermann, Hagmann, & Eye, 2015; Wiedermann et al., 2013). Under normality of the "true" errors e_y, skewness and kurtosis of the errors in the mis-specified model (e_x) can be expressed as functions of third and fourth moments of the true predictor x, i.e. $\gamma_{e_x} = (1 - \rho_{xy}^2)^{3/2}\gamma_x$ and $\kappa_{e_x} - 3 = (1 - \rho_{xy}^2)^2(\kappa_x - 3)$. When normality of the "true" error is assumed, one obtains $\gamma_{e_y} = \kappa_{e_y} - 3 = 0$. Thus, the "true" errors will always be closer to the normal distribution than the errors of the mis-specified model as long as $\rho_{xy}^2 \neq 1$ and x deviates from normality.

The same set of tests proposed to compare distributional properties of observed variables can also be used to make statements about distributional properties of errors. In addition, asymptotic significance tests for skewness and kurtosis differences, i.e. $\Delta(\gamma_e) = |\gamma_{e_x}| - |\gamma_{e_y}|$ and $\Delta(\kappa_e) = |\kappa_{e_x}| - |\kappa_{e_y}|$ have been proposed (Wiedermann, 2015; Wiedermann & von Eye, 2015b). These tests again assume normality of the "true" error term. Bootstrap CIs are preferable whenever the normality assumption of the "true" error is violated.

14.1.3 Independence of Error Terms with Predictor Variable

Although, in the "true" model $x \to y$, errors are assumed to be stochastically independent from model predictors ($x \perp e_y$), ordinary least squares (OLS) regression residuals cannot be used to evaluate this crucial assumption. The reason for this is that estimated OLS residuals are always linearly uncorrelated with model predictors. This uncorrelatedness will also hold in a directionally mis-specified model. In addition, uncorrelatedness does not necessarily imply independence. The former indicates "independence" only in linear terms while the latter indicates stochastic independence in any possible way. However, when the "true" predictor x is non-normal, the error term and the predictor of the mis-specified model (y and e_x), will be stochastically non-independent (Wiedermann et al., 2017). Since the independence assumption will be preserved in the correctly specified model, there is asymmetry between the causally competing models with respect to the predictor-error independence. Again, this asymmetry can be used to differentiate between causally competing models.

Several procedures have been suggested to evaluate predictor-error independence beyond first-order (linear) uncorrelatedness. First, homoscedasticity tests have been suggested such as the Breusch–Pagan test (Breusch & Pagan, 1979) and the robust Breusch–Pagan test (Koenker, 1981; Koenker & Bassett Jr, 1982). Although these tests are conventionally used to identify heteroscedasticity in linear regression, the tests are also useful to identify causal model mis-specifications when the "true" predictor is non-normal (Wiedermann et al., 2017). To find empirical support for the model $x \to y$, homoscedasticity should be observed for $x \to y$ and heteroscedasticity should be present in the causally reversed model $y \to x$. Second, nonlinear correlation (NLC) tests (Hyvärinen, Karhunen, & Oja, 2001), i.e. Pearson correlation tests applied to nonlinearly transformed variates can also be used to detect asymmetry in predictor-error independence. Here, the square function proved particularly useful due to its connection to higher moments of the "true" predictor. For the three possible nonlinear covariances one obtains $\text{cov}(y, e_x^2) = \rho_{xy}(1 - \rho_{xy}^2)^2 \gamma_x$ (as the basis for NLC_{12}), $\text{cov}(y^2, e_x) = \rho_{xy}^2(1 - \rho_{xy}^2)\gamma_x$ (as the basis for NLC_{21}), and $\text{cov}(y^2, e_x^2) = \rho_{xy}^2(1 - \rho_{xy}^2)^2(\kappa_x - 3)$ (as the basis for NLC_{22}; see Wiedermann & Li, 2018).

Alternatively, a universal independence test, the Hilbert–Schmidt Independence Criterion (HSIC; Gretton et al., 2008) can also be used to perform DDA model selection. The HSIC evaluates the independence of functions of random variables based on Euclidian distance matrices of variables and is provably omnibus in detecting any dependence between two random variables as $n \to \infty$. Sen and Sen (2014) introduced the HSIC in identifying the independence of predictors and error terms of linear regression models and proposed a bootstrap approach to approximate the null distribution of the test statistic.

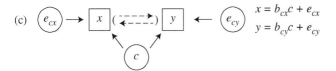

Figure 14.2 Conceptual diagram of model (c) with an unmeasured confounder together with the model equations. Here, e_{cy} and b_{cy} are errors and slope co-efficient of c in the regression with y as the dependent variable, e_{cx} and b_{cx} are the errors and slope co-efficient of c in the regression with x as the dependent variable. The dashed arrows in parentheses indicate that either $x \rightarrow y$ or $y \rightarrow x$ can be subject to confounding.

14.1.4 DDA in Confounded Models

The DDA framework is also applicable when an unmeasured confounder c is present (see Figure 14.2). Although asymmetry properties of variable and error distributions are affected by the magnitude of the confounder effect (b_{cx} and b_{cy}) and non-normality of c, DDA's independence component can indicate the presence of meaningful confounding. In the presence of c, symmetric (instead of asymmetric) patterns will occur when one evaluates the independence of residuals and predictors in the two competing models. Specifically, in the presence of c, independence is likely to be rejected in both candidate models. Further, the presence of c can alter the distributions of variables in a way that they are too close to normality to perform independence tests with sufficient statistical power. Therefore, a confounder is also likely to be present when predictor-error independence is retained in both models. Taken together, the two causally competing models in Figure 14.1 and the confounded model in Figure 14.2 can be uniquely identified based on model-specific DDA patterns. Table 14.1 summarizes these patterns for each candidate model. For more details, see Wiedermann and Li (2018) and Li and Wiedermann (2019).

14.1.5 DDA in Multiple Linear Regression Models

In the previous section, we focused on the simple linear regression model to give an overview of asymmetry properties of causally competing models under non-normality of variables. We now turn to the multiple linear regression setting. Asymmetry properties can be easily extended to multiple regression scenarios by making use of the Frisch–Waugh–Lovell (FWL) theorem (Filoso, 2013; Lovell, 2008). The FWL theorem provides a mathematical basis that allows one to dissect a multiple linear regression model into several simple linear regression models while preserving the original residual term and the original regression coefficients. This transformation can be achieved by separating the independent variables of the multiple regression model into one focal predictor and a set of

Table 14.1 Summary of DDA components and model-specific DDA patterns.

Properties	Tests	Model (a)$x \rightarrow y$	Model (b)$y \rightarrow x$	Model (c)$x \leftarrow c \rightarrow y$
Variable skewness	D'Agostino test	$\|\gamma_x\| > 0$ $\|\gamma_y\| = 0$	$\|\gamma_x\| = 0$ $\|\gamma_y\| > 0$	if $\|\rho_{xc}\| > \|\rho_{yc}\|, \|\gamma_y\| < \|\gamma_x\|$; if $\|\rho_{xc}\| < \|\rho_{yc}\|, \|\gamma_y\| > \|\gamma_x\|$
Variable kurtosis	Anscombe–Glynn test	$\|\kappa_x\| \neq 3$ $\|\kappa_y\| = 3$	$\|\kappa_x\| = 3$ $\|\kappa_y\| \neq 3$	if $\|\rho_{xc}\| > \|\rho_{yc}\|, \|\kappa_y\| < \|\kappa_x\|$; if $\|\rho_{xc}\| < \|\rho_{yc}\|, \|\kappa_y\| > \|\kappa_x\|$
Variable higher moment differences	bootstrap CIs	$\|\gamma_y\| < \|\gamma_x\|$ $\|\kappa_y\| < \|\kappa_x\|$	$\|\gamma_y\| > \|\gamma_x\|$ $\|\kappa_y\| > \|\kappa_x\|$	if $\|\rho_{xc}\| > \|\rho_{yc}\|,$ $\|\gamma_y\| < \|\gamma_x\|$ and $\|\kappa_y\| < \|\kappa_x\|$; if $\|\rho_{xc}\| < \|\rho_{yc}\|,$ $\|\gamma_y\| > \|\gamma_x\|$ and $\|\kappa_y\| > \|\kappa_x\|$
HOC differences	Bootstrap ΔHOC_s CI	$\Delta HOC_s > 0$	$\Delta HOC_s < 0$	if $\|\rho_{xc}\| > \|\rho_{yc}\|, \Delta HOC_s > 0$; if $\|\rho_{xc}\| < \|\rho_{yc}\|, \Delta HOC_s < 0$;
	Bootstrap ΔHOC_c CI	$\Delta HOC_c > 0$	$\Delta HOC_c < 0$	$\Delta HOC_c > 0$ unless κ_c is extremely large and $\|\rho_{xc}\|$ is close to 1
Residual skewness	D'Agostino test	$\|\gamma_{e_x}\| > 0$ $\|\gamma_{e_y}\| = 0$	$\|\gamma_{e_x}\| = 0$ $\|\gamma_{e_y}\| > 0$	$\|\gamma_{e_x}\| > 0$ $\|\gamma_{e_y}\| > 0$
Residual kurtosis	Anscombe–Glynn test	$\|\kappa_{e_x} - 3\| > 0$ $\|\kappa_{e_y} - 3\| = 0$	$\|\kappa_{e_x} - 3\| = 0$ $\|\kappa_{e_y} - 3\| > 0$	$\|\kappa_{e_x} - 3\| > 0$ $\|\kappa_{e_y} - 3\| > 0$

Table 14.1 (Continued)

Properties	Tests	Model (a)x → y	Model (b)y → x	Model (c)x ← c → y
Residual higher moment differences	Bootstrap CIs	$\|\gamma_{e_y}\| < \|\gamma_{e_x}\|$ $\|\kappa_{e_y}\| < \|\kappa_{e_x}\|$	$\|\gamma_{e_y}\| > \|\gamma_{e_x}\|$ $\|\kappa_{e_y}\| > \|\kappa_{e_x}\|$	if $\|\rho_{y(c\|x)}\| < \|\rho_{x(c\|y)}\|$, $\|\gamma_{e_y}\| < \|\gamma_{e_x}\|$ and $\|\kappa_{e_y}\| < \|\kappa_{e_x}\|$; if $\|\rho_{x(c\|y)}\| < \|\rho_{y(c\|x)}\|$, $\|\gamma_{e_x}\| < \|\gamma_{e_y}\|$ and $\|\kappa_{e_x}\| < \|\kappa_{e_y}\|$
	Asymptotic significance test			
Nonlinear correlations	(t-test)	NLCs = 0	NLCs ≠ 0	NLCs ≠ 0 for both models (a) and (b)
Homoscedasticity	Breusch–Pagan test	Non-significant	Significant	Significant for both models (a) and (b)
Omnibus independence test	HSIC test	HSIC = 0	HSIC ≠ 0	HSIC = 0 or HSIC ≠ 0 for both models

ρ_{xc} and ρ_{yc} are the correlation between the confounder and the focal variables.
CI = confidence interval, HOC = Higher order correlation, HSIC = Hilbert–Schmidt Independence Criterion, NLC = nonlinear correlation.

(remaining) covariates. The covariates are then regressed on the focal predictor and the focal outcome. The estimated regression residuals of these auxiliary regressions constitute covariate-adjusted ("purified") versions of x and y which are subsequently used in DDA. To give an example, consider the case of two covariates z_1 and z_2 (covariates may either be categorical or continuous). The two auxiliary regression models $y = b_{z_1} z_1 + b_{z_2} z_2 + e_y^P$ and $x = b'_{z_1} z_1 + b'_{z_2} z_2 + e_x^P$ are used to partial out the effects of the covariates and the extracted residuals e_y^P and e_x^P represent "purified" versions of y and x. Because the regression coefficient and residuals of the model $e_x^P \rightarrow e_y^P$ will be identical to the ones obtained for the multivariate model $\{x, z_1, z_2\} \rightarrow y$, applying DDA on e_y^P and e_x^P allows covariate-adjusted decisions on the causal direction of effect of x and y (see also the related Chapter 4 by Wiedermann in this volume). In the following sections, we apply the presented DDA principles to test the causal direction of the association between intrinsic motivation and academic achievement using SPSS.

14.2 The Causal Relation of Intrinsic Motivation and Academic Achievement

Although the association between intrinsic motivation and academic achievement has long been discussed among researchers, it still remains a controversial topic in education. People are intrinsically motivated when they perform behaviors out of interest and for which the primary reward is the feelings of joy as their basic psychological needs are satisfied, i.e. the needs for *autonomy, competence* and *relatedness* (Ryan & Deci, 2017). In the educational context, *academic self-concept* (ASC), *academic self-efficacy* (ASE) and *interest* are often considered three key factors for intrinsic motivation (Hannula et al., 2016). With subtle conceptual differences, both ASC and ASE concern *competence,* i.e. a basic need to feel able to operate effectively within a specific context (Bong & Skaalvik, 2003; Ryan & Deci, 2017). ASC refers to the individuals' perception of their capabilities in academic domains, and ASE represents individuals' beliefs of their successful performance in specific academic tasks (Bong & Skaalvik, 2003). In school contexts, since it is very difficult to track and record learning behaviors of individual students, the association between intrinsic motivation and academic achievement is often examined to explore motivational mechanisms. Although a positive association between intrinsic motivation and academic achievement is well recognized among educational researchers, there is considerable disagreement on the causal direction of effect between the two concepts. Many studies support the notion that academic achievement is the outcome of a motivational status, including personal interest, self-concept, and self-efficacy (Alivernini & Lucidi, 2011; Areepattamannil, Freeman, & Klinger, 2011; Hornstra, van der Veen, Peetsma, &

Volman, 2013; Liu & Hou, 2018; Murayama, Pekrun, Lichtenfeld, & vom Hofe, 2013; Wang & Eccles, 2013), i.e. *motivation* → *achievement*. In contrast, some researchers support a bi-directional association (Corpus, McClintic-Gilbert, & Hayenga, 2009; Luo, Kovas, Haworth, & Plomin, 2011; Schöber, Schütte, Köller, McElvany, & Gebauer, 2018; Williams & Williams, 2010) or an association with reversed causal direction of effect, i.e. *achievement* → *motivation* (Garon-Carrier et al., 2016; Skaalvik & Valås, 1999). A theoretical explanation for the causally reversed model is that these motivational factors predict learning behaviors and at the same time are influenced by experiences within social environment, such as academic achievement which provides resources for social comparison and external feedback (Hannula et al., 2016; Ryan & Deci, 2017).

Previous examinations of the causal direction of effect of intrinsic motivation and academic achievement have mainly relied on longitudinal information using cross-lagged panel models (CLPMs). The CLPM is a longitudinal structural equation model (SEM) that allows researchers to test causal predominance by comparing cross-lagged relations across multiple time points while controlling for both associations within time points and construct stability across time (Kearney, 2017; Newsom, 2015). Applying the CLPM implies tracking individuals and collecting a large set of data across multiple time points, which can be expensive, time-consuming, and sometimes less practical. More importantly, the CLPM relies on three important (potentially unrealistic) assumptions: (i) there is no between-subject difference in stability over time (Kearney, 2017; Selig & Little, 2012), (ii) there is no unconsidered confounder present (Hamaker, Kuiper, & Grasman, 2015; Selig & Little, 2012), and (iii) the causal direction of effect is not conditional on time (Li & Wiedermann, 2019; Selig & Little, 2012). DDA does not require precedence of time and, therefore, does not rely on some of the time-related assumptions (Note that based on cross-sectional data, DDA implicitly relies on the equilibrium assumption, i.e. cross-sectional variances and covariances do not depend on the time of measurement). In addition, the existence of influential unconsidered confounders may lead to inconclusive results in DDA (i.e. neither causal model finds empirical support), but may cause CLPM to return spurious results (Garon-Carrier et al., 2016; Hamaker et al., 2015; Newsom, 2015). Therefore, in the present study, we re-examine the causal direction of the association between intrinsic motivation and academic achievement using principles of DDA.

14.2.1 High School Longitudinal Study 2009

We used public data of the High School Longitudinal Study 2009 (HSLS: 09) collected by the National Center for Education Statistics (NCES). The sample includes fall-term 9th-graders in more than 900 public and private high schools in the United States (Ingels et al., 2011). Considering that cultural background can be a crucial factor when examining the association between motivation and achievement, the present study focuses on Asian students as the target sample. Asian students are a widely researched ethnic group in education who often outperform other ethnic groups in mathematics tests and shows different motivational mechanisms, where intrinsic motivation may not be an important driving force for learning efforts (Chen & Stevenson, 1995; Ng, 2003). The analysis sample consisted of $n = 674$ students who identified themselves as Asian (328 males and 346 females).

Students took a 72-item mathematics assessment in algebraic reasoning and an online survey covering various topics, including educational experiences, sociodemographic background, expectancies, and values for science and mathematics. Students' parents, teachers, and principals took their surveys online or via phone (for study details see Ingels et al. (2011)). Norm-referenced IRT scale scores were used to measure mathematics achievement (MATH). Intrinsic motivation in mathematics (IM) was measures using sum scores of standardized mathematics self-identity/concept, self-efficacy, and interest scales. Higher motivation scores indicated that students were more intrinsically motivated as they felt more interested, competent, and confident in mathematics. Mathematics self-identity/concept was based on students' responses to two survey items asking whether they and others viewed themselves as a "math person." Mathematics self-efficacy was measured by four items asking how confident they were to excel in mathematics tests, understand mathematics textbooks, possess mathematics skills, and excel in mathematics assignments. Mathematics interest was based on six items asking how students were interested in the 2009 mathematics course.

We also include covariates that are potentially relevant for both focal variables, i.e. student gender (SEX; 0 = male, 1 = female), Socio-economic status (SES; a composite score based on parental education, occupation, and family income), average parents' education (PEDU; calculated as the mean of parents' highest degree, which are coded from 1 to 7 with higher scores reflecting higher degrees), math teachers' expectation (EXP; a standardized score based on principal components analysis of eight items), region of the United States (dummy variable per region with RGN1 = Northeast, RGN2 = Midwest, RGN3 = South, RGN4 = West; Northeast served as the reference region), and variables indicating the type of school (PUBL; 0 = private school, 1 = public school) and the students' first language (EN1ST; 1 = English, 0 = other). Descriptive statistics and bivariate correlations of all variables are given in Table 14.2. As expected, mathematics achievement is positively correlated with intrinsic motivation ($\hat{\rho} = 0.38$), SES ($\hat{\rho} = 0.36$), and parents' education ($\hat{\rho} = 0.39$). Correlations between mathematics achievement and the remaining covariates are rather low (ranging from -0.12 to 0.08). Other pairwise correlations involving intrinsic motivation tend to be low as well (ranging from -0.11 to 0.17).

Table 14.2 Bivariate Pearson correlation coefficients and descriptive measures of all variables.

Variables	(2)	(3)	(4)	(5)	(6)	(7)	(8)	(9)	(10)	(11)	M	SD	γ	κ
(1) Mathematics score (MATH)	0.38	−0.09	0.36	0.39	−0.01	0.06	0.08	−0.12	−0.04	−0.10	0.97	0.96	−0.29	−0.23
(2) Intrinsic Motivation (IM)	—	−0.09	0.16	0.17	0.03	0.05	0.00	−0.11	0.03	−0.06	0.45	0.69	−0.08	−0.45
(3) Gender (SEX, 0 = male, 1 = female)		—	0.03	0.01	0.03	−0.01	−0.06	−0.04	0.01	0.01	1.51	0.50	−0.05	−2.00
(4) Socio-economic status (SES)			—	0.92	0.04	0.09	−0.02	−0.12	0.21	0.08	0.60	0.91	−0.14	−0.73
(5) Parents' Education (PEDU)				—	0.06	0.09	−0.01	−0.12	0.13	0.02	3.79	1.48	0.10	−0.69
(6) Math teachers' expectation (EXP)					—	0.04	−0.03	−0.06	0.07	−0.03	−0.72	2.93	−2.25	3.70
(7) Midwest (RGN2)						—	−0.49	−0.24	0.03	0.03	0.26	0.44	1.10	−0.80
(8) South (RGN3)							—	−0.34	−0.06	−0.01	0.41	0.49	0.38	−1.87
(9) West (RGN4)								—	−0.04	−0.05	0.14	0.35	2.05	2.21
(10) Public school (PUBL, 0 = private, 1 = public)									—	0.08	1.24	0.43	1.24	−0.47
(11) First language (EN1ST; 0 = other, 1= English)										—	0.66	0.47	−0.70	−1.52

M = mean, SD = standard deviation, γ = skewness, κ = kurtosis.

14.3 Direction Dependence Analysis Using SPSS

Because the majority of previous studies suggest that intrinsic motivation is more likely to affect achievement (and not the other way around), we use IM → MATH as the causal target model. Thus, we start with the assumption that math achievement is the result of intrinsic motivation. In contrast, MATH → IM constitutes the (causally reversed) alternative model. In other words, when IM → MATH constitutes the one model that better describes the underlying causal flow of the two variables, we expect that (i) MATH is closer to normality than IM, (ii) the error term of IM → MATH is closer to normality than that of MATH → IM, and (iii) predictor-error independence holds for IM → MATH and is violated for MATH → IM. In contrast, when the alternative model is more likely to approximate the causal mechanism (i) IM is closer to normality than MATH, (ii) the error of MATH → IM are closer to normality than IM → MATH, and (iii) independence holds for MATH → IM but not for IM → MATH. The regression results for the two competing models are provided in Table 14.3. Though some of the covariates in the models are not statistically significant for the target model, we decided to keep them in the models due to theoretical relevance.

In the following paragraphs, we provide a step-by-step tutorial on using DDA in SPSS. After installing the SCDs at the desired dropdown menu location (for installation details see https://ddaproject.com), one can launch the DDA main dialogue box in SPSS (see Figure 14.3). One has to specify a working directory before running DDA procedures (note that PC users are assumed to be authorized to read and write in the designated folder). To check the distributional requirements of DDA, we start with evaluating distributional characteristics of observed variables through selecting DDA Tests:Variable Distribution in the upper left corner of the main dialogue box.

14.3.1 Variable Distributions and Assumption Checks

Before running the DDA procedures, the assumption of non-normality of variables was checked for both focal variables, MATH and IM. The Shapiro–Wilk tests rejected the null hypothesis of normality for both measures ($p < 0.001$). Following Wiedermann and Li (2018), we ruled out the presence of outliers and floor/ceiling effects using visual diagnostics. To continue the analysis of asymmetry properties of variable distributions, we select the "Options" dialogue box (upper right corner of the main dialogue box) to specify the confidence level of bootstrap CIs (here 0.95), number of resamples (here 1000), sidedness of significance tests ("Two Sided," "Less," or "Greater"), and type of missing value treatment (cf. Figure 14.4). In case of missing values, "Listwise deletion" (the default) and "Pairwise deletion" are available. The option "Save Residuals"

Table 14.3 Results of the competing regression models.

Variables	*b*	SE	*t*-Value	*p*-Value
Target model: IM → MATH ($R^2 = 0.279$)				
Intercept	0.552	0.250	2.205	0.028
Intrinsic motivation	0.428	0.045	9.471	<0.001
Gender	−0.094	0.065	−1.445	0.149
Socio-economic status	0.153	0.092	1.669	0.096
Parents' education	0.142	0.056	2.539	0.011
Math teachers' expectation	−0.011	0.011	−0.982	0.326
US Region: Midwest	0.099	0.097	1.014	0.311
South	0.165	0.090	1.831	0.068
West	−0.057	0.115	−0.497	0.619
Public school	−0.209	0.078	−2.663	0.008
English as first language	−0.157	0.069	−2.274	0.023
Alternative model: MATH → IM ($R^2 = 0.156$)				
Intercept	0.332	0.201	1.649	0.100
Mathematics score	0.275	0.029	9.471	<0.001
Gender	−0.082	0.052	−1.574	0.116
Socio-economic status	0.003	0.074	0.046	0.964
Parents' education	−0.005	0.045	−0.106	0.916
Math teachers' expectation	0.006	0.009	0.627	0.531
US Region: Midwest	−0.050	0.078	−0.639	0.523
South	−0.096	0.073	−1.320	0.187
West	−0.189	0.092	−2.061	0.040
Public school	0.078	0.063	1.234	0.218
English as first language	−0.056	0.055	−1.005	0.315

b = regression weight, IM = intrinsic motivation, MATH = math achievement, SE = standard error.

can be used to save the covariate-adjusted ("purified") focal variables for further analyses. In the present application, we keep all the default settings shown in Figure 14.4 and click "Continue" to return to the main menu.

After specifying dependent and independent variables (MATH and IM) of the target model, we use "OK" to run variable distribution-based DDA tests, i.e. the D'Agostino and Anscombe–Glynn tests and bootstrap CIs on higher moment

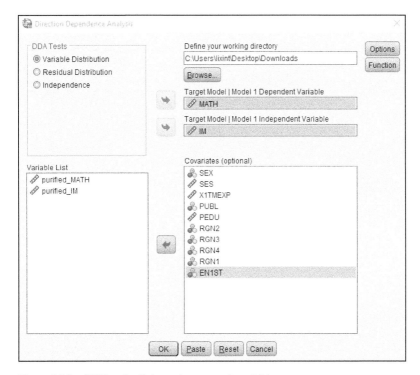

Figure 14.3 SPSS main dialogue box to perform DDA.

Figure 14.4 DDA options dialogue box.

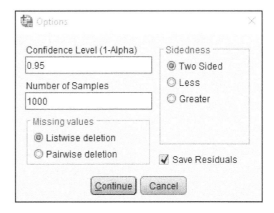

differences. When using the "Save Residuals" option, purified mathematics scores (here p_MATH) and intrinsic motivation scores (p_IM) are automatically added to the data file. Please note that the purified variables have to be renamed as desired manually – the default names are "ry" and "rx" representing the purified

```
Skewness and Excess Kurtosis Tests
        p_MATH  z-value      Sig    p_IM    z-value      Sig
Skew    -0.3241 -3.3892   0.0007  -0.0121  -0.1300   0.8965
Exkurt  -0.2537 -1.4217   0.1551  -0.3814  -2.3965   0.0166


Bootstrap Confidence Intervals (Skewness and Kurtosis Difference)
        BootCILo BootCIUp
Skew     -0.4111  -0.1132
Exkurt   -0.4137   0.1463


Higher-Order Correlation Difference Test
         Estimate BootCILo  BootCIUp
Delta.sq   -0.0051  -0.0328    0.0146
Delta.cu    0.1122  -0.0308    0.2286
```

Figure 14.5   SPSS output of DDA variable distribution tests.

dependent and independent variables of the target model. The SPSS output of DDA variable distribution tests is given in Figure 14.5. Scatterplots and univariate histograms of the purified variables are shown in Figure 14.6 for illustrative purposes (not included in the current DDA macro version). D'Agostino z-tests and 95% bootstrap CIs suggest that the "purified" MATH variable (p_MATH) is more skewed than the "purified" IM scores (p_IM; see also Figure 14.6). The excess kurtosis (ExKurt) of p_IM is significantly smaller than zero in the z-test, indicating a platykurtic distribution. However, the 95% bootstrap CIs suggests for kurtosis differences indicates that the kurtosis values do not differ from each other. No statements can be made based on the HOC tests (Delta.sq and Delta.cu; here values significantly larger than zero provide evidence for the target model).

14.3.2   Residual Distributions

Overall, four significance procedures were implemented to evaluate asymmetry properties of error distributions in the two competing models: The D'Agostino test evaluates symmetry of residual distributions, the Anscombe–Glynn test evaluates the kurtosis of distributions and bootstrap CIs and asymptotic significance tests are provided to evaluate differences in higher moments (see Table 14.1). After launching the DDA main dialogue box, we select "DDA tests: Residual Distribution" in the DDA tests section. In addition, we use the "Options" settings to specify details of the significance procedures (see above). One needs to untick the "Save Residuals" option if "purified" variables should not be saved in the data file. The corresponding SPSS output is given in Figure 14.7. Here, r_MATH represents the residual term of the target model (IM → MATH), and r_IM represents the residual term of the alternative model (MATH → IM). The upper part of the

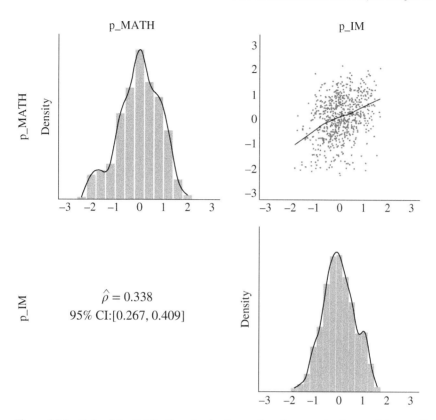

Figure 14.6 Univariate distributions (main diagonal) and scatterplot (upper right panel) of "purified" mathematics scores (p_MATH) and intrinsic motivation scores (p_IM), including the bivariate LOESS smoothed line for 674 9th graders.

output summarizes the results of separate higher moments tests. The lower part summarizes results on the differences in higher moments. In the present application, all significance procedures focusing on the symmetry of residual distributions (i.e. D'Agostino tests as well as asymptotic and resampling-based skewness difference procedures) suggest that residuals of the target model (IM → MATH) are more skewed than the residuals of the alternative model (MATH → IM). In contrast, the excess kurtosis significantly deviates from zero for r_IM but does not for r_MATH. Kurtosis difference procedures, however, suggest that residual distributions for the two models are equally kurtotic. Overall, assuming error normality in the one model that better approximates the underlying causal mechanism, we again find empirical support for the alternative model MATH → IM.

```
Skewness and Excess Kurtosis Tests
          r_MATH   z-value    Sig      r_IM    z-value     Sig
Skew     -0.2963   -3.1108  0.0019   0.0428   0.4582    0.6468
Exkurt   -0.1195   -0.5444  0.5862  -0.3696  -2.2993    0.0215

Skewness and Excess Kurtosis Difference Tests
          Diff.    z-value    Sig    BootCILo   BootCIUp
Skew     -0.2535   -2.4762  0.0133  -0.3981    -0.0400
Exkurt    0.2501    1.2491  0.2116  -0.5896     0.0980
```

Figure 14.7   SPSS output of residual distribution tests.

14.3.3   Independence Properties

To evaluate predictor-residual independence in the two competing models, non-linear correlation tests, Breusch–Pagan homoscedasticity tests, and the HSIC test were included in the DDA SCDs. When selecting "Independence" in the DDA main dialogue box, Breusch–Pagan homoscedasticity tests (cf. Wiedermann et al., 2017) are performed for the two candidate models as a default. In addition, we can use "Function" (upper right corner of the main dialogue box) to select nonlinear functions ("Square," "Cubic," "Tangent Hyperbolic," "Other power function") for nonlinear correlation procedures (Figure 14.8). For each nonlinear function, three tests are performed: One in which the nonlinear transformation is only applied to the predictor, one where only residuals are transformed, and a third test were both, predictor and residuals, are nonlinearly transformed. For example, selecting "Square" results in testing cor(predictor^2, errors), cor(predictor, errors^2), and cor(predictor^2, errors^2). When selecting the Hilbert Schmidt Independence Criterion ("HSIC"), specifying the number of resamples (the default value is 200) is required. Since the HSIC is computationally demanding, for large sample sizes (e.g. $n > 800$), we recommend that nonlinear correlation tests (using the square function for skewed data and the hyperbolic tangent function for symmetric non-normal data) are used first to find out whether severe dependencies are observed in both models, indicating the presence of confounders (Wiedermann & Li, 2018). Note that, in the present data example with $n = 674$, it takes a standard desktop PC about 10 minutes to perform the HSIC test with 200 resamples.

The results of DDA independence tests are given in Figure 14.9. First, the upper panel summarizes results for the Breusch–Pagan and the robust Breusch–Pagan test. "Model 1" refers to the target model, "Model 2" refers to the alternative model. Breusch–Pagan tests indicate heteroscedasticity in the target model and homoscedasticity in the alternative model. Next, the middle section of the SPSS output summarizes the results of nonlinear correlation tests (using the square function). No decision can be made based on these tests because all tests retain the null hypothesis. Finally, the lower panel of the output gives the results of

Figure 14.8 DDA function dialogue box.

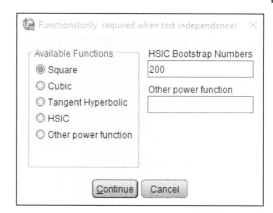

the HSIC tests. Again, results indicate that the null hypothesis of predictor-error independence is rejected in the target model (IM → MATH) and retained in the alternative model (MATH → IM).

14.3.4 Summary of DDA Results

To better illustrate the patterns of how competing causal models were supported by DDA, we provide a summary of the results in Table 14.4. Recall that, in the "true" causal model (with "true" normal errors), the outcome variable is supposed to be closer to the normal distribution than the predictor, and the residuals are supposed to be closer to the normal distribution than those observed for the causally mis-specified model. Further, the predictor and residuals should be stochastically independent in the "true" model and dependent in the mis-specified model. In the present example, the following patterns can be observed: First, two out of three approaches to evaluate the independence assumption (homoscedasticity and HSIC tests) point at the alternative model MATH → IM; non-linear correlation tests give inconclusive results. Second, all DDA measures based on third higher moments (i.e. separate D'Agostino tests and skewness difference procedures for observed variables and regression residuals) also favor a causal flow of the form MATH → IM. Third, separate Anscombe–Glynn tests suggest that the reversed model IM → MATH better fits the data. However, bootstrap CIs of kurtosis differences suggest that variates of the two competing models (i.e. residuals and observed variables) are equally kurtotic. Thus, overall, we can conclude that the model MATH → IM seems to be better suited to approximate the underlying data generating mechanism than IM → MATH. That is, in the present sample, the intrinsic motivation of Asian 9th graders is more likely to be the result of their mathematics achievement instead of vice versa. One possible explanation for this

```
Tests of Heteroscedasticity: Model 1
               Chisq          df         Sig
BP-test      20.3168     10.0000      0.0264
robustBP     21.6076     10.0000      0.0172

Tests of Heteroscedasticity: Model 2
               Chisq          df         Sig
BP-test       6.0718     10.0000      0.8092
robustBP      7.4482     10.0000      0.6826
```

```
MODEL 1----Predictor_Residual Non-linear Correlation Test
           est.corr    t-value          df          Sig
IM2_R       -0.0397    -1.0311    672.0000       0.3029
IM_R2       -0.0505    -1.3115    672.0000       0.1902
IM2_R2      -0.0219    -0.5675    672.0000       0.5706

MODEL 2----Predictor_Residual Non-linear Correlation Test
           est.corr    t-value          df          Sig
MA2_R        0.0031     0.0800    672.0000       0.9362
MA_R2       -0.0630    -1.6376    672.0000       0.1020
MA2_R2      -0.0365    -0.9457    672.0000       0.3446
```

```
MODEL 1----Predictor_Residual Hilbert-Schmidt Independence Criterion
              HSIC         Sig
IM_R         0.1726      0.0250

MODEL 2----Predictor_Residual Hilbert-Schmidt Independence Criterion
              HSIC         Sig
MA_R         0.0328      0.8050
```

Figure 14.9 SPSS outputs of DDA independence tests.

finding is that self-benefiting intrinsic motivation is not as valued in Asian cultures as in western cultures (Ng, 2003; Zhu & Leung, 2011), and therefore, the underlying motivational mechanisms may also be different between Asian and western cultures. Conducting DDA on other ethnic groups in the US is a topic for future research; however, extensive discussion is off the illustrative purpose of the presented empirical example.

Further, cultural differences may also explain the contradicting directional results observed for fourth-moment based DDA tests. Specifically, the so-called middle response bias (i.e. fewer self-reported extreme IM scores) commonly identified among Asian Americans (Harzing, 2006; Johnson, Shavitt, & Holbrook, 2011; Wang, Hempton, Dugan, & Komives, 2008) may serve as a potential explanation for the disagreement in distribution-based DDA tests.

Table 14.4 Summary of DDA decisions.

DDA property	DDA procedure	Causal decision
Variable distributions	D'Agostino skewness test	MATH → IM
	Bootstrap CI of skewness differences	MATH → IM
	Anscombe–Glynn kurtosis test	IM → MATH
	Bootstrap CI kurtosis differences	Inconclusive
	Higher-order correlations	Inconclusive
Residual distributions	D'Agostino skewness test	MATH → IM
	Bootstrap CI of skewness differences	MATH → IM
	Asymptotic skewness difference test	MATH → IM
	Anscombe–Glynn kurtosis test	IM → MATH
	Bootstrap CI kurtosis differences	Inconclusive
	Asymptotic kurtosis difference test	Inconclusive
Independence	Non-linear correlation tests[a]	Inconclusive
	Breusch–Pagan Heteroscedasticity test	MATH → IM
	Hilbert Schmidt Independence Criterion	MATH → IM

IM = intrinsic motivation, MATH = mathematics achievement.
a) Non-linear correlation tests are based on squared residuals.

14.4 Conclusions

The DDA framework provides an innovative perspective to examine the causal direction of effects in theoretically reversible linear associations. Since DDA does not require precedence of time to aid causal inference, fewer time related assumptions are required. However, DDA is capable of examining whether the causal direction of contemporaneous effects is consistent over time. More importantly, with DDA, the causal direction of effect can be evaluated in data situations where experiments are not feasible, or when accurate longitudinal data are very difficult to collect. In addition, unlike covariance-based methods such as the CLPM, DDA is able to identify spurious effects due to confounding. Although principles of DDA have been used in the educational, developmental, and behavioral sciences (see, e.g. von Eye & DeShon, 2012; Wiedermann, Dong, & von Eye, 2019), DDA has not yet drawn enough attention from applied researchers in the social sciences. This is partly rooted in the lack of accessible software implementations. The present introduction of DDA SCDs addresses this limitation by providing a GUI add-on for SPSS. As was shown in the data example, when researchers have familiarized themselves with the DDA principles and model selection criteria, the application of DDA SCDs is intuitive and in line with other SPSS operations.

14.4.1 Extensions and Future Work

Several extensions of DDA SCDs are planned in the future. First, we plan to extend DDA SCDs to evaluate the impact of moderators (i.e. testing the causal direction of conditional effects; cf. Li & Wiedermann, 2019) and to handle nonlinear variable relationships (i.e. testing the causal direction of effects in polynomial regression models). Second, we plan to incorporate options for data visualization designed to detect outliers and influential data points in non-normal variables (see, e.g. Walker, Dovoedo, Chakraborti, & Hilton, 2018). Third, to evaluate the stability/robustness of DDA decisions, we plan to implement bootstrap aggregating/bagging techniques (Breiman, 1996) and Monte-Carlo based sensitivity analyses (for examples see Wiedermann and Sebastian, 2019, and the related Chapter 11 by Rosenström and García-Velázquez, in this volume). Version updates and extensions will be posted on the DDA portal https://ddaproject.com.

References

Alivernini, F., & Lucidi, F. (2011). Relationship between social context, self-efficacy, motivation, academic achievement, and intention to drop out of high school: A longitudinal study. *The Journal of Educational Research, 104*(4), 241–252. doi:10.1080/00220671003728062

Anscombe, F. J., & Glynn, W. J. (1983). Distribution of the kurtosis statistic b 2 for normal samples. *Biometrika, 70*(1), 227–234.

Areepattamannil, S., Freeman, J. G., & Klinger, D. A. (2011). Intrinsic motivation, extrinsic motivation, and academic achievement among Indian adolescents in Canada and India. *Social Psychology of Education, 14*(3), 427. doi:10.1007/s11218-011-9155-1

Bong, M., & Skaalvik, E. (2003). Academic self-concept and self-efficacy: How different are they really? *Educational Psychology Review, 15*(1), 1–40. doi:10.1023/a:1021302408382

Breiman, L. (1996). Bagging predictors. *Machine Learning, 24*(2), 123–140. doi:10.1023/A:1018054314350

Breusch, T. S., & Pagan, A. R. (1979). A simple test for heteroscedasticity and random coefficient variation. *Econometrica: Journal of the Econometric Society, 47,* 1287–1294.

Chen, C., & Stevenson, H. W. (1995). Motivation and mathematics achievement: A comparative study of Asian-American, Caucasian-American, and East Asian high school students. *Child Development, 66*(4), 1215–1234. doi:10.2307/1131808

Corpus, J. H., McClintic-Gilbert, M. S., & Hayenga, A. O. (2009). Within-year changes in children's intrinsic and extrinsic motivational orientations: Contextual predictors and academic outcomes. *Contemporary Educational Psychology, 34*(2), 154–166. doi:10.1016/j.cedpsych.2009.01.001

D'Agostino, R. B. (1971). An omnibus test of normality for moderate and large size samples. *Biometrika, 58*(2), 341–348.

Deci, E. (1975). *Intrinsic motivation.* New York, NY: Plenum Press.

Deci, E., & Ryan, R. (1980). The empirical exploration of intrinsic motivational processes. In L. Berkowitz (Ed.), *Advances in experimental social psychology* (Vol. *13,* pp. 39–80). Cambridge, MA: Academic Press.

Dodge, Y., & Rousson, V. (2000). Direction dependence in a regression line. *Communications in Statistics-Theory and Methods, 29*(9–10), 1957–1972.

Dodge, Y., & Rousson, V. (2001). On asymmetric properties of the correlation coeffcient in the regression setting. *The American Statistician, 55*(1), 51–54.

Dodge, Y., & Yadegari, I. (2010). On direction of dependence. *Metrika, 72*(1), 139–150.

Filoso, V. (2013). Regression anatomy, revealed. *Stata Journal, 13*(1), 92–106.

Garon-Carrier, G., Boivin, M., Guay, F., Kovas, Y., Dionne, G., Lemelin, J. P., … Tremblay, R. E. (2016). Intrinsic motivation and achievement in mathematics in elementary school: A longitudinal investigation of their association. *Child Development, 87*(1), 165–175. doi:10.1111/cdev.12458

Gretton, A., Fukumizu, K., Teo, C. H., Song, L., Schölkopf, B., & Smola, A. J. (2008). A kernel statistical test of independence. In J. C. Platt, D. Koller, Y. Singer, & S. T. Roweis (Eds.), *Advances in neural information processing systems 20* (pp. 585–592). Vancouver, BC, Canada: Curran Associates, Inc.

Hamaker, E. L., Kuiper, R. M., & Grasman, R. P. P. P. (2015). A critique of the cross-lagged panel model. *Psychological Methods, 20*(1), 102–116. doi:10.1037/a0038889

Hannula, M. S., Di Martino, P., Pantziara, M., Zhang, Q., Morselli, F., Heyd-Metzuyanim, E., … Goldin, G. A. (2016). Attitudes, beliefs, motivation, and identity in mathematics education. In *Attitudes, beliefs, motivation and identity in mathematics education: An overview of the field and future directions* (pp. 1–35). Cham, Switzerland: Springer International Publishing.

Harzing, A.-W. (2006). Response styles in cross-national survey research: A 26-country study. *International Journal of Cross Cultural Management, 6*(2), 243–266. doi:10.1177/1470595806066332

Hornstra, L., van der Veen, I., Peetsma, T., & Volman, M. (2013). Developments in motivation and achievement during primary school: A longitudinal study on group-specific differences. *Learning and Individual Differences, 23*, 195–204. doi:10.1016/j.lindif.2012.09.004

Hyvärinen, A., Karhunen, J., & Oja, E. (2001). *Independent components analysis.* New York, NY: Wiley & Sons.

IBM Corp. (2017). *IBM SPSS statistics for windows, version 25.0.* Armonk, NY: Author.

Imbens, G., & Angrist, J. (1994). Identification and estimation of local average treatment effects. *Econometrica, 62*, 467–476. doi:10.3386/t0118

Ingels, S. J., Pratt, D. J., Herget, D. R., Burns, L. J., Dever, J. A., Ottem, R., … Leinwand, S. (2011). *High school longitudinal study of 2009 (HSLS: 09): Base-year data file documentation.* Washington, DC: U.S. Dept. of Education, Institute of Education Sciences, National Center for Education Statistics.

Johnson, T. P., Shavitt, S., & Holbrook, A. L. (2011). Survey response styles across cultures. In *Cross-cultural research methods in psychology* (pp. 130–175). New York, NY: Cambridge University Press.

Kearney, M. W. (2017). Cross lagged panel analysis. *The SAGE encyclopedia of communication research methods,* 1–6.

Koenker, R. (1981). A note on studentizing a test for heteroscedasticity. *Journal of Econometrics, 17*(1), 107–112.

Koenker, R., & Bassett, G., Jr., (1982). Robust tests for heteroscedasticity based on regression quantiles. *Econometrica: Journal of the Econometric Society, 50*, 43–61.

Li, X., & Wiedermann, W. (2019). Conditional direction dependence analysis: Evaluating the causal direction of effects in linear models with interaction terms. *Multivariate Behavioral Research.* doi:10.1080/00273171.2019.1687276

Liu, Y., & Hou, S. (2018). Potential reciprocal relationship between motivation and achievement: A longitudinal study. *School Psychology International, 39*(1), 38–55. doi:10.1177/0143034317710574

Lovell, M. C. (2008). A simple proof of the FWL theorem. *The Journal of Economic Education, 39*(1), 88–91.

Luo, Y. L., Kovas, Y., Haworth, C. M., & Plomin, R. (2011). The etiology of mathematical self-evaluation and mathematics achievement: Understanding the relationship using a cross-lagged twin study from age 9 to 12. *Learning and Individual Differences, 21*(6), 710–718. doi:10.1016/j.lindif.2011.09.001

Muenchen, R. A. (2017). The popularity of data science software. Retrieved from http://r4stats.com/articles/popularity/

Murayama, K., Pekrun, R., Lichtenfeld, S., & vom Hofe, R. (2013). Predicting long-term growth in students' mathematics achievement: The unique contributions of motivation and cognitive strategies. *Child Development, 84*(4), 1475–1490. doi:10.1111/cdev.12036

Newsom, J. T. (2015). *Longitudinal structural equation modeling: A comprehensive introduction*. New York, NY: Routledge.

Ng, C.-H. (2003). Re-conceptualizing achievement goals from a cultural perspective. Paper presented at the the Joint Conference of NZARE & AARE, Auckland, New Zealand.

Pornprasertmanit, S., & Little, T. D. (2012). Determining directional dependency in causal associations. *International Journal of Behavioral Development*, *36*(4), 313–322.

Rosenbaum, P. R., & Rubin, D. B. (1983). The central role of the propensity score in observational studies for causal effects. *Biometrika*, *70*, 41–55. doi:10.1017/cbo9780511810725.016

Ryan, R. M., & Deci, E. L. (2017). *Self-determination theory: Basic psychological needs in motivation, development, and wellness*. New York, NY: The Guilford Press.

Schöber, C., Schütte, K., Köller, O., McElvany, N., & Gebauer, M. M. (2018). Reciprocal effects between self-efficacy and achievement in mathematics and reading. *Learning and Individual Differences*, *63*, 1–11. doi:10.1016/j.lindif.2018.01.008

Selig, J. P., & Little, T. D. (2012). Autoregressive and cross-lagged panel analysis for longitudinal data. In *Handbook of developmental research methods* (pp. 265–278). New York, NY: Guilford Press.

Sen, A., & Sen, B. (2014). Testing independence and goodness-of-fit in linear models. *Biometrika*, *101*(4), 927–942.

Shimizu, S., Inazumi, T., Sogawa, Y., Hyvärinen, A., Kawahara, Y., Washio, T., … Bollen, K. (2011). DirectLiNGAM: A direct method for learning a linear non-Gaussian structural equation model. *Journal of Machine Learning Research*, *12*(Apr), 1225–1248. Retrieved from http://www.jmlr.org/papers/v12/shimizu11a.html

Skaalvik, E., & Valås, H. (1999). Relations among achievement, self-concept, and motivation in mathematics and language arts: A longitudinal study. *The Journal of Experimental Education*, *67*(2), 135–149. doi:10.1080/00220979909598349

Thistlewaite, D., & Campbell, D. (1960). Regression-discontinuity analysis: An alternative to the ex post facto experiment. *Journal of Educational Psychology*, *51*, 309–317. doi:10.1037/h0044319

von Eye, A., & DeShon, R. P. (2008). Characteristics of measures of directional dependence-Monte Carlo studies. *Interstat*, *14*, 1–33.

von Eye, A., & DeShon, R. P. (2012). Directional dependence in developmental research. *International Journal of Behavioral Development*, *36*(4), 303–312. doi:10.1177/0165025412439968

Walker, M. L., Dovoedo, Y. H., Chakraborti, S., & Hilton, C. W. (2018). An improved boxplot for univariate data. *The American Statistician*, *72*(4), 348–353. doi:10.1080/00031305.2018.1448891

Wang, M.-T., & Eccles, J. S. (2013). School context, achievement motivation, and academic engagement: A longitudinal study of school engagement using a multidimensional perspective. *Learning and Instruction, 28*, 12–23. doi:10.1016/j.learninstruc.2013.04.002

Wang, R., Hempton, B., Dugan, J. P., & Komives, S. R. (2008). Cultural differences: Why do Asians avoid extreme responses? *Survey Practice, 1*(3). doi:10.29115/SP-2008-0011

Wiedermann, W. (2015). Decisions concerning the direction of effects in linear regression models using fourth central moments. In *Dependent data in social sciences research* (pp. 149–169). Cham, Switzerland: Springer.

Wiedermann, W. (2017). A note on fourth moment-based direction dependence measures when regression errors are non normal. *Communications in Statistics-Theory and Methods, 47*(21), 1–10.

Wiedermann, W., Artner, R., & von Eye, A. (2017). Heteroscedasticity as a basis of direction dependence in reversible linear regression models. *Multivariate Behavioral Research, 52*(2), 222–241.

Wiedermann, W., Dong, N., & von Eye, A. (2019). Advances in statistical methods for causal inference in prevention science: Introduction to the special section. *Prevention Science, 20*(3), 390–393. doi:10.1007/s11121-019-0978-x

Wiedermann, W., & Hagmann, M. (2016). Asymmetric properties of the Pearson correlation coefficient: Correlation as the negative association between linear regression residuals. *Communications in Statistics-Theory and Methods, 45*(21), 6263–6283.

Wiedermann, W., Hagmann, M., & Eye, A. (2015). Significance tests to determine the direction of effects in linear regression models. *British Journal of Mathematical and Statistical Psychology, 68*(1), 116–141.

Wiedermann, W., Hagmann, M., Kossmeier, M., & von Eye, A. (2013). Resampling techniques to determine direction of effects in linear regression models. In R. G. Graf (Ed.), *Interstat*. Blacksburg, VA: Virginia Tech.

Wiedermann, W., & Li, X. (2018). Direction dependence analysis: A framework to test the direction of effects in linear models with an implementation in SPSS. *Behavior Research Methods, 50*(4), 1581–1601. doi:10.3758/s13428-018-1031-x

Wiedermann, W., & Li, X. (2019). Direction dependence analysis in R. Retrieved from http://www.ddaproject.com

Wiedermann, W., Li, X., & von Eye, A. (2019). Testing the direction of mediation effects in randomized intervention studies. *Prevention Science, 20*(3), 419–430.

Wiedermann, W., & Sebastian, J. (2019). Sensitivity analysis and extensions of testing the causal direction of dependence: A rejoinder to thoemmes. *Multivariate Behavioral Research*. doi:10.1080/00273171.2019.1659127

Wiedermann, W., & von Eye, A. (2015a). Direction-dependence analysis: A confirmatory approach for testing directional theories. *International Journal of Behavioral Development, 39*(6), 570–580.

Wiedermann, W., & von Eye, A. (2015b). Direction of effects in multiple linear regression models. *Multivariate Behavioral Research, 50*(1), 23–40.

Williams, T., & Williams, K. (2010). Self-efficacy and performance in mathematics: Reciprocal determinism in 33 nations. *Journal of Educational Psychology, 102*(2), 453–466. doi:10.1037/a0017271

Zhu, Y., & Leung, F. K. (2011). Motivation and achievement: Is there an East Asian model? *International Journal of Science and Mathematics Education, 9*(5), 1189–1212.

Author Index

Direction Dependence in Statistical Modeling: Methods of Analysis, First Edition.
Edited by Wolfgang Wiedermann, Daeyoung Kim, Engin A. Sungur, and Alexander von Eye.
© 2021 John Wiley & Sons, Inc. Published 2021 by John Wiley & Sons, Inc.

Subject Index

Direction Dependence in Statistical Modeling: Methods of Analysis, First Edition.
Edited by Wolfgang Wiedermann, Daeyoung Kim, Engin A. Sungur, and Alexander von Eye.
© 2021 John Wiley & Sons, Inc. Published 2021 by John Wiley & Sons, Inc.